EFCE Event No. 303

First European Conference on

Coal Liquid Mixtures

Organised by the Institution of Chemical Engineers
in association with the Institute of Energy and
the UK Department of Energy, and held at the
Golden Valley Hotel, Cheltenham, UK,
5–6 October 1983.

Organising Committee

G. O. Davies (Chairman)	Coal Research Establishment
J. Highley	Coal Research Establishment
R. L. Jones	Consultant, Gloucester
D. Merrick	Institute of Energy
D. T. Ponsford	Department of Industry
W. Prescott	ISC Chemicals Ltd
M. Smith	Coal Research Establishment

INSTITUTION OF CHEMICAL ENGINEERS
SYMPOSIUM SERIES No. 83

ISBN 0 85295 162 0

PUBLISHED BY THE INSTITUTION OF CHEMICAL ENGINEERS

Copyright © 1983 The Institution of Chemical Engineers

First edition 1983 – ISBN 0 85295 162 0

MEMBERS OF THE INSTITUTION OF CHEMICAL ENGINEERS (Worldwide)
SHOULD ORDER DIRECT FROM THE INSTITUTION

Geo. E. Davis Building, 165–171 Railway Terrace, Rugby, Warks CV21 3HQ.

Australian orders to:

R. M. Wood, School of Chemical Engineering and Industrial Chemistry, University of New South Wales, PO Box 1, Kensington, NSW, Australia 2033.

Distributed throughout the world (excluding Australia) by Pergamon Press Ltd, except to IChemE members.

U.K.	Pergamon Press Ltd., Headington Hill Hall, Oxford OX3 0BW, England
U.S.A.	Pergamon Press Inc., Maxwell House, Fairview Park, Elmsford, New York 10523, U.S.A.
CANADA	Pergamon Press Canada Ltd., Suite 104, 150 Consumers Rd., Willowdale, Ontario M2J 1P9, Canada
FRANCE	Pergamon Press SARL, 24 rue des Ecoles, 75240 Paris, Cedex 05, France
FEDERAL REPUBLIC OF GERMANY	Pergamon Press GmbH, 6242 Kronberg-Taunus, Hammerweg 6, Federal Republic of Germany

---

British Library Cataloguing in Publication Data

European Conference on Coal Liquid Mixtures (1st: 1983: Cheltenham)
Coal liquid mixtures – (Institution of Chemical Engineers symposium series, ISSN 0307'0492;83)
1. Coal liquefaction – Congresses
I. Title   II. Series
662'.662   TP352

ISBN 0-08-031397-3

Printed & bound in Great Britain by A. Wheaton & Co. Ltd., Exeter

Preface

Following the escalation of fuel oil prices the market for coal in industry is expanding and with this expansion there is a growing interest in new methods for the combustion of coal. The use of coal as a fluid mixture with oil, water or alcohol has great potential in terms of user convenience and environmental acceptability and considerable attention is being paid in many parts of the world to the development of the fuel as a direct replacement for heavy petroleum oil. As with all new technologies many problems have to be solved before the fuel is accepted commercially and the authors of papers on all aspects of the subject have not been reluctant in directing attention to the problem areas. Nevertheless a considerable amount of confidence in the new fuels was also expressed and these problems are being tackled with enthusiasm.

The Organisers intend that this will be the first of many conferences on the topic and hope to hold the Second European Conference in September 1985.

Contents

Commercial Applications and Economics

* Three late papers appear out of sequence at the end of the volume.

HISTORY AND DEVELOPMENT OF COAL-LIQUID MIXTURES

Daniel Bienstock[1] and Oliver K. Foo[2]

Coal-liquid mixtures (CLM) has emerged as a viable alternative fuel to displace fuel oil and/or gas in both the utility and industrial sectors. Although fuel oil costs are presently low with oil supplies abundant, the concern for depending on a fuel with an uncertain future in price and availability, especially for importing countries, makes the CLM technology very attractive. The history and development of CLM technology is presented, focusing on coal-oil and coal-water developments. The status of CLM technology is assessed and the major CLM activities in various countries are described.

INTRODUCTION

Coal-liquid mixtures (CLM), particularly coal-water mixture (CWM), has emerged as a viable alternative fuel to displace fuel oil and/or gas in both the utility and industrial sectors. Although fuel oil costs are presently low with oil supplies abundant, the concern for depending on a fuel with an uncertain future in price and availability, especially for importing countries, makes the CLM technology very attractive.

This paper presents the history and development of CLM technology. Although it focuses on coal-oil mixture (COM) and CWM developments, CLM technology has been broadened to include coal-methanol mixture, solvent refined coal-oil mixture, petroleum coke-oil mixture, and other solid fuel-liquid mixtures. The status of CLM technology is assessed and the major CLM activities in various countries are described.

HISTORY OF COM

The history of COM development was reviewed by Demeter,[1] Morrison,[2] and in 1981 by Bienstock and Jamgochian.[3] Significant events in COM development are briefly described.

Significant Early Events

There is a tendency for CLM researchers and engineers to focus on COM developments in the last ten years. However, it should be noted that the recent rapid progress on COM was made possible because significant COM research had been done in the past century.

[1]Pittsburgh Energy Technology Center; [2]The Mitre Corporation

The earliest patent on COM is about 100 years old. It was recognized at that time that oil could be mixed with small coal particles to form a collodial fuel which would retain many advantages of oil, such as ease of storage, handling, and combustion. During World War I, this coal-oil fuel was evaluated for use in submarines. The work was discontinued when the war ended.

During the 1930s, development of coal-oil fuel was renewed in England, Germany, and Japan. The English conducted successful tests in locomotives, and in a Cunard liner on a round-trip between Liverpool and New York. This early research was reviewed by Manning and Taylor[4] and by Schroeder.[5]

COM research started in the United States in the 1940s. The reason for the revival of interest in COM research was the realization of dwindling domestic petroleum resources at that time. The technical investigations, such as coal grinding, mixing, stability, hydraulic transport, atomization, and combustion were summarized by Jonnard.[6] COM technology in the 1940s reached an advanced level and the technical data generated at that time on COM rheology, stability, additive evaluation, and combustion are still useful today. As the cost of the oil was only 20 percent higher than that of coal on a caloric basis, it was not economically profitable to justify the use of COM.

In the 1960s, COM was evaluated as a method for reducing blast furnace iron-making costs. Successful short-term COM tests were conducted by the major industrial nations.[7-9] However, these efforts were aborted when low-cost, low-sulfur, Middle East oil became available. One interesting development by Rudzki[7] is worth mentioning. Rudzki used a fluid energy mill to grind the coal to an average 10-micron diameter. Good COM fluidity and stability were achieved by mixing these ultrafine coal particles with oil without using chemical additives. The high grinding cost was a drawback for this ultrafine coal process.

In 1974, one year after the oil crisis of 1973, the General Motors Corporation (GM) initiated a COM project which was funded and supported by 26 U.S. companies.[10] The U.S. Department of Energy (DOE) and the Electric Power Research Institute (EPRI) joined in the support of this project in 1975. The most significant achievement was the discovery of low-cost but effective chemical additives which made COM conversion economically attractive to users. The 750-hour COM firing in an oil-designed package boiler reaffirmed that COM was a viable fuel. Since then, COM programs spread with revitalized interest worldwide.

In 1976, the DOE solicited a competitive procurement from industry for large-scale COM demonstrations in utility and industrial applications, on a cost-shared basis. As a result, New England Power Service Company (NEPSCO) was selected to convert an 80-MW oil-fired but coal-designed utility boiler to COM-firing, and Interlake Inc. was selected to demonstrate COM injection in a 1,200-tons of hot metal/day blast furnace. In addition, an extensive base technology program on COM fuel preparation, atomization, combustion, and pollution control was carried out at the DOE in-house research facilities at the Pittsburgh Energy Technology Center (PETC) and by DOE contractors.

In 1972, the Canadian Combustion Research Laboratory conducted a limited evaluation of the combustion characteristics of a coal in a No. 2 fuel oil suspension in a pilot-scale test furnace. The results of this work were presented at an industry/government seminar to stimulate interest in COM technology.[11] At that time, there was little interest in Canada and the COM program was subsequently discontinued. After the oil crisis a COM program was initiated in 1976 to encourage the use of COM in place of premium fuels.

Energy, Mines and Resources of Canada and the New Brunswick Electric Power Commission jointly funded a three-phase, COM test program. The initial phase of the program, which began in October 1977, was a short test firing of a low-coal concentration COM in a 12.5-MW utility boiler originally designed for coal firing. Three other projects were funded to study physical properties of COM, COM combustion, and equipment development.

After the oil embargo, the Japanese government decided to secure diverse energy supplies to reduce its dependence on petroleum. Under this new policy, a government-subsidized COM program was initiated in 1976. The Electric Power Development Company (EPDC) was the lead organization in charge of a nationwide COM program which converted base technology research and full-scale COM demonstrations in blast furnaces and utility boilers. EPDC designed, constructed, and operated a COM pilot plant and provided technical guidance to industrial participants.

In May 1978, DOE sponsored the First International Symposium on Coal-Oil Mixture Combustion. The purpose of this symposium was to bring together the international community of users, manufacturers, and researchers interested in and active in COM to facilitate the transfer of technology and know-how necessary to hasten COM commercialization. In five years, the annual COM symposium had been expanded to include coal-water mixtures, coal-alcohol mixtures, and other forms of slurry fuels. The fifth symposium, in April of this year, attracted more than 700 attendees from major industrial and developing nations. The convening of the First European Conference on Coal-Liquid Mixture shows that there is a need for a new forum for the exchange of ideas due to growing interest in CLM in Europe.

Low-Cost Additives in the 1970s

Two different approaches had been used to achieve reasonable COM stability when GM's demonstration program was started in 1976. One approach, which was to grind the coal to micron size to reduce the coal settling rate, required a very-high energy consumption. The other approach was to add a small amount of stabilizers, such as an emulsifier, a gel-forming agent, or a surfactant to the mixture. The stabilizers were sensitive to the coal types and oil characteristics and were expensive. In 1977, Metzger indicated that it was impractical and unnecessary to produce 100-percent stability by using expensive stabilizers.[12] It was only necessary to produce adequate stability for handling and storing by using a low-cost chemical additive with the aid of mechanical agitation to maintain an homogeneous mixture.

Additive costs should not be greater than the potential fuel cost savings from COM conversion. COM replaces only a portion of the energy which is originally provided by oil. A 50 wt% of coal in COM replaces only about 40 percent of the energy; the remaining 60 percent of the energy is still supplied by oil. Therefore, the fuel price differential between COM and oil is about 40 percent of the price differential between coal and oil. In 1977, the costs of U.S. coal and oil averaged $1.30/10^6$ Btu and $2.20, respectively. The price differential between COM and oil was only $0.36/$10^6$ Btu, which had to cover the COM processing cost, the additional transportation cost, the incremental boiler operating and maintenance costs, capital charges for conversion, and a profitable return on investment. In 1976, additives used during early COM testing were quoted at $1.00/$10^6$ Btu. Obviously, it was imperative to reduce the cost of additives to a small fraction of that amount.

An intensive search for low-cost additives was undertaken by COM researchers.[13] In 1977, GM selected a $0.09/$10^6$ Btu additive for the 500-hour COM demonstration test. COM stability was monitored by measuring coal content of randomly collected samples from the storage tank and the burner line. A small amount of coal settling in the storage tank was measured. However, this instability did not cause any major operational problems. The era of low-cost additive thus began and a major hurdle to COM commercialization was finally overcome.

COM EXPERIENCE

COM experience in utility and industrial boilers and blast furnace fuel injection is summarized in the following sections. Experience now being acquired with COM in process heaters is usually not available in the open literature because of the competitive nature of the process industries.

Utility Boilers

Table 1 summarizes experience with COM combustion in utility boilers. An important observation revealed by this table is that, despite the potentially more favorable economics of coal-water mixture, there are still several important projects aimed at converting utility boilers to COM. In Japan, the Electric Power Development Co. (EPDC) plans to convert an oil-designed 350 MWe boiler to COM during the 1984 to 1985 time period,[14] and Tokyo Electric Power plans to convert at Yokosuka two 265 MWe units to COM in 1984.[14] Whether these full scale COM projects will continue to multiply will depend on COM and CWM technology development in the next few years. Potential trends in COM fuel development may make it a more competitive fuel for utility boiler application. These are the possibilities for higher coal loading, 60 wt%, and the use of coal cleaning to reduce ash. For example, EPDC in Japan plans to burn a deashed COM in an oil-designed boiler.

A three-phase COM demonstration project was conducted at a 10-MWe utility boiler at Chatham Station, New Brunswick, from 1977 to 1980 by the New Brunswick Electric Power Commission (Canada) to find a way to burn local high-sulfur (7 percent) and high-ash (20 to 30 percent) coals economically in existing oil-fired boilers.[20,21,22] Significant wear on existing pumps and burner tips was encountered during the initial phase. However, effective COM burning in an utility boiler was successfully demonstrated. During the second and third phases, coal concentration was increased to 40-50 percent by weight. Meanwhile, a spherical oil agglomeration process was used to reduce ash and sulfur in coal to 65 percent and 50 percent, respectively. More than 1,000 hours of COM firing were conducted during Phase II and Phase III test periods. Satisfactory boiler performance on COM was demonstrated. Specially designed burner tips and hardened materials overcame erosion problems. The work at Chatham Station had been expected to lead to a COM demonstration in a larger utility boiler

TABLE 1

COM UTILITY BOILER DEMONSTRATIONS

Country	Power Station	MWe	Design Fuel	Test Period	Reference
Canada	Chatham Station	10	coal	1977 - 1980	20, 21, 22
China	Yang Shupu Station	20	oil/coal	1981 - 1983	23
	Anshan Iron and Steel Company	20	coal	1981 - 1983	41
Japan	Takehara Unit 1	250	coal	1981 - 1982	14, 24
	Takehara Unit 2	350	oil	planned	24
	Yokosuka (2 units)	265	N.A.	planned	14
Spain	Almeria Plant	30	coal	1981 - 1982	25
Sweden	Uppsala, Kraftvarma, Cogeneration Plant (315 MW thermal for heat)	200	oil	1980	26
United Kingdom	Padiham Power Station	120	coal	1981	19
United States	Crystal River Station	383	coal	1977 - 1978	17
	Salem Harbor Station	80	coal	1979 - 1981	18
	Sanford Station	400	oil	1980 - 1981	16
	P.L. Bartow Unit 1	120	oil/coal	1982 - current	15

N.A. = Not available.

in the 100-MWe category, but rapid development of coal-water mixture (CWM) technology and deflated oil prices shifted the Canadian research direction from COM to CWM.

The Chinese have made significant progress in COM development since their COM program was started in 1979. In 1981, COM demonstrations were conducted in two 20-MWe utility boilers.[23,41] Boiler performance, burners, pumps, and instruments were carefully evaluated during a year-long test period. Burner plugging was a major problem, which was attributed to the high-fibrous content of the coal.

The Japanese industries, under the leadership from EPDC, have carried out a comprehensive COM R&D program since 1976.[14,24] COM demonstration tests in a 250-MWe boiler at Takehara Station in 1982 signified that COM technology in Japan had advanced to commercialization. Valuable technical data on COM preparation plants and boiler tests were accumulated during an 18-month test period. These data will be utilized in the design and construction of a 900,000-ton/yr COM plant at Onahama scheduled for operation in November 1984 and a deashed COM demonstration in a 350-MWe oil-designed boiler at Takehara Station.[24] Tokyo Electric Company plans to convert two utility boilers (265 MWe each) from fuel oil to COM in 1984.

In a response to the Spanish government's demand to reduce drastically the oil consumption in utility boilers, Spanish utilities began in 1980 to evaluate COM as one of the alternate fuels to displace oil.[25] A 30-MWe coal-designed boiler at Almeria Station was selected for COM demonstration tests and the only boiler modification needed was the installation of four Verloop burners. Approximately 200,000 gallons of COM, 23 to 50 wt% of coal, were mixed and burned. No chemical additives were used, as the COM was always burned immediately after mixing. Data on COM preparation systems and boiler operations showed that COM can definitely and realistically be extended to large utility boiler applications.

The main market for COM in Sweden is in small- and medium-size industrial and district-heating boilers.[26] A 200-MWe cogeneration plant designed for oil was selected for investigation of the technical and economic feasibility of COM conversion. An on-site preparation plant with 3 ton/hr capacity produced a COM with a moderately cleaned coal. A series of 8-hour tests was run in 1980, burning 1,000 M^3 of COM with 35 percent coal. Boiler load was limited to 50 percent, because of unforeseen, low-load demand.

In 1981, the Central Electricity Generating Board, in cooperation with the British Petroleum Company (BP), carried out a four-day combustion trial to evaluate the performance of BP's Coal-Oil Dispersion (COD) in a 120-MW boiler rebuilt for oil-firing.[19] It represented the first COM trial in an operational power station in Europe. The existing fuel oil handling and firing equipment was not modified for COD firing. The short trials showed that the use of BP's COD may be feasible in large water-tube boilers.

In 1980, the Florida Power and Light Company (FPLC) constructed an on-site COM preparation facility and converted a 400-MWe oil-designed boiler to COM firing in six months.[16] Long-term, continuous boiler operation with virtually no derating was achieved with a 42-percent-coal COM. The significance of this project was that it was the first COM conversion of an oil-designed utility boiler, which is the largest segment of the potential

COM market in the United States. The FPLC project, which was funded by the utility, signified the beginning of an era for COM commercialization.

The first permanent COM conversion of a utility boiler is now about to enter its second year operation. Florida Power Corporation (FPC) converted the 120-MWe Bartow Unit No. 1 to COM, under a five-year COM supply agreement with COMCO.[15] The burning of COM in the unit has been highly successful. The experience gained from this conversion will aid the possible conversion of other large units in Florida Power Corporation's system.

Two earlier COM utility boiler conversions in the United States were conducted by FPC in 1977[17] and by New England Power Service Co. (NEPSCO) in 1981.[18] Satisfactory COM experience in these two coal-designed boilers has been reported.

Blast Furnace

Recent tests on COM fuel injection into blast furnaces are summarized in Table 2. In Japan, Sumitomo Metal Industries, Ltd. operated a 20-TPH COM preparation plant and injected COM into all tuyeres of a large modern blast furnace continuously for three months.[27] In the United Kingdom, British Steel Corporation conducted a short feasibility trial of COM injection in 1979.[28] In the United States, Interlake, Inc., under a cost-sharing contract from DOE, showed that COM injection into a blast furnace can save fuel oil without affecting the hot metal quality.[29] Republic Steel Corporation pursued with its own funds the benefits of COM injection.[30]

The COM that was used for blast furnace injection was significantly different from the COM used in boiler testing. Instead of a utility grind, (70 percent less than 200 mesh), a coarse grind (top size ¼-inch) was used, because of high-intensity combustion and longer residence times in the combustion zone of a blast furnace. The lance that was used to inject the COM into the tuyeres had a large internal diameter (3/8-inch) which allowed the use of a COM that would clog most burners.

Results from these tests were decisively favorable to COM conversion. Blast furnace operations and hot metal quality were not affected by COM injection; replacement ratio of COM to coke was about 80 percent of that for fuel oil. Uniform COM injection into all tuyeres was achieved with minimum modification of the existing oil-injection system. The steel industry has demonstrated that COM is a technically viable alternative to the injection of fuel oil in the blast furnace and may decide to commercialize this technology when there is an upturn in steel demand and fuel oil prices start to escalate.

Industrial Boilers

The uncertainty in fuel oil supplies and high costs in the late 1970's caused many industries to consider options for converting oil-fired boilers to coal or to COM. COM tests were conducted to determine economic and technical feasibility by COM suppliers and their potential customers. The technical and cost data were usually not available to the public because of the highly competitive COM market. Given the current economic conditions and the prices of coal and oil, COM conversion appears attractive only to a limited number of industrial users. COM was burned successfully in

TABLE 2

COM BLAST FURNACE APPLICATIONS

Country	Blast Furnace	COM Injection	Test Period	Reference
Japan	Kashima Works			
	No. 3 Blast Furnace 10,100 THM/D	COM injection through all tuyeres	1981 – 1982	27
United Kingdom	British Steel Corp. Middlesbrough No. 3 Blast Furnace 2,000 THM/D	One out of 20 tuyeres	1979	28
United States	Interlake Inc. Blast Furnace B 1,300 THM/D	3 out or 9 tuyeres 6 out of 9 tuyeres at 13 GPM (85 lb/NTHM)	1980 – 1982	29
	Republic Steel Cleveland District No. 5 Blast Furnace	18 tuyeres 27 GPM	1980 – 1982	30

THM/D = Tons of Hot Metal Per Day

8

industrial boilers, as shown in Table 3, and these projects form the basis for future commercial applications.[10,28,31-40]

Boiler efficiencies achieved with COM firing in industrial boilers approach that obtained in residual oil firing. Tests performed by PETC in a 700-Hp industrial boiler and by Ontario Research Foundation in a package hot water boiler have demonstrated successful COM burning.[36,31] Heat losses due to slightly lower carbon utilization relative to residual oil are apparently compensated by improved heat transfer when firing COM, provided that tube surfaces are kept clean by sootblowers.

PETC found during one 500-hour test in the 700-Hp boiler was that ash accumulated inside the furnace and slagging occurring and flame impingement on the accumulated ash layer.[35,36] This problem was overcome by using a pair of retractable sootblowers inserted from two rear viewports of the furnace to entrain the ash into the flue gas stream entering the convective section. The convective section had a sootblower installed in its manufacture.

Boiler derating needs to be addressed further because it is an important area which has insufficient data. Occidental Petroleum Corp. limited its 135,000 pph oil-designed boiler to 79 percent load due to deposits on the superheater screen tubes.[38] The furnace exit gas temperature was 2240°F which was 80°F higher than the initial deformation temperature of the parent coal (2160°F) in the COM. A coal with a higher ash-fusion temperature will eliminate fouling and therefore reduce boiler derating. PETC used a Pittsburgh seam coal with a high initial deformation temperature of 2350°F.[35] No boiler derating was attributed to slag formation during PETC COM combustion trials. However, this does not imply there is no boiler derating when firing COM with a high ash-fusion coal. Boilers will be derated to avoid tube erosion/corrosion by high-flue gas velocity and high-ash content.

COM PREPARATION PLANTS

Over the years, a variety of COM preparation processes have been developed to produce COM, usually with locally available coals and oils. There are significant differences in composition, coal particle size distribution, and use of additives. Nevertheless, these preparations have demonstrated adequate stability and good burning characteristics.

Major COM preparation plants in the world are listed in Table 4, grouped together according to their grinding and mixing process design: dry-grinding and mixing, wet-grinding, ultrafine coal grinding, beneficiated coal, and coal-oil dispersion. Special features and advantages of each process are briefly discussed.

Dry-Grinding and Mixing

A dry-grinding and mixing COM preparation process involves mixing of pulverized coal in heated oil, usually with additives. This approach is sufficiently simple so that Florida Power and Light Company (FPLC) was able to design, build, and start to operate a 10,000 barrels per day COM preparation plant in six months.[16] The dry-grinding and mixing process was adopted by Yang Shupu Power Plant and Anshan Iron and Steel Company in China, Blanzy Station in France, Almeria Power Station in Spain, and NEPSCO, PETC, FPLC, COM Energy Inc., Coaliquids Inc., and Allied Oil

TABLE 3

COM INDUSTRIAL BOILER EXPERIENCE

Country	Project	Capacity	Design Fuel	Test Period	Reference
Canada	Ontario Research Foundation	1.2 MWth	oil	1980 – 1981	31
United Kingdom	British Petroleum Inc.	Various boilers in U.K. and and other countries	Mostly oil	1979 – 1981	32
	Shell Research Ltd.	60,000 pph	coal	1976 – 1977	28
United States	PETC/DOE	100 Hp 700 Hp	oil oil	1976 – 1981 1977 – 1981	1,34 35,36
	Coaliquids, Inc. General Motors Occidental Petroleum Ohio Univ. at Athens	65,000 pph 120,000 pph 135,000 pph 150 Hp	coal oil oil oil	1980 1977 1981 Ongoing Since 1981	37 10 38 39
	Adelphi Energy Center	350 Hp	oil	1978 – 1981	40

TABLE 4

COM PREPARATION PLANTS

Process	Country	Project	Capacity	Additive	Reference
Dry-Grinding and Mixing	China	Yang Shupu Power Plant	10 TPH	No	23
	France	Anshan Iron and Steel Co.	12 TPH	Yes/No	41
	Spain	Bianzy Station	8 TPH	Yes	42
		Almeria Power Station	N.A.	No	25
	U.S.	New England Power Service	6,440 bpd	Yes	18
		PETC	100 bpd	Yes	34
		Florida Power and Light Co.	10,000 bpd	Yes	16
		COM Energy Inc.	3,000 bpd	Yes	38
		Coaliquids Inc.	1,200 bpd	Yes	37
		Allied Oil Co.*	1,500 bpd	Yes	39
Wet-Grinding	Japan	EPDC - Takehara Unit 1	10 TPH	Yes	24
		Sumitomo Metal Industries	20 TPH	Yes	27
	U.K.	Shell International Pet.	N.A.	No	28
	U.S.	Interlake Inc.	1,540 bpd	Yes	29
		Allied Oil Co.*	1,500 bpd	Yes	39
Ultrafine Coal Grinding	Canada	General Comminution Inc.	3 TPH	No	43
	U.S.	COMCO	3,500 bpd	No	44
		ERGON	5,000 bpd	No	45
Beneficiated Coal	Canada	Scotia Liquicoal Ltd.	3 TPH	No	31
	Japan	EPDC - Takehara Unit 2	10 TPH	N.A.	14
	Sweden	Uppsala Kraftvarme AB	3 TPH	No	26
	U.S.	Gulf & Western Ind.	1.5 TPH	No	46
Coal-Oil Dispersion	U.K.	British Petroleum	100,000 TPY (planned)	No	19

*Both dry-grinding and mixing and wet-grinding can be done at Allied Oil Co. COM preparation plant.
Nomenclature: TPH = tons per hour, N.A. = not available, bpd = barrels per day, TPY = tons per year.

Company in the United States.[23,41,42,25,18,34,16,38,37,39] The differences among these COM preparation processes are with regard to the selection of different types of pulverizers, agitators, or additives.

Coaliquids, Inc., produces a COM by mixing pulverized coal, from three cage mills in series, oil and water into a slurry, which then passes through an ultrasonic dispersion system.[37]

Wet-Grinding

The technique of wet-grinding was used by EPDC at Takahara Station Unit 1 and Sumitomo Metal Industries Ltd. in Japan, Shell International Petroleum Co. in the United Kingdom, and Interlake Inc. and Allied Oil Company in the United States.[24,27,29,39]

A wet-grinding process developed by the Electric Power Development Co. (EPDC) in a pilot plant involves coal and oil pulverized and mixed simultaneously in a ball mill. An additive, less than 1 percent by weight, is introduced after pulverization and is further mixed by a homogenizer. It has been successfully scaled to a 10-ton/hr plant at the Takehara Power Station and a 20 ton/hr plant at Sumitomo Metal Industries, Ltd. A 900,000-ton/yr plant by Japan COM Company, scheduled for 1984 start-up, is also based on the EPDC wet-grinding process design.

Shell International Petroleum Co. has developed the COLLOIL process to suspend finely ground coal in oil. The crushed coal is mixed with surfactant type additives and 10 to 20 wt% water. The wetted coal and additives are pulverized in a ball mill and a circular vibrating sieve is used to remove oversize particles.

A wet-grinding circuit developed by Interlake Inc. consisted of a disperser followed by a grinding pump for size reduction of large particles that escape from the disperser. Both disperser and grinding pump experienced wear. A COM with coarse coal particles at 70 percent minus 25 mesh was produced for blast furnace injection.

Wet-grinding combines the grinding and mixing steps into one operation such that handling and storing of pulverized coal is not required. Currently, wet-grinding of coal is regarded as safer than dry-grinding.

Ultrafine Coal Grinding

Ultrafine coal with an average coal particle size of 10 microns was used to make COM by General Communition Inc. in Canada and COMCO and ERGON in the United States.[43,44,45] Advantages of ultrafine coal are stability without the use of additives, improved combustion efficiency, and less ash deposition in the boiler. These advantages help to offset the higher cost of ultrafine coal grinding.

General Communition, Inc. operates a Szego mill which grinds coal in oil to very small size, rapidly and efficiently, with a relatively narrow range of particle size distribution. Coal loadings of 55 percent and 51 percent were achieved in No. 2 oil and No. 4 oil, respectively.

COMCO's patented COM process consists of a two-stage crushing and pulverizing circuit in which coal is reduced to 95 percent minus 325 mesh. The pulverized coal in oil is then metered and homogenized in a dispersion tank. ERGON uses a fluid energy mill to pulverize coal to ultrafine coal particles of 98 percent minus 20 microns. The pulverized coal passes through a venturi section and mixes with oil in a mixing tank. COM with 50 percent coal loading was produced.

Beneficiated Coal

In the preparation of beneficiated COM, additional steps are required for the removal of ash and pyritic sulfur. The use of beneficiated coal will add a premium to the price of COM, but will reduce downstream costs when the cleaned COM is burned in a boiler originally designed for oil-firing. Several organizations in different countries are currently developing integrated oil agglomeration and COM preparation processes. A Spherical Agglomeration Process is being developed by the National Research Council of Canada and the University of Waterloo.[31]

Scotia Liquicoal Ltd. operates a 6 ton/hr COM plant in Dartmouth, Nova Scotia and utilizes two developing Canadian technologies. Coal is beneficiated by using the spherical agglomeration process after being pulverized to 15 microns mean diameter in a Szego mill. Ultrasonic emulsification is used to stabilize the mixture.

EPDC constructed a 2-ton/hr COM oil-agglomeration pilot plant.[47] The deashing ratio (percent of ash removal, initially at 38 percent), is increased to 50 percent by various techniques such as changing agitator speeds and additive use. A 10-ton/hr COM oil-agglomeration plant will be built at Takehara Station for an 18-month test in an oil-designed boiler.

A flotation process to remove coal ash before final mixing of coal with fuel oil was adopted by Uppsala Kraftvarme AB (Sweden) and achieved a 33 percent ash reduction.

Gulf & Western Industries Inc. (G&W) has developed a clean coal COM or CWM preparation process.[46] An automated 1-ton/hr coal pilot plant has been operated for almost three years. G&W employs a patented coal beneficiation process in which chemicals are used to condition the coal before it enters the beneficiation cells. The diluted slurry is sprayed through a nozzle onto the surface of a beneficiation tank. The force of the spraying action applies intense shear to the flocculates and separates the mineral matter from the treated coal particles. The freed coal particles are forced through the air and stay on the surface. The hydrophilic and heavier mineral matter travels down into the water and stays there. The beneficiated coal is de-watered. Water, chemicals, and de-watered coal are blended in the first mixing tank. Chemicals are added into the second and third mixing tanks in series to achieve thickening and stabilization. Coal loading up to 65 and 75 percent by weight COM and CWM were produced and burned successfully in large test furnaces.

Coal-Oil Dispersion

The British Petroleum Company (BP) has developed a patented coal-oil dispersion (COD) process, and has conducted ten or more successful short-term COD handling and combustion tests.[19] A 100,000-ton/year COD development plant at West Thurrock had been planned for late 1981 but is now delayed.

In the BP approach crushed coals are pulverized in a conventional dry pulverizer to 80 percent through 200 mesh. The pulverized coal is then introduced to a second milling operation together with fuel oil under carefully controlled conditions. During this second phase which is patented, the coal is further reduced in size to below 20 microns. The BP fuel needs no agitiation in storage nor the addition of chemical additives. A wide range of coals from the United Kingdom and other countries was used to produce a stable COD.

STATUS OF COM TECHNOLOGY

The review of the current status of COM technology focuses on three major areas: preparation, combustion and boiler performance.

COM Preparation

The major operations in COM processing are fuel preparation, grinding, mixing, and stabilization. Washed, crushed coals are delivered to the COM processing facility to minimize on-site coal-refuse handling. Bituminous coals with high ash-fusion temperatures have been used in large scale COM processing plants. Sub-bituminous coals, lignites, and anthracites mixed with oil have been tested in laboratories or small-scale test facilities.[36,48]

Residual oil (or No. 6 oil) is commonly used for COM processing because of its low cost. A major difficulty with No. 6 oil for COM is the lack of uniformity among different oils of this grade. The high viscosity of residual oil has limited coal loading in COM primarily to the 50 percent level although higher concentrations have been achieved in some cases. No. 2 or No. 4 oils were used for special COM fuels despite their high costs.[49]

In almost all processes, an additive is considered necessary to improve stability and prevent hard compaction of coal particles. Different types of additives are needed for different coals and oils. In Japan, thousands of additives were evaluated together with various mechanical mixing and stabilizing equipment.[50] In Canada, COM rheological properties and stability testing were studied by Trass of Toronto University and Al Taweel of Technical University of Nova Scotia.[49,51] In the United States, Ekmann of PETC measured extensively stability, settling characteristics, and rheological behaviors of a large number of COMs made of different coals and additives.[13] Rowell designed a sedimentation column which became a measuring standard for COM static stability.[52] Researchers in China, France, Spain, and Israel also performed additive screening and stability evaluation of COMs suited for locally available coals and oils.[53,54]

In grinding and mixing, standard off-the-shelf equipment has been operated with satisfactory results. The characteristics of five different grinding and mixing schemes were discussed in the previous section.

With some reservations, the selected slurry pumps have operated satisfactorily, provided conservative design principles and proper operating and maintenance procedures were implemented and followed. The materials presently used by manufacturers are sometimes subject to erosion. For rotary pumps which can trap coal particles between a moving member and a stationary member, it has been found that one member should be constructed of a resilient elastomer and the other should be constructed of a very hard material. Nevertheless, the expected life of slurry pumps in COM applications may not be as long as can be expected for clean liquid pumps in process industries. Maintenance costs obviously will be higher. Additional pump improvements and testing are necessary to improve performance and achieve high reliability.

COM Combustion

COM combustion tests in the past decade have addressed several important questions on atomization and nozzle erosion. The most encouraging fact is that existing commercial oil burners, with slight modifications, can be used to burn COM in both industrial and utility boilers, and to achieve good atomization and flame stability. Service life of a burner nozzle modified for COM firing is about 1000 hours or longer, which is approaching the normal service life of two to eight months for oil-firing.

There is no clear indication of which type of atomizer is most suitable for COM firing. The high-pressure mechanical atomized burner will have to be replaced by an air-atomized or steam-atomized burner for COM firing.

The external air-atomized burners such as the Forney Verloop burner and the Vortometric burner have large fuel passages and low pressure drop.[18,31] However, the nozzle which imparts swirl to the fuel stream is still subject to erosion caused by coal particles in the fuel. NEPSCO's experience with the Forney Verloop burner showed that nozzle erosion could be minimized by using selected materials.[18]

High-pressure (about 100 psig), steam-atomized burners have been widely used to fire COM. External mixing type burners made by I.C.L. (U.K.) were used at the FPLC Sanford Station COM demonstration tests in 1981.[55] Through a series of unique field modifications and use of more exotic materials, wear rate was reduced. Nozzle tips made of steel with a sprayed application of tungsten carbide chip in a nickel matrix had a service life of more than 500 hours.

Internal mixing such as single orifice type, Y-jet and T-jet nozzles have been tested with varying degrees of success. Y-jet nozzles which are widely used in both utility and industrial boilers for oil-firing were tested for COM firing in Canada, Japan, and the U.S.[31,24,56] Results showed that severe erosion, which occurred at the junction where atomizing steam interacted with COM, can be minimized by using hardened materials. In Canada wear-rate data showed that the use of harder materials such as tungsten carbide significantly reduced burner tip erosion.[32] In Japan, no wear was observed for tips with ceramic junction after continuous burning of 6000 tons of COM for about 1000 hours.[24] This tip is expected to operate for several thousand hours.

At PETC, an off-the-shelf, high-pressure, inside-mix steam-atomized burner was tested.[36] COM viscosity at the burner was 600 SSU while the design value was 100 to 200 SSU. By adjusting atomizing steam flows, pressures, and firing temperatures, satisfactory atomization and combustion were obtained. Burner nozzle wear rate data on 45 wt% coal COM are given in Table 5. By using tungsten-carbide inserts, the wear rate was greatly reduced.

TABLE 5

WEAR RATES OF COM BURNER NOZZLE AT PETC

Boiler/Burner	Coal Slurry Composition	Coal Slurry Flow Rate, Lb/Min	Burner Nozzle Material	Wear Rate Inch/Hour
700-hp/Coen 8-5/32" Holes	COM 40% Pittsburgh seam coal (80% minus 200 mesh)	28.7	Hardened Tool Steel Cap	0.00034
		28.7	Case-Hardened Stainless Steel (1440C SS) cap	0.00035
		28.7	Carbon Steel Cap with Inconel X-750 inserts	0.00056
		28.7	Carbon Steel Cap with Tungsten Carbide inserts	0.00000 (40 hours of operation)
	COM 40% Pittsburgh seam coal (90% minus 325 mesh to 60% minus 200 mesh)	27-29	Carbon Steel Cap with Tungsten Carbide inserts	0.000035

Reference 36.

16

In addition, a high-pressure, steam-atomized internal-mix nozzle can achieve good combustion and carbon burnout with a turndown ratio of 5:1. Atomizer plugging problems can be controlled by using large orifice openings and operating at a high COM temperature.

Other burners having potential for COM firing are the ultrasonic and vortometric types. Organizations in Canada, China, and Germany reported success with the ultrasonic type burners.[32,57,58] The ultrasonic burner developed by DUMAG, Vienna, Austria, was used to burn COM in a rotary kiln in Germany.[58] The fundamental principle is based on the generation of an ultrasonic field by an atomizing gas which passes through a resonance chamber at high velocity. The fuel flows out of the nozzle and is atomized in the ultrasonic field. Because of the low fuel velocity, good flame stability is achieved. The advantages of an ultrasonic burner are numerous. It can handle a high-viscosity fuel and the required fuel pressure and pressure drop are very low. When using a T-jet nozzle, clogging and nozzle erosion was not observed for COM combustion. The flame was stable even at partial loads. The shortcomings of the ultrasonic burner is that a high volume of combustion air is required.

Preliminary tests results on the vortometric burner reported by the Ontario Research Foundation, showed that wear did not appear to be progressive after a few hours of operation.[32]

BOILER PERFORMANCE

Boiler Efficiency and Combustion Efficiency

At the Sanford Station COM demonstration, FPLC showed that the COM boiler efficiencies tracked the oil-fired efficiencies over the load range, being highest at 100 MW and lowest at 400 MWe.[59] The boiler efficiencies obtained with COM were 0.7 to 1.5 percent lower than those obtained with 100 percent oil. The slightly lower efficiencies were attributed to higher combustible losses ranging from 0.3 percent to 0.8 percent over the load range, with the higher excess-air requirements to burn COM. Similar findings were observed in other COM demonstrations in both utility and industrial boilers. In general, good boiler thermal efficiencies and combustion efficiencies were achieved even though atomization with the burner used was not considered optimum. The slightly lower efficiencies were not considered by utilities to present an operational limit on the boiler.

Ash Deposition and Slagging

FPLC found that ash deposition on the furnace walls became significant after 30 wt% coal concentration in the COM was reached. To overcome the accumulation of ash, wall blowers were added to the furnace walls. The ash distribution was about 20 percent bottom ash and 80 percent fly ash. The nature of the ash varied, depending on the location within the boiler. In the furnace, ash varied from a light powdery deposit to heavy layers of two to five inches thick. These heavy layers resulted from incomplete combustion prior to deposition on the furnace wall. Incomplete combustion was attributed to limited capacity of combustion air at the higher coal concentrations. Pressure loss measurements in the convection passes indicated little additional fouling in comparison to oil firing.[59]

FPC inspected the boiler at two and seven months after commencing COM firing and found slag on the front wall around the burner.[60]   Ash deposits were found on the secondary superheater and all tubes in the convection pass.   Slagging and fouling can be controlled by the sootblowers.   Sootblowing is scheduled to minimize large swings in furnace exit gas temperatures and main and reheat steam temperatures.

Ash deposits on test probes and superheaters during tests conducted by CEGB indicated that fluxing of coal ash by oil ash contributed significantly to fusion of ash particles on the superheater tubes.[19]   It was concluded that frequent sootblowing would be required to prevent serious fouling of the superheater tubes.

Derating   •

Derating will vary with the type of oil-designed boiler. Derating of boilers with coal-firing capability is very small, as indicated by the FPC experience at Bartow Station.[60]   The maximum load carrying capacity of this unit was essentially unchanged.   The rating before conversion was 115-MWe which has been attained repeatedly after COM conversion and even exceeded at times.

The only COM experience with an oil-designed boiler was obtained by FPLC on Sanford Station Unit No. 4.[59]   As the percentage of coal was increased to 50 percent, handling of COM became more difficult.   It was sometimes necessary to derate the unit from 400 MWe to 355 MWe to keep temperatures within the desired limit, due to inadequate fan capacity for combustion air. However, no derating was required with a 42 percent coal in COM.

There is no COM experience with a compact-type oil-designed utility boiler. Engineering analyses performed by B&W indicated that a derating of 34 percent from maximum continuous rating (MCR) could result for the particular coal assumed.[61]   No major internal modification was required although sootblowers would have to be added in the furnace, convection bank, and air heater. The major constraints to increasing the load of this boiler were the furnace exit gas temperature as it related to the slagging potential of the fuel and the secondary superheater tube spacings which limit flue gas velocity. Modification of heat-transfer surfaces and use of high ash-fusion coal would allow higher loads to be reached, or a lower derating. More conservative estimates of derating have been made for this type of boiler indicating differences in designs and engineering estimates. In the final analysis, testing may be required to resolve questions of derating with regard to specific boilers.

Emissions

FPLC measured the quantity of fly ash and confirmed that the distribution of COM fly ash is similar to the distribution normal to pulverized coal firing.[59]   The approximate split was 20 percent to furnace ash hoppers with the balance of 80 percent to fly ash. NEPSCO measured particulate loadings and opacity and showed that both quantities increased with increased coal concentration in the COM.[18]   Measurements of ash loadings in the 700-Hp industrial boiler at PETC showed similar results. Flue gas and ash analyses by PETC (Table 6), showed that particulate loadings are higher with a coarser coal particle distribution in the COM.[36]   A

TABLE 6

EMISSIONS AND FURNACE ASH DEPOSITION,
PETC 700-HP BOILER COAL-OIL MIXTURE TESTS

Coal Type	Montana Subbituminous	Pittsburgh Seam		Illinois No. 6
Test Number	1	2	3	4
Flue Gas Analysis				
O_2 (%)	3.0	3.0	3.0	2.9
CO_2 (%)	14.2	14.1	14.2	14.1
CO (ppm)	52	446	272	32
SO_2 (ppm)	426	977	920	1,490
(lb/MBtu)	0.80	1.80	1.72	2.81
NO_x (ppm)	555	396	460	447
(lb/MBtu)	0.73	0.52	0.62	0.60
THC (ppm)	0.3	0.1	0.0	0.0
Particulate Emissions (lb/hr)	62	132	69	63
Avg. Particle Size (μ)	29.8	51.5	24.7	44.7
Analysis (%)				
Carbon	5.8	36.2	19.7	8.6
Sulfur	0.6	0.7	0.6	0.8
Ash	91.1	60.9	78.6	90.1
Furnace Deposits (lb/hr)	6.5	12.8	8.8	8.6
Analysis (%)				
Carbon	3.7	13.4	12.7	33.9
Sulfur	1.8	0.7	0.4	1.9
Ash	96.1	84.2	87.0	59.2

*Total hydrocarbons.

Reference 36.

baghouse was installed at PETC to collect particulates from its 700-Hp boiler and has operated satisfactorily on COM and CWM.

In 1980, a pilot electrostatic precipitator was installed at FPLC Sanford Station and 93 performance tests were conducted. Test data demonstrated that the COM ash can be effectively collected in an electrostatic precipitator. To meet the particulate emissions regulations, a new electrostatic precipitator designed for COM firing was installed at the FPC Bartow Station.[63]

Table 7 shows that ash collection efficiencies of 98 percent or higher are achieveable and particulate emissions and opacity are within the regulation limits.

Low-sulfur COM can be made to meet SO_2 emission regulations with use of low-sulfur fuel oil and low-sulfur coal. An alternative approach is to use a low-cost, flue gas desulfurization system. PETC investigated the technical feasibility of injecting powdered sorbent into the flue gas stream just prior to the baghouse as a means of removing SO_2.[36] SO_2 removal efficiencies increased with Na_2/S ratio from 1.0 to 2.5 and with baghouse temperatures of 350 to 475°F. Removal efficiencies as high as 90 percent were achieved at the higher Na_2/S ratio but with a lower sorbent utilization. PETC plans to develop techniques to improve sorbent utilization, evaluate sorbents other than bicarbonate, and determine process economics.

Emissions of NO_x from boilers firing COM are influenced by nitrogen content of the fuels, percent coal, and, possibly the presence of water in the COM. Data from the FPLC COM demonstration indicate that NO_x emissions from COM combustion are higher than those from oil combustion.[59] NO_x control technology was evaluated in Japan for COM applications. The effect of low-excess-air operation on NO_x emissions from COM and oil-only combustion is described.[64] Flue gas recirculation was effective in reducing NO_x emissions from COM combustion, particularly when it was used in combination with staged combustion.[65]

FUTURE COM DEVELOPMENT

Three major areas are identified for future COM developments: beneficiated COM, a compact oil-designed boiler demonstration test, and a highly loaded coal COM. A low-ash, low-sulfur COM is still an attractive fuel for utility and industrial users. Oil-agglomeration (Australia, Canada, and Japan) and froth frotation (Sweden and U.S.) processes have been incorporated into the COM preparation process. Preliminary results from the pilot scale tests are promising. However, there is a need for long-term operation of scaled up facilities to improve the effectiveness of beneficiated techniques and to obtain reliability and economic data.

Compact oil-designed boilers are a large share in the potential COM market.[66] COM conversion of this type of boiler has been considered impractical because of high conversion costs and boiler derating. Burning beneficiated COM in a compact, oil-designed boiler may be a solution to these problems. However, test burns are required to prove this concept.

In the past, coal loading in COM usually has been limited to 50 percent by weight because COM viscosity increases rapidly beyond that concentration. Recently, there have been indications that a highly loaded

TABLE 7

PERFORMANCE OF AN ELECTROSTATIC PRECIPITATOR ON COM AT FLORIDA POWER CORPORATION B.L. BARTOW STATION

ESP Mode	Unit Load, MWe	Test	Gr/ACF Inlet	Gr/ACF Outlet	Eff. %	Opacity %
Full Power	112	1	0.3704	0.0040	98.92	2.3
	110	2	0.3704	0.0030	99.19	2.1
	101	3	0.3876	0.0137*	96.47*	2.4 – 12.0
	57	6	0.2956	0.0060	97.97	3.5
	56	7	0.2827	0.0052	98.16	3.0

*60% of collecting area inadvertently tripped during test.

Reference 63.

COM may be possible. During the FPC/Dravo (the predecessor of COMCO) COM tests at the Crystal River Station in 1977, 60 percent by weight of coal in the COM with reasonable viscosity was produced and pumped to the storage barge where it was diluted for combustion. Currently, G&W produces a 65 percent or higher of coal in COM and COWM for small boiler testing.[67] A low-sulfur, low-ash, highly loaded coal COM could become an economical alternative fuel for the future.

HISTORY OF COAL-WATER-MIXTURE DEVELOPMENT

Significant Early Events

The European and U.S. experience in coal-water-mixture (CWM) combustion to 1970 was reviewed by Marnell, Table 8.[68]

In 1961, a full-scale coal-slurry demonstration program was conducted at the Werner Station of the Jersey Central Power and Light Company.[69] The boiler was fired by two 9-ft. cyclone furnaces using coal as a base fuel with oil as a secondary fuel. More than eight thousand tons of coal in a water slurry at 67 to 68 percent solids was fired in the demonstration program. Coal slurry was fired a total of 443 hours with a continuous firing of more than 11 days with the boiler operating on system demand. The boiler operated at 86.5 percent efficiency, a decrease of a four percentage points from coal firing. This decrease in boiler efficiency was due to the water content in the slurry and the higher excess air used during the test. Cyclone furnace turndown with the slurry was comparable to the use of dry coal. Despite this successful CWM demonstration, virtually no further work on CWM combustion was carried out in the U.S. until Atlantic Research Corporation burned CWM in a 1 million Btu per hour test furnace in 1979.

Major early CWM combustion tests were conducted by Germans and Russians in both pilot plants and large boilers. In Germany, rod mills were used for wet-grinding of coal, using the experience of the ore dressing and cement industries as a guide.[70,71] A 60 wt% coal in the CWM, having a coal particle size distribution similar to that required for conventional pulverized coal combustion was prepared, stored, and distributed without using a stabilizer. Mechanical agitators and circulating pumps were used to keep coal particles in suspension.

During a CWM demonstration test in a 200,000 pph coal-designed boiler, stable CWM flame without supplementary oil-firing was achieved after several burner modifications. However, excessive wear rates of rotary atomizers and coal sedimentation in a heat exchanger were experienced. Otherwise, boiler operation on CWM was as smooth as when it was on fuel oil. Later, steam atomizers were used to replace worn-out rotary atomizers and the wear rate was reduced.

The Russian work comprised both theoretical and experimental studies of CWM combustion and rheology.[72-74] Conventional ball mills were used to produce CWM and additives were not used for CWM handling, storage, and combustion. Consequently, the Russian CWM exhibited high viscosity when coal loading was increased to the 60 percent level. At least six known CWM combustion tests in stoker, pulverized coal, and cyclone fired boilers were conducted. Carbon burnouts at 98 percent and higher and boiler efficiencies of 80 percent with excess-air levels at 7 to 12 percent were

TABLE 8

MAJOR COAL-WATER COMBUSTION EXPERIENCE UP TO 1970[a]

Fuel Type	% Water	Slurry Flow, Metric Ton/Hr	Application	Steam Rate, Metric Ton/Hr	Plant/Location
Coal Beneficiation Wastes	50	40	Power Plant Boiler	170	Magnitogorsk Metallur- gical Combine/USSR
Coal Beneficiation Wastes	45	8	LMZ Boiler	40	Anzherskaya Central Power Plant/USSR
Coal Beneficiation Wastes	45	1.5	Spray Above Stoker Fired LMZ Boiler	0.6	Anzherskaya Central Power Plant/USSR
Wet Coal	15-50	2	DKV Boiler	--	Lutuginskaya-Severnaya Mine/USSR
Coal-Water Mixture	40	2	Horizontal Cyclone	--	Zhilevskaya Plant Moscow Oblast/USSR
Coal Waste	40	1	Cyclone	--	Khigmavt/USSR
Coal-Water Mixture	30	29	Twin Cyclone Boiler	220	Werner, South Amboy, New Jersey/USA
Coal-Water Mixture	40	20	Pulverized Coal Boiler	100	Kellerman, Lunar/ Germany

[a]P. Marnell, Reference 68.

obtained; however, the air-preheat temperature had to be maintained at 650° to 800°F to sustain the CWM flame.

The early CWM development demonstrated the technical feasibility of producing a fuel which could be handled and burned like a liquid fuel, in spite of the high-water content (40 percent by weight) as compared with the normal moisture level (10-15 percent by weight) for pulverized coal firing. These early achievements constituted the foundation on which recent CWM research and development activities have evolved.

Highly Loaded Coal in CWM

During the early CWM development, optimization of grinding techniques to produce a controlled coal particle size distribution for maximum coal loading and minimum viscosity was not emphasized. The importance of using various dispersants and stabilizers to reduce viscosity and retard coal settling was generally overlooked. The drive, in recent years, to obtain a highly loaded CWM with acceptable viscosity and stability became the focal point in recent CWM research and development activities.

In 1977, Atlantic Research Corporation (ARC) was contracted to develop, evaluate, and test burn highly loaded coal slurries under the DOE COM program.[75] ARC developed a two-stage grinding technique which employed a dry-grinding hammer mill or ball mill and a wet-grinding ball mill to produce coarse and fine coal particles, respectively. The optimum particle-size distribution was achieved by blending appropriate proportions of coarse and fine coal particles. ARC screened various additives in combination with different sources of coals. Additives were used during grinding and mixing to lower visocsity, promote stability, control slurry acidity, and to prevent bio-degradation. CWM containing up to 70 percent finely pulverized bituminous coals were burned for the first time in the United States in 1979 at ARC's 1.3×10^6 Btu/hr test furnace. Carbon burnouts in the mid-90 percent range were achieved and the combustion efficiencies were comparable to that achieved with pulverized coal. The successful CWM test burn by ARC marked the beginning of an intensive CWM program in the United States.

In Sweden, the Carbogel concept started under government funding at Scaniainventor in 1974.[26] Carbogel is a coal-water dispersion containing 70 percent by weight of upgraded coal (less than 4-percent ash) and 30-percent water and is stabilized with organic additives to create a low-viscosity, storable fuel. Boliden AB and AB Scaniainventor formed a joint company Carbogel AB to develop and market this fuel. A unique feature of the Carbogel process is the inclusion of a modified froth flotation process to remove ash and sulfur. In doing so, the Carbogel process is applicable to a wide range of coals of various ranks and the beneficiated CWM can be used in several types of combustors.

CWM Growth

CWM is a near-term technology intended for the conversion of oil-designed boilers and furnaces to coal. The fuel could be applicable to utility, industrial, and large commercial boilers plus many types of industrial furnaces. The goals of the CWM program are not very much different from those previously established for CWM. Technologies developed for COM are transferrable to CWM in most cases. Consequently, COM programs in Canada, Japan, United Kingdom, and United States were expanded to include CWM and other forms of coal slurry fuels. In 1981, a CWM workshop was coordinated by Pittsburgh Energy Technology Center of U.S. DOE to invite inputs from industries which had been active in COM and CWM as to the direction and R&D needs in CWM. Based on these inputs, a comprehensive CWM program was formulated by DOE. Many new CWM projects on combustion, rheology, burner developments, and instrumentation were initiated.

CWM development was emphasized early by the Swedish government and industries; there are more large-scale CWM projects in Sweden than in any other country on the European continent. Swedish companies have signed license agreements of their CWM processes with major corporations in Canada and the United States and are actively marketing CWM worldwide. Other countries such as China, Italy and Spain have rapidly expanded CWM R&D activities, independently or through international cooperation.

CWM developments have experienced a very rapid growth in the last two years and the projected growth rate is even greater. Pilot CWM plants in Canada, Japan, Italy, Sweden, and the United States, can be readily scaled up for commercial production. CWM combustion tests in both utility and industrial boilers have been conducted and long-term tests are scheduled for the near future.

In 1982, the annual COM symposium in the United States was renamed as the International Symposium on Coal Slurry Combustion to accommodate the increasing interest in CWM with its growing number of publications which exceeded that of COM publications in this conference as well as in the Fifth Symposium in April 1983.

CWM EXPERIENCE

Recent Combustion Trials

Recent combustion trials in Japan, Spain, Sweden, United Kingdom, and the United States are summarized in Table 9.

In Japan, CWM R&D which started in 1980 under the direction of EPDC includes domestic efforts and joint developments between private industries in Japan and other countries. At the beginning, EPDC operated the existing COM pilot plant which was modified slightly to produce batches of CWM to gain experience and data for designing a new CWM pilot plant.[76] CWM was burned in a 1.3 MWt cylindrical water-cooled furnace. Combustion and emissions characteristics of CWM were compared with that of the parent coal. Data generated from the pilot-scale test facilities were analyzed to prepare a large-scale combustion test scheduled in 1984.

In 1981, the Fuyo group which consists of Marubeni Corp., Nippon Kokan K.K., Nippon Oil and Fats Co., and Hitachi Ltd. joined a consortium to fund the research and development activities for Co-Al, a low-viscosity and highly loaded CWM.[77] One hundred tons of Co-Al was transported to Japan and burned in an 80 million Btu/hr vertical, water-cooled test furnace. This was the first large-scale CWM combustion test in Japan and demonstrated that CWM can be burned in a large furnace without supplemental oil or gas firing. CWM handling and combustion characteristics were studied at the same time.

TABLE 9

RECENT CWM COMBUSTION TRIALS

Country	Organization	Test Furnace	Test Period	Reference
Japan	Babcock Hitachi K.K.	80 x 10^6 Btu/hr test furnace	1982	77
	Electric Power Development Co. Ltd.	1.3-MMt Btu/hr Cylinderical test furnace	1982	76
Spain	UNESA	4-MW test furnace	1983 - 1985	78
Sweden	AB Carbogel	Various prototype burners with capacities up to 3.5-MW have been tested	From 1979	26
United States	Atlantic Research	1.3 x 10^6 Btu/hr furnace at ARC, and 4.0 x 10^6 Btu/hr test furnace at Babcock & Wilcox Alliance Research Center	From 1979	80,81
	Slurrytech, Inc.	4.0 x 10^6 Btu/hr test furnace at R&W Alliance Research Center	1980	82
	Gulf & Western Industries, Inc.	80 x 10^6 Btu/hr test furnace at Combustion Engineering, Inc. 70 x 10^6 Btu/hr test furnace at Forney Eng. Co.	1982	83
United Kingdom	NEI International Combustion, Ltd.	5 x 10^6 Btu/hr test furnace	1982 - 1983	79

The Spanish Ministry of Industry has delegated to UNESA, an association of major utility companies, the responsibilities of CLM development.[78] After successfully completing COM demonstration firing at a 30-MW boiler at Almeria, UNESA started its own CWM program. The CWM program includes CWM preparation and combustion. CENIT, a burner manufacturer, is responsible for developing a CWM burner and conducting combustion tests in a 4-MW furnace. CENIT will be complemented by other institutions working on CWM preparation and coal beneficiation. Data will be carefully evaluated prior to full-scale CWM tests in a 40-MW industrial boiler and a 150-MW utility boiler.

Carbogel (Sweden) has been working with burner suppliers to develop burners for its CWM. Various burner prototypes to 3.5-MW capacity have been tested in open-air demonstration firings, as well as in a flame tunnel, smelter furnace, and water tube boilers. Work done so far indicates a very high degree of combustion at low-excess air, low-flame temperatures and low-NO_x levels in the exhaust gas. However, data or burner design is not available in the open literature.

NEI International Combustion Ltd. (U.K.) set up a small (5 x 10^6 Btu/hr) and a large (300 x 10^6 Btu/hr) test rig for CWM combustion in 1983.[79] The early work on the small rig was concentrated on CWM properties and their influence on handling equipment and burner design. The large test rig with CWM firing rate of 2-5 tons/hr will be used for burner development.

Atlantic Research Corporation (ARC) first burned its CWM in a 1.3 x 10^6 Btu/hr refractory-lined sheet tube furnace in September 1979. In early 1980, a stable CWM flame, without natural gas firing, was obtained when the refractory was heated.[80] In 1981, over 4 tons of ARC-Coal was fired in a Babcock & Wilcox 4 x 10^6 Btu/hr test furnace.[81] A commercial circular burner was set at maximum setting for swirl and turbulent mixing. The furnace was preheated to operating temperature with combustion air at 600°F. Measured combustion efficiencies were a few percentage points higher than those obtained in the ARC small test furnace.

In 1980, Co-Al, a CWM produced by Slurrytech, Inc., was successfully fired in the same B&W test furnace.[82] The tests demonstrated the feasibility of burning Co-Al directly in conventional liquid-fuel handling and combustion equipment. Only minor burner modifications were made to improve performance and to achieve stable combustion. Four blends of Co-Al prepared from a high-volatile and a low-volatile bituminous coal were tested to determine the influence of coal volatile content on the combustion air required to achieve ignition stability. Carbon conversion efficiencies were comparable to those generally experienced for coals of similar volatile content when burned in the same test furnace.

In 1982, CWM prepared by Gulf and Western Industries, Inc. (G&W), was burned in a Combustion Engineering Inc. (CE) test furnace of 80 x 10^6 Btu/hr capacity. Approximately 116 tons of coals cleaned at EPRI's coal cleaning facility were processed by G&W into a nominal 70 percent solids CWM. A burner-atomizer combination was successfully developed by CE during the project. G&W also supplied CWM to Forney Engineering Company for their burner development tests in 1982.[84]

Results from these combustion test trials demonstrated that CWM could be handled and burned like a liquid fuel. Encouraged by these results, building of large scale CWM preparation plants was initiated in anticipation of CWM conversions in industrial and utility boilers.

Utility Boilers

In 1982, the Italian Electricity Generating Board (ENEL) conducted a CWM test burn in two 125-MWe boilers originally designed for lignite and fuel oil at the S. Barbara Station.[85] Snamprogetti, an engineering firm operating a CWM pilot plant at Fano, produced CWM with 68-72 percent by weight of coal. Two existing burners at two elevations (each rated 14 MW$_t$) fired CWM while the remaining burners were operated on oil. Good qualitative results on flame ignition, stability, and atomization were achieved and will be applied to a planned CWM test in an industrial boiler at the Livorno Power Station in 1983. The Italian experience is the latest CWM firing in utility boilers while tests in Canada and the United States are scheduled, Table 10.

In 1982, Energy, Mines and Resources Canada (EMR), the New Brunswick Electric Power Commission, and Cape Breton Development Corporation entered into a collaborative agreement to demonstrate the preparation of CWM and its utilization in utility boilers.[86] The major objectives of the joint project are to build a pilot plant to produce 6000 tons of CWM for burner evaluations to be undertaken at two coal-designed utility boilers at Chatham Station. The boiler tests will include boiler efficiency and flue gas composition measurements, establishment of derating levels, and furnace modifications. CWM characteristics might be modified to achieve proper atomization and reduce erosion of burner nozzles. Preliminary tests for burner development at Forney Engineering Co. and Combustion Engineering, Inc. burned about 500 tons of CWM produced by Carbogel AB in Sweden.

The last phase of the project includes scale up and fabrication of proven burners and demonstration of firing in an oil-designed boiler in the 50-150-MWe range in 1984. If the project is successful, it has significant implications both for the CWM suppliers and the utilities in Eastern Canada with 5-GW of potentially convertible oil-fired generating capacity.

In the United States, the Electric Power Research Institute plans to sponsor a demonstration one-year burn at an as-yet-unspecified utility.[87] Thirty utility boilers are currently being considered for the project and a decision on the test site will be made in the next 18 months. Burns and Roe, Inc. was contracted by PETC to perform a study to assess the technical feasibility and economics of converting boilers from oil to either pulverized coal (PC) or to CWM firing.[88] Results showed that boilers can be retrofitted to either PC or CWM firing with acceptable derating while achieving considerable economic benefits. Utility companies which own oil-fired boilers also commissioned such feasibility studies to evaluate CWM conversion and other options to reduce oil consumption. In most cases CWM appears to have technical and economical advantages over COM, and to be competitive with PC.[88]

Industrial Boilers

The major CWM market in Sweden is the industrial and district-heating boiler sector which is mostly less than 20 years old and burns imported oil. Several large-scale demonstrations in industrial and district-heating boilers, shown in Table 11, were conducted to assess CWM conversion potential. Earlier CWM boiler tests were not very successful because

TABLE 10

CWM UTILITY BOILER DEMONSTRATIONS

Country	Power Station	MWe	Design Fuel	Test Period	Reference
Canada	Chatham Station	12	coal	1983	86
		22	coal	1983	86
		100-150	oil	1984	86
Italy	S. Barbara Station	125	Lignite/oil	1982	85
United States		To Be Determined		1984 - 1985	87

TABLE 11

CWM INDUSTRIAL BOILER DEMONSTRATIONS

Country	Organization	Size	Design Fuel	Test Period	Reference
Sweden	NYCOL AB	In Several Industrial Boilers		From 1980	26
	Svenska Fluidcarbon	5 MW	oil	1981	90
	Thermal Engineering Research Foundation	50,000 pph	oil	1983	91
United States	PETC	100-Hp	oil	From 1981	35,36
		700-Hp	oil		
	DuPont Co.*	65,000 pph	oil	1983	87
	Occidental Research Corp.	130,000 pph	oil	1984	92

*Co-sponsored by U.S. DOE and EPRI.

existing oil burners gave unstable flames.[26] After a concerted effort on burner development by CWM suppliers, government, and industries, optimistic results were obtained. For example, Svenska Fluidcarbon AB reported that CWM tests in a 5-MW hot water heating boiler at the Malmo plant of the Swedish National Tobacco Company had few operational problems with overall boiler efficiency at 86 percent.[90] To obtain more data on boiler performance, pollutant emissions, and burner performance, a 50,000 pounds of steam-per-hour package boiler is scheduled to operate on CWM, supplied by NYCOL, for 90 days in 1983.[91]

At PETC, both 100-Hp boiler and 700-Hp boiler test facilities which had been used for COM test firings were retrofitted for combustion of CWM and other slurries. The 100-Hp test facility was designated to perform short-term, 100-hour tests for preliminary evaluations of fuels whereas the 700-Hp test facility was employed to conduct a comprehensive test program to establish CWM combustion characteristics such as carbon burnout, boiler efficiency, pollutant emissions, ash deposition, and corrosion/erosion of boiler tubes. The 700-Hp facility was also used to screen commercially available burners and to determine the effects of beneficiated coals on CWM utilization. Data and results have been reported in a series of papers by Pan and will be discussed in later sections.[35,36]

DuPont Co., EPRI, DOE and other organizations will sponsor a 30-day trial burn of CWM in 65,000 pounds of steam-per-hour boiler in August 1983.[87] About 2,500 tons of CWM will be supplied mainly by Atlantic Research Corporation, with the remainder coming from Slurrytech, Inc. This represents the first, large-scale, long-term CWM test in the United States. Occidental Research Corporation has planned to convert a larger boiler at 130,000 pounds of steam per hour to CWM firing in 1984.[92] The boiler was retrofitted to COM firing in 1982 and the changeover to CWM is expected to be minimum.

Data from CWM industrial boiler test firings have shown that CWM can be burned in oil-designed boilers with modest boiler and facility modifications. However, long-term operation on CWM is needed to demonstrate its reliability and to obtain operating and maintenance costs.

Industrial Applications

Recently, CWM has been tested as an injection fuel for a blast furnace, a feedstock for a gasifier, an alternate means of coal feeding into a fluidized bed combustor, and a replacement fuel for oil in a process heater (Table 12). CWM made with beneficiated coal has an even greater potential. Heat engines such as a slow-speed stationary diesel engine, gas turbine, and Stirling engine are the potential targets.

Disposal of large amounts of combustible tailings from a coal cleaning plant poses a difficult problem. EMR Canada intends to explore the feasibility of burning a slurry made of coal tailings in a fluidized bed combustor as an alternate means of disposal.[91] B&W was contracted to conduct parametric tests in its 1 ft. x 1 ft. FBC pilot combustor to determine whether stable and efficient combustion can be achieved.

In Italy, Centro Sperimentale Metallurgico prepared CWM in its laboratory and evaluated the effects of CWM injection on blast furnace performance through mathematical modeling and engineering calculations.[93] The estimated coke replacement ratio is one pound of coal in a 70/30 CWM

TABLE 12

CWM INDUSTRIAL APPLICATIONS

Country	Project	Application	Test Period	Reference
Canada	EMR/B&W	Fluidized Bed Combustor	1983	91
Italy	Centro Sperimentale Metallurgico	Blast Furnace Fuel Injection	Planned	93
Sweden	Carbogel AB	Gasifier	N.A.	26
United Kingdom	National Coal Board, Coal Utilization Research Laboratory	Pressurized Fluidized Bed	1982	94
United States	Standard Havens Research Corporation	Rotary Rock Dryer	1983	95

for one pound of coke. When the data are evaluated, the company plans to inject CWM into one tuyere and build a pilot scale CWM plant.

CWM was investigated as a feedstock for a gasifier which is being used by Boliden AB for the production of ammonia.[26] Synthesis gas has been produced with Carbogel and was found to be excellent for ammonia production. Less soot was produced than that with heavy fuel oil. Boliden AB is planning a conversion in the future.

Hoy reported a CWM trial in a pressurized fluidized bed combustor of 12-inch diameter burning a low-sulfur, high-volatile coal in a CWM produced by Carbogel AB of Sweden.[94] The combustor response was somewhat faster than with an equivalent crushed oil, indicating that the water was not presenting any hindrance to the combustion. High combustion efficiency (99%), low NO_x emissions, and good fuel distribution were achieved. Inspection of the interior of the combustor after the test revealed no evidence of any sintering, clinkering or other agglomeration. As expected for the finely ground coal in the CWM, all the ash was elutriated. CWM with large coal particles could be used to minimize this problem. The test demonstrated that a CWM is technically attractive in that it would eliminate coal drying and the need to provide an inert gas for coal pressurizing. CWM also offers the possibility of simpler coal feed control and distribution.

In the United States, Standard Havens Research Corporation plans to build regional CWM plants to service a network of asphalt manufacturers, who burn lots of fuel in heating their asphalt mixes.[95] Two preliminary combustion tests were successfully concluded. The commercialization plan includes a full-scale CWM preparation facility to be installed at Gallagher Asphalt Co. in Chicago in the fall of 1983. CWM produced by this facility will be used to fire the company's hot-mix asphalt plant rock dryers.

CWM PREPARATION PLANTS

The CWM preparation process is more sophisticated and controlled than COM preparation so as to achieve high coal loading with minimum viscosity. There are not much differences in coal particle size distributions and solid loadings among CWMs for boiler applications. But CWM developed for special applications such as fluidized bed combustion or blast furnace injection may use coarser coal particles while heat engines may use finer coal particles.

Major CWM preparation plants (Table 13) are listed in two groups according to whether the coal benefication process is an integral part of the CWM preparation process. Special features and advantages of each process will be briefly discussed.

Beneficiated Coal CWM

Coal beneficiation is relatively more important to CWM than to COM. It is conceivable that an ultra-clean CWM could be burned in an oil-designed boiler without installing expensive particulate removal and flue gas desulfurization systems. The potential also exists that ultra-clean CWM could improve boiler performance by reducing erosion and ash slagging/fouling.

TABLE 13

CWM PREPARATION PLANTS

Process	Country	Company	Capacity (TPH)	Status	Reference
Beneficiation Included	Canada	New Brunswick Electric Power Commission, Cape Brenton Development Corp., Licensee of Carbogel (Sweden)	?	Start up in in June 1983	86
	Japan	Electric Power Development Corp.	1.5	Start up in mid 1983	76
	Sweden	AB Carbogel NYCOL AB NYCOL AB	7-10 3(1) 100	Operational Operational Start up in late 1983	91 91 91
		Svenska Fluidcarbon AB Svenska Fluidcarbon AB	5(2) 44(3)	Operational Start up in late 1984	97
	United States	Gulf & Western Industries	1.5	Operational	46,96
Beneficiation Currently Not Included	Italy	Snamprogetti	N.A.	Operational	85
	United States	Atlantic Research Corp. Slurrytech Inc. Occidental Research Corp.	4(4) 5(5) 15	Operational Operational Start up in late 1983	98 87 92
		Standard Havens Research Corp.	N.A.	Start up in late 1983	95
		Pittsburgh Energy Technology Center	0.5	Operational	36

(1) Capacity is given as 25 tons per day.
(2) Capacity is given as 30,000 tons per year.
(3) Capacity is given as 250,000 tons per year.
(4) Capacity is given as 600 barrels per day.
(5) Capacity is given as 75 tons per day.

Most physical coal cleaning processes utilize water as the medium by which pyrite and coal particles are separated. The water remaining on the coal particles is dewatered or thermally dried. The cost of dewatering increases dramatically as the particle size decreases. Therefore, fine coal cleaning is sometimes too costly although the ash separation potential is high. However, for CWM applications, it may be economically advantageous to beneficiate fine coal particles because expensive dewatering can be avoided.

Variations of multi-stage froth flotation have been incorporated into Carbogel, G&W, EPDC, and Fluidcarbon AB processes.(26,91,96,76,97) Other coal cleaning processes, including chemical deashing, have been considered for future production of ultra-clean CWM.

Boliden Carbogel/AB Carbogel has operated a 7- to 10-tons/hr plant at Saxerget since 1980. The Carbogel plant has tested a wide range of coals. A controlled grinding system which includes two wet-grinding mills is used to obtain desired coal size distribution. The ground coals are beneficiated in a multi-stage froth flotation system. The beneficiated coals have less than 4 percent ash, with up to 95 percent of the inorganic sulfur removed. Chemical additives are used to enhance yield and selectivity in the separation process. Three tenths (0.3) wt% organic additives are added to the dewatered coal concentrates to obtain a stable CWM containing 70 wt% coal. Carbogel AB has signed license agreements with Cape Breton Development Commission (CBDC) in Canada and Foster Wheeler Corporation in the United States. CBDC is building a 7-ton/hr CWM pilot plant at Sydney, Nova Scotia and will produce and sell CWM with exclusive rights to Eastern Canada.

The development of the Fluidcarbon fuel and system has been underway for three years in Sweden. In 1982, Sonesson of the Volvo group and other large Swedish chemical and construction companies, as well as Allis Chalmers of the United States acquired equity interests in Svenska Fluidcarbon AB. The Fluidcarbon fuel and system have been demonstrated on a pilot scale since 1981. The first commercial operation started late 1982 at an annual production rate of 30,000 tons. The plant capacity will be increased to 250,000 tons per year in October 1984. Fluidcarbon has a coal loading of 70-75 wt% with 1 wt% additive and 2 to 5 wt% ash.

In the EPDC CWM pilot plant two-stage wet grinding is used to obtain the desired coal-particle-size distribution. Additives are introduced during the first stage of wet-grinding. EPDC plans to use a ship simulator to study the changes of CWM properties due to vibration and swing during ship transportation. A pump test loop will be used to collect and evaluate CWM hydraulic data and pump performance.(76)

G&W (U.S.) has operated an automated 1.5-ton/hr pilot plant for more than 2 years.(46,96) NYCOL AB of Sweden has constructed a 25-ton/day CWM pilot plant under a licensing agreement with G&W. G&W has developed a patented coal beneficiation process to remove ash and sulfur from coal. Chemicals are used to condition the coal before it enters the beneficiation cells. The diluted slurry is sprayed through a nozzle onto the surface of a beneficiated tank. The force of spraying action applies intense shear to the flocculates and separates the mineral matter from the treated coal particles. The beneficiated coals are dewatered and then blended together with water and chemicals. More chemicals are added into the second and third tanks to achieve thickening and stabilizing effects.

Unbeneficiated Coal CWM

Snamprogetti in Italy, Atlantic Research Corporation (ARC), Slurrytech, Inc., Occidental Research Corporation, Standard Havens Research Corporation, and Pittsburgh Energy Technology Center (PETC) in the United States have CWM preparation plants which currently do not include a coal benefication process, but are capable of doing so.[85,98,87,92,95,36]

Data on the Snamprogetti process and Occidental Research Corporation are not available and ARC's process was described earlier. At PETC, the COM preparation facility was modified to produce CWM for burning in its test boilers. Pulverized coal with a size distribution of 70-90 percent minus 200 mesh is mixed with water and additivies by a mechanical mixer. Sixty-three percent by weight coal in CWM has been produced. PETC prepares CWM only for its own needs.

CWM produced by Standard Havens Research Corporation contains only 50-60% wt% solids as a trade-off for process simplicity and reliability to suit customer's needs. The penalty of the high-water content in CWM is not a major concern in rock drying. ARC has evaluated various coal benefi- ciation processes which can be best fitted into its CWM process. Currently, ARC uses low-sulfur and low-ash coal for special applications.

STATUS OF CWM TECHNOLOGY

The review of the current status of CWM technology focuses on three major areas: preparation, combustion, and boiler performance.

CWM Preparation

Coal loading in a CWM is a function of the properties of coal, water content, and additives. The chemical and physical properties of the coal have a major influence on the amount of water needed to achieve a slurry with desired flow characteristics. Japanese researchers reported that CWM made of bituminous coals have higher coal loadings than CWM made with sub- bituminous coals or lignites.[99] High-volatile bituminous coals are also more desirable for making CWM because they give favorable ignition and com- bustion characteristics.

Ball mills were selected for early CWM tests because of the abundance of data on wet grinding of coals and minerals, their reliability, and economics. Other types of communition equipment were used to produce coarser or finer coal CWM. ARC and Carbogel favored a two-stage grinding technique to produce a bimodal distribution of coal particles. Recently, both ARC and Slurrytech have demonstrated that for a number of coals, satisfactory CWM can be produced by one-step grinding technique in a ball mill. This improvement will simplify the CWM process and reduce production costs.

Dispersants and stabilizers are added at various points of the CWM preparation process. An acceptable dispersant should have the following features: (1) non-foaming, (2) a structure with both a hydrophobic portion and a hydrophillic portion, (3) water solubility, (4) compatibility with a stabilizer, and (5) effectiveness at low concentrations. It must also be low in cost and commercially available in adequate quantities. An effective dispersant will produce CWM with a high-solids loading and minimum viscosity.

Anionic and non-ionic type dispersants have been used for making CWM. An ionic dispersant content required for fluidity can range from a few tenths of a part per hundred of coal for a coal size distribution of 60 percent minus 200 mesh to over one pph for 100 percent minus 200 mesh. The anionic type dispersants are in the class of sodium organosulfonates, e.g. Lomar-D and Tamol SN and Marasperse CBO-3, which is a sodium lignosulfonate, a wood derivative.

Non-ionic dispersants containing long chains of polyoxyethylene are used at a concentration of less than 0.4 wt% to lower the surface tension of water, keep the coal particles from flocculating, promote suspension, and provide good flow properties. The selection of anionic and non-ionic type dispersants depends on the process design, stabilizer, coal characteristics, and other factors which CWM producers keep proprietary.

CWM stability is a strong function of the type and quantity of the stabilizer. Without the stabilizer in a CWM, coal settling is dependent on dispersant concentration, coarse/fine ratio, solids loading, and storage conditions. Gums, salts, clays and other chemicals have been used as stabilizers.

CWM has rheology and hydraulic characteristics similar to that of COM. Currently, there is limited experience on CWM pumps. Progressive cavity pumps which were also used in COM applications are now being used by PETC, Atlantic Research Corporation, Adelphi University, B&W, and others. PETC reported that the rotary gear pump, which was effective for continuous slurry recirculation or transfer service with COM, experienced total flow blockage when operated on CWM.[36] The pump cavities tended to foul with coal particles. Compressed air-driven diaphragm pumps have been installed to replace the rotary gear pumps.

Wear rates of a pump used for CWM are expected to be higher than those of COM pumps operating under the same conditions. There are two possible reasons: (1) CWM has high solids loading and therefore is more abrasive, and (2) the fuel oil in COM provides lubrication during pumping and, thus, reduces the impact of solid particles on the surfaces of a pump. However, CWM pump data are presently insufficient to draw final conclusions.

The U.S. DOE has funded a project to evaluate major equipment and instrumentation.[100] Various types of pumps will be installed in a test loop in which short-term tests and a 200-hour test will be performed. Test data will be correlated to the physical properties, such as density and viscosity, of each slurry tested. Changes in CWM hydraulic characteristics or pump performance will also be evaluated. The CWM hydraulic data will be correlated to known fluid flow models which will be used to predict fluid flow characteristics of CWM. Records on the operation and maintenance of the pumps will be collected to evaluate pump performance. Erosion/corrosion coupons or wear plugs will be installed to evaluate the wear rate and to predict the useful service life of various materials and components.

Canada, Japan, Spain, and Sweden all have plans to evaluate slurry pump performance during their CWM demonstration tests.[86,76,85,31]

CWM Combustion

CWM burner development has been identified as one of the major areas for investigation. Bench scale and full-scale test programs are now being conducted to evaluate CWM burner performance. COM burner experience is helpful, but not totally useful for CWM development because of large differences in fuel composition and combustion characteristics. Early CWM data have been obtained in small laboratory facilities of 1 to 4×10^6 Btu/hr and, in general, results were poor, i.e., test furnaces required extensive preheat, burner turndown was limited, and carbon conversion efficiencies were in the high 80 to mid-95 percent range.[81,101,102]

Since 1981, PETC has burned CWM with 63 wt% coal loading in a 100-Hp firetube boiler and a 700-Hp watertube boiler.[36] A Peabody nozzle with six 5/32-inch holes was used in the 100-Hp boiler tests. Atomizing air and CWM were mixed inside the cap before injection into the boiler. A swirl vane air register was used to provide better air distribution. The COEN burner used with the 700-Hp watertube boiler was modified for CWM firing. Nozzle holes were enlarged to double the area to reduce fuel velocity and pressure drop. The spray angle of the nozzle cap was reduced from 75° to 60° so that the burner gun and the diffuser could be retracted. This modification exposed the CWM to a hot environment for a longer time, and aided CWM combustion. The dual-air-zone register was also modified. The inner and outer air streams were separated by a sliding shroud to create a high swirl flow at the outer air stream. These modifications brought recirculating combustion gas close to the burner nozzle by means of the high-swirl combustion air to maintain a stable CWM flame. With 500°F preheated combustion air, stable CWM combustion was achieved without the use of supplemental fuel.

Wear rate data for various material inserts with both the Peabody and COEN burners are shown in Table 14. When the holes were enlarged from 7/32-inch to 15/64-inch, the wear rate was reduced even further for a carbon steel cap with carbide inserts.

A joint program between Combustion Engineering (C.E.) and the Electric Power Research Institute (EPRI) was organized to develop and demonstrate a commercial CWM burner.[83] A Y-jet type atomizer was targeted to produce the same atomization quality for CWM as that for residual fuel oil. The C.E. Atomizer Test Facility (ATF), which is equipped with optical diagnostic instrumentation, was used to characterize the quality of the CWM atomization. The CWM fuel spray droplet size distribution and velocity/trajectory were measured in a non-combustion mode. Results showed that the atomizer effectively atomized the high-viscosity CWM (up to 2800 cps). Atomization quality and ratio of atomizing fluid to CWM were similar to those for heavy fuel oil.

Forney Engineering Company (FECO) designed and fabricated two new atomizers.[84] The first design is a conical Y-jet atomizer which features a large annulus instead of individual orifices. Plugging problems were eliminated by using large openings, and erosion is reduced by gradual changes in direction of fuel flow. The fuel exit velocity was reduced to allow combustion air to entrain the coal quickly. The second design used conical internal mixing and incorporated the same considerations as the conical Y-jet atomizer, except that the fuel flowed through a right angle turn causing higher impact and atomizing medium consumption.

In addition to newly developed atomizers, commercial oil burner configurations were also modified for CWM firing. The major features of a CWM burner are refractory-lined divergent throat and a combustor swirl. The purpose of the refractory-lined throat is to increase the mass

TABLE 14

WEAR RATES ON CWM BURNER NOZZLES AT PETC*

Boiler/Burner	Coal Slurry Composition	Coal Slurry Flow Rate Lb/Min	Burner Nozzle Material	Wear Rate Inch/Hour
100-Hp/Peabody 6-5/32" holes 50° spray angle	CWM 60 wt% Pittsburgh seam coal (90% minus 200 mesh)	9.8	Hardened Tool Steel Cap	0.001246
100-Hp/Peabody 6-0.159" holes 50° spray angle	CWM 60 wt% Pittsburgh seam coal (90% minus 200 mesh)	6.0	Low-Carbon Steel Cap with Tungsten-carbide overlay	0.00400
		5.2	Low-Carbon Steel Cap	0.00450
	CWM 60 wt% Pittsburgh seam coal (90% minus 200 mesh) 32% methanol, 8% water	7.2	Stainless Steel Cap	0.00550
	CWM 55 wt% Pittsburgh seam coal (90% minus 200 mesh) 45 wt% methanol and water	8.6-9.6		0.00005
700-Hp/COEN 8-7/32" holes 60° spray angle	CWM 60 wt% Pittsburgh seam coal (90% minus 200 mesh)	47.0	Hardened Tool Steel Cap	0.00660
		38.5	Hardened Tool Steel Cap with Ferrotic Inserts	0.00039
700-Hp/COEN 8-15/64" holes 60° spray angle		61.8	Carbon Steel Cap with Carbide Inserts	0.00011

*Y.S. Pan et al., Reference 36.

recirculation ratio and thus to stabilize the flame both aerodynamically and thermally. The swirl combustion air stabilizes the flame and contributes to high combustion efficiency.

Tests showed that, with the proper combination of burner and atomizer design, CWM can be successfully burned with carbon conversion efficiencies in the range of 96 to 99+ percent. CWM was reliably ignited in a cold furnace using conventional ignitors and low air-preheat temperatures (250°F). A turndown ratio of 4:1 was achieved and CWM flame stability was acceptable.

Both C.E. and FECO will supply prototype CWM burners for boiler testing at the Chatham Station for the New Brunswick Electric Power Commission in 1983. Performance of burners and atomizers will be evaluated. C.E. under a DOE contract, will conduct CWM combustion evaluation/atomization and burner testing.[103] The purpose of this work is to demonstrate the reliability of firing CWM using commercially available products and technology. Standards and practices for CWM burner design and operation will be developed, and CWM characteristics that affect burner performance will be identified. A Peabody burner and a COEN burner, in addition to two C.E. burners, will be evaluated based on their performance on ignition, turndown, carbon utilization, atomization, and atomizer tip life. This effort is scheduled to be completed in 1985.

Boiler Performance

The testing of CWM industrial boilers is at an early stage while it is only at a planning stage for utility boiler testing. In the next two years more boiler data will be available.

Boiler Efficiency and Combustion Efficiency

The combustion performance of the 700-Hp boiler at PETC during four tests using CWM is compared to No. 6 oil in Table 15.[36] The CWM consisted of 60 wt% Pittsburgh Seam coal in water, with 0.5 wt% Lomar-D to reduce viscosity. The combustion air was heated to 500°F. At full load the boiler efficiency was 75 percent with a carbon conversion efficiency of 96 percent. Boiler and combustion efficiencies varied slightly with conditions of excess air and load level.

The boiler efficiency with CWM was approximately seven percent lower than that with oil firing. Although not specifically analyzed, this lower boiler efficiency appears to be due to differences in heat losses caused by water and hydrogen in the fuels, heat losses due to carbon conversion, and differences in combustion-air temperature. The combustion-air temperature for oil was approximately 80°F compared to about 500°F for the CWM.

Svenska Fluidcarbon AB tested CWM in a 5-MWt hot water boiler. The water content of the CWM caused a reduction of approximately 1.8 percent in combustion efficiency relative to coal burned at a nominal 10 percent moisture content. The overall boiler efficiency was about 86 percent during the Fluidcarbon firing.

PETC also conducted a series of parametric tests to evaluate the effects of coal ash content on boiler performance.[104] With CWM prepared from three coals of 11.2, 6.8 and 2.8 percentage ash levels, carbon conversion efficiencies increased slightly with the increase in excess air.

TABLE 15

COMBUSTION PERFORMANCE OF PETC 700-HP BOILER ON CWM*

	Coal-Water Mixtures				No. 6 Oil
	1	2	3	4	5
Fuel Characteristics					
Weight-Percent Coal	59.6	60.2	60.6	58.6	----
Coal Size Consist (% minus 200 mesh)	90	90	91	90	----
Heating Value (Btu/lb)	7,729	7,807	6,938	7,599	18,695
Boiler Operating Conditions					
Load	Full	Full	Full	Half	Full
Flue Gas O_2 (%)	2.5	2.9	3.3	3.7	2.8
Combustion Air Temp. (°F)	493	492	501	508	81
Boiler Performance					
Flue Gas Temp. (°F)	570	592	599	532	584
Carbon Conversion Eff. (%)	95.5	96.5	96.9	98.8	100
Boiler Efficiency (%) (Heat-Loss Method)	75.0	74.9	74.1	76.6	----

*Y.S. Pan et al., Reference 36.

No apparent correlation exists between coal ash levels and carbon conversion efficiencies, but boiler efficiency increased with decreasing ash levels under the same operating conditions.

CWM burner and atomization are still in the developmental stage. Boiler efficiency and carbon conversion efficiency will improve when the CWM burner is optimized.

Ash Deposition and Slagging

CWM tests at PETC showed that measured ash deposition in the furnace section and calculated deposition in the convective section increased monotonically with ash level in the coal. There was significant ash deposition even when burning CWM with 2.6 percent ash. With high-ash level of 11.2 percent in coal, the deposition rate increased drastically, resulting in higher flue gas temperature and particulate emissions. Ash was accumulated because sootblowers were intentionally idled during the tests. CWM ash was found to be friable and easily removed by sootblower action.

Prediction of slagging and fouling from the ash chemistry of coal is still very much an art. Because of the heterogeneity of the coal and the variability of the boiler designs, varying deposit rates and in situ sintering strength will yield different slagging and fouling problems in boilers. Current data on these areas are still insufficient to draw any meaningful conclusions.

Derating

Full load on CWM was reached at PETC's 700-Hp boiler after burner and air register modifications.[104] Engineering studies were conducted by B&W on three site specific industrial boilers of different types ranging in size from 60,000 pounds of steam per hour (ppH) to 200,000 ppH.[105] The smallest unit of 60,000 lbs/hr was a saturated steam B&W Integral Furnace Type FJ boiler designed to fire natural gas and No. 6 oil. It was determined that this unit could achieve the original maximum continous rating (MCR) with CWM. The Type FJ boiler has a standard furnace hopper design which can be retrofitted to this boiler at a reasonable cost if necessary. However, the need for a furnace hopper would not be decided until active tests are conducted to determine effects of ash deposition in the furnace. In addition, one sootblower would be required in front of the furnace outlet screen tubes and a second sootblower in the convection bank. Modification may also be required in the economizer to avoid plugging in that area.

The second unit evaluated by B&W was a 200,000 lbs/hr Integral Furnace Type FH boiler designed to fire oil and acid sludge tower bottoms. It was determined that, with minor modifications, this unit could achieve 90 percent of MCR. This unit does not have a furnace ash hopper but could be modified to include one. Four sootblowerrs would have to be added in front of the furnace screen tubes and superheater tubes to remove deposited ash. If the sootblowers in the superheated area do not prove to be effective, an increase in the superheater tube clear-side spacing may be required. The flue gas velocities in the convection pass also would have to be reduced by modification of the baffling arrangement to reduce potential fly ash erosion to acceptable levels.

The third unit considered by B&W was a 150,000 lb/hr two-drum Stirling power boiler designed to fire oil with future coal capability. This unit would reach 100 percent MCR with the addition of furnace wall sootblowers and a furnace ash hopper. Sootblowers would also have to be added to the convection pass area. This unit could also be converted to a coal-fired stoker, but, it would be less costly to convert to CWM firing.

Derating could occur when CWM is fired in an oil-designed utility boiler. The amount of boiler derating depends on original boiler design and fuel characteristics. In addition, the boiler can be modified, such as enlarging furnace volume, and adding heat-transfer surfaces to maintain original rating. Estimates on boiler derating and capital costs of boiler modifications have been reported by Burns & Roe and other engineering companies.[88,101,106] Nevertheless, derating is still a site specific problem which requires actual boiler testing.

Emissions

Flue gas emissions data from CWM combustion in test furnaces and small boilers have been collected. The SO_2 emissions were found to correspond to the sulfur content in CWM.[36] The NO_x emissions were higher for CWM combustion than for No. 6 oil combustion, apparently due to the higher nitrogen content in the coal than oil. The particulate emission were found to be higher for a coarse coal particle size consist or a high-ash level coal.

In Japan, NO_x emissions data from CWM and pulverized coal in a 80×10^6 Btu/hr test furnace were compared.[77] The measured NO_x emissions from CWM were about 100 ppm lower than that from the parent coal at the same oxygen level. The principal factor for NO_x reduction is the lower CWM flame temperature. The Japanese researchers extended the test by measuring NO_x emissions when a common NO_x reduction technique, staged combustion, was applied. It was found that NO_x emissions from CWM was reduced even further.

The current control technology for SO_2 and particulate emissions appears to be adequate. However, the effect of high-moisture level in flue gas from CWM combustion on flue gas desulfurization and baghouse/electrostatic precipitator may need to be determined prior to the installation of full-scale equipment.

FUTURE CWM DEVELOPMENT

At this time, CWM technology has not achieved commercial status and more R&D is needed.

There is a strong need to expand the CWM technology base to incorporate new developments in coal cleaning and coal slurry transportation, and to continue the technical data exchange among interested countries. The future CWM development needs for preparation, major equipment, combustion, environmental control, transportation, and handling are identified in Table 16 and briefly reviewed as follows:

1. Advanced physical coal cleaning and chemical coal cleaning processes should be evaluated to determine feasibility and cost for CWM applications.

TABLE 16

FUTURE CWM DEVELOPMENT NEEDS

Category	Need	Accomplished	Benefit
Preparation	Advanced physical coal cleaning and chemical coal cleaning to reduce sulfur and ash.	Laboratory scale test; ash reduction not significant; cost too high.	Low ash and sulfur fuel.
Major Equipment	Improved reliability of CWM handling equipment and instrumentation.	Test loops; short-term tests.	Reduce O&M cost; automate process control.
	Better burner design for CWM firing.	Short-term tests; insufficient data.	Improved boiler performance and reliability.
Combustion	Long-term full scale combustion tests in boilers representing major markets to determine reliability and derating.	Test furnace firing; insufficient data.	Remove CWM conversion uncertainties.
Environmental Control	Emissions control technology.	Very little data available.	Required by law for long-term CWM firing. Cost reduction possible.
Handling and Transportation	CWM characteristics during pumping.	Very little data available.	Data needed for CWM stability and handling.
	Long distance transport of CWM.	Very little data available.	A complete mine-mouth to user CWM preparation/transport system.

2. More R&D for CWM major equipment is desirable. CWM burners and pumps have been developed by private companies. However, more information is needed on durability and performance. For instance, the change of CWM characteristics during pumping should be investigated. Also, the use of additives in CWM to reduce SO_2 emissions by direct capture of sulfur is also worthy of investigation.

3. Long-term CWM combustion tests in oil-designed utility and industrial boilers are needed to determine boiler derating and reliability. Short term test data are insufficient to convince users to accept CWM. The selected boiler should be representative of a major CWM conversion market segment and the tests should be planned such that the data can be applied by other utility/industrial users.

4. More R&D activities are needed in the environmental control area. The trade-offs of using low-ash and low-sulfur CWM versus emissions control equipment should be evaluated.

5. Handling and Transportation of CWM, particularly long distance transportation between the fuel preparation site and the user, should be evaluated.

SUMMARY REMARKS

COM technology has become a mature technology in the last decade while CWM technology is approaching commercialization rapidly. CLM will compete with other coal utilization and coal conversion technologies such as coal liquefaction, medium Btu gasification, and direct coal firing, for combustor retrofit and new combustor markets. CWM has a competitive edge on cost over COM. CWM also has a potentially lower cost than liquefaction and gasification due to the lower capital and operating costs. Most of all, CWM has the advantage of earlier availability. Finally, CWM, as a liquid fuel, is easier to transport, store, and handle, than solid coal, and may be the preferred way to get coal to the user, particularly in industrial plants.

Commercial acceptance of CWM technology hinges on the magnitude of the price differential between the cost of oil and cost of coal, and the projected long term availability of oil. Market acceptance will also require proven operating reliability which is the focal point of ongoing and future CWM projects. Additional research and development opportunities exist for further expanding CWM market penetration into new areas such as transportation and heat engines. The importance of coal benefication in CWM to an expanding market is paramount. A low-ash and low-sulfur CWM needs to be developed to create a greater demand for CWM fuel.

REFERENCES

1. Demeter, J.J., C.R., McCann, G.T. Bellas, J.M. Ekmann, and D. Bienstock, "Combustion of Coal-Oil Slurry in a 100-Hp Firetube Boiler," PERC RI 77/85, May 1977.

2. Morrison, G.F., "Conversion to Coal and Coal/Oil Firing," ICTIS/TR07, IEA Coal Research, December 1979.

3. Bienstock, D., and E.M. Jamgochian, "Coal-Oil Mixture Technology in the U.S.," Fuel, Volume 60, p. 851, September 1981.

4. Manning, A.B., and R.A.A. Taylor, "Collodial Fuel," Trans. Inst. Chem. Engr., Vol. 14, p. 44 (1936).

5. Schroeder, W.C., "Use of Mixture of Oil and Coal in Boiler Furnaces," Mech. Engr. 64(11), p. 793, 1942.

6. Jonnard, A., "Collodial Fuel Development for Industrial Use," Bulletin No. 48, Kansas State College Engineering Experiment Station, Manhattan, Kansas, 1946.

7. Rudzki, E.M., B.K. Pease and T.H. Weidner, "Use of Coal-in-Oil Mixtures to Improve Open-Hearth Furnace Performance," Journal of the Institute of Fuel, Vol. 38, No. 291, p. 154, April 1965.

8. Morinaga, K., and Y. Jonoto, "Coal Slurry Injection into Blast Furnace," Proceedings of Coke in Iron and Steel Industry, International Congress, 1966.

9. Agarwal, J.C., "Technical Considerations of Fuel Injection in Blast Furnace," ASME paper 62-WA-286, November 1962.

10. Brown, Jr., A (ed)., "Final Report of the General Motors Corporation Powdered Coal-Oil Mixture (COM) Program," FE-2267-2, August 1977.

11. Canadian Combustion Research Laboratory, "Proceedings of the Coal-in-Oil Seminar, May 17, 1972," Fuel Research Center Divisional Report FRC 72/95-CCRL, 1972.

12. Metzger, G.W., "Coal-Oil Mixture Additives and Process," Proceedings of the Coal-Oil Mixture Combustion Technology Exchange Workshop, Energy Research and Development Administration, CONF-761019, October 1976.

13. Ekmann, J.M., and D. Bienstock, "Stability of Coal-Oil Mixtures," Proceedings of First International Symposium on Coal-Oil Mixture Combustion, May 1978.

14. Nakabayashi, Y., "The Overview of the R&D Status on CLM in Japan," Proceedings of Fourth International Symposium on Coal Slurry Combustion, May 1982.

15. Higgins, M.E., "Commercial Operation of the Paul L. Bartow Unit 1 Coal-Oil Mixture," Proceedings of Fifth International Symposium on Coal Slurry Combustion and Technology, April 1983.

16. Cook, M.C., "Operating Florida Power and Light Company's Sanford Plant on Coal-Oil Mixture," presented at the Joint Power Generation Conference, October 1981.

17. Rodriguez, L., F. Sell, "Florida Power Corporation/DRAVO Corporation Coal/Oil Composite Fuel Program," Proceedings of First International Symposium on Coal-Oil Mixture Combustion, May 1978.

18. New England Power Service Company, "Demonstration Program for Coal-Oil Mixture in a Utility Boiler," Final Report (Draft), 1981.

19. Conolly, R., P. Gadbury, M.T. Jacques and K.J. Matthews, "The Combustion of Coal in Oil Dispersion - A Direct Comparison with the Firing of Residual Fuel Oil in a 120 MWe Utility Boiler." Proceedings of Fourth International Symposium on Coal Slurry Combustion, May 1982.

20. Whaley, H., "Coal-Oil Mixture Projects in Canada," Proceedings of First International Symposium on Coal-Oil Mixture Combustion, May 1978.

21. Whalen, P.J., F.W. Davies, L.K. Lee, M. Natarajan and R.L. Wang, "Coal-Oil Mixture Combustion Applied to Canadian Utility Boilers," Proceedings of Second International Symposium on Coal-Oil Mixture Combustion, November 1979.

22. Whaley, H., "Overview of the Canadian Program," Proceedings of Third International Symposium on Coal-Oil Mixture Combustion, April 1981.

23. Zhao, X., E.H. Hou, and F.J. Xu, "COM Test Program at Yang Shupu Power Plant," Proceedings of Fifth International Symposium on Coal Slurry Combustion and Technology, April 1983.

24. Matsuura, Y., Y. Sugino, K. Yamaguchi, I. Koyama, T. Shinohara, and M. Nishizawa, "Test Results of COM at an Actual Utility Boiler," Proceedings of Fifth International Symposium on Coal Slurry Combustion and Technology, April 1983.

25. Forte, J.P., R.H. Hickman and R.S. Folgueras, "COM Test Program - Almeria Power Plant," Proceedings of Fourth International Symposium on Coal Slurry Combustion, May 1982.

26. Borgne, K.C., "Coal Combustion Technology, Featuring the Use of Coal-Liquid Mixtures and Related Environmental Aspects," presented at the Chinese-Swedish Energy Symposium, September 1981.

27. Kurashige, I., T. Iba, T. Miyazaki, K. Satoh, M. Kojima and Y. Aminaga, "COM Injection into All Tuyeres of Kashima No. 3 Blast Furnace," Proceedings of Fifth International Symposium on Coal Slurry Combustion and Technology, April 1983.

28. Clayfield, E.J., W.R. Dorresteijn, E.C. Lumb, K.J. Wilbraham and D.J. Barratt, "COLLOIL Manufacture and Application," Fuel, Volume 60, p. 865, September 1981.

29. Jansto, S.G., A. Mertdogan, L.A. Martin and V.D. Beaucaire, "Coal-Oil Mixture Combustion Program Injection Into a Blast Furnace," Final Report, DOE/ET/10387-T1, April 1982.

30. Ivey, R.T., and B. Schiller, "Utilization of a Coal-in-Oil Mixture as a Replacement for No. 6 Fuel Oil at a Republic Steel Blast Furnace," Proceedings of Fourth International Symposium on Coal Slurry Combustion, May 1982.

31. Bruno, L., A.S. Desphande, and H. Whaley, "Coal-Oil Slurry Combustion and Tribology -- A Canadian Experience," Proceedings of Fourth International Sympoisum on Coal Slurry Combustion, May 1982.

32. Centre for Energy Studies, Technical University of Nova Scotia, "Evaluation of Combustion Burners for Coal Cleaning Containing Liquids," February 1982.

33. Dinham, E., D.R. Wall and D.M. Whitehead, "The Handling and Combustion Characteristics of Stable Coal/Fuel Oil Dispersions," Proceedings of Fourth International Symposium on Coal Slurry Combustion, May 1982.

34. Pan, Y.S., G.T. Bellas, M.P. Mathur, J.I. Joubert, and D. Bienstock, "Recent Coal-Oil Mixture Combustion Tests at PETC," DOE/PETC/TR-80/5, Pittsburgh Energy Technology Center, June 1980.

35. Pan, Y.S., G.T. Bellas and J.I. Joubert, "Coal-Liquid Mixture Combustion Tests in Oil-Designed Boilers," ASME paper 82-IPC-FU-2 presented at the 1982 ASME Industrial Power Conference, New Orleans, La., October 1982.

36. Pan, Y.S., G.T. Bellas, R.B. Snedden, and J.I. Joubert, "Coal Slurries as Alternate Fuels for Oil-Designed Boilers," Proceedings of International Symposium on Conversion to Solid Fuels, October 1982.

37. Page, L.E., "Case Study of Conversion to Coal-Oil Mixture," presented at 8th Energy Technology Conference, Washington, D.C., February 1981.

38. Adams, D.R., G.S. Kapp, G.F. Gresham and E.T. Flanigan, "Commercial Demonstration of Coal-Oil Mixtures in Industrial Applications," Proceedings of Fifth International Symposium on Coal Slurry Combustion and Technology, April 1983.

39. Savage, R.L., W.R. Warfield, S.N. Singh, G.D. Craig, W.E. Krauss, and J.E. Ross, "Coal-Oil Mixture Utilization in Small Industrial and Commercial Boilers," Proceedings of Fourth International Symposium on Coal Slurry Combustion, May 1982.

40. Dooher, J., R. Genberg, S. Moon, D. Wright, S. Jakatt, B. Gilmartin, J. Skura, and J. Lepore, "The Feasibility of Using Coal/Water/Oil as a Clean Liquid Fuel." Proceedings of First International Symposium on Coal-Oil Mixture Combustion, May 1978.

41. Xu, X.Y., K.F. Cen, G.B. Shi and P.S. Wu, "The Test of COM in a 100 ton/hr Power Station Boiler in Anshan Iron and Steel Company," Proceedings of Fourth International Symposium on Coal Slurry Combustion, May 1982.

42. Dreuilhe, J., "Blanzy, The First Coal-Oil French Installation," Proceedings of Fourth International Symposium on Coal Slurry Combustion, May 1982.

43. Whaley, H., "Canadian Initiatives in Coal-Liquid Mixture Combustion and Development," Proceedings of Fourth International Symposium on Coal Slurry Combustion, May 1982.

44. Wilda, J.C., "COMCO's Composite Fuels Preparation Plant, Port Sutton, Tampa, Florida," Proceedings of Fifth International Symposium on Coal Slurry Combustion and Technology, April 1983.

45. Hartness, J.L. ERGON, Inc., personal communication, 1982.

46. Burgess, L., P. McGarry and D. Herman, "Spray Flotation, A New Way to Clean Coal," Proceedings of Fifth International Symposium on Coal Slurry Combustion and Technology, April 1983.

47. Kurihara, M., Y. Nakamura, M. Yamamura, Y. Kiyonaga, A. Naka, and Y. Ogura, "Deashing of Coal by Oil Agglomeration," Proceedings of Fourth International Symposium on Coal Slurry Combustion, May 1982.

48. Cundy, V.A., and D. Maples, "Fuel Mixtures of Low Rank Lignite Coal with Fuel Oils," Fuel, Vol. 61, p. 1277, December 1982.

49. Trass, O., V.R. Koka, G. Papachristodoulou, and E.R. Vasquez, "Grinding of Coal-Oil Slurries with the Szego Mill--Effects of Solids Concentration," Proceedings of Fourth International Symposium on Coal Slurry Combustion, May 1982.

50. Naka, A., "The Use of Additives to Stabilize Coal-Oil Mixture," Proceedings of Second International Symposium on Coal-Oil Mixture Combustion, November 1979.

51. Al Taweel, A.M., H. Farag, and K. Langille, "An Accelerated Method for Testing COM Stability," Proceedings of Third International Symposium on Coal-Oil Mixture Combustion, April 1981.

52. Rowell, R.L., "Structure and Stability of Coal-Oil Mixtures and Coal-Oil Mixtures," CS-1695. EPRI Final Report, February 1981.

53. Hu, C., Y. Zeng, L. Qian and H. Zhang, "The Study of the Additives of COM Fuel," Proceedings of the Fourth International Symposium on Coal Slurry Combustion, May 1982.

54. Bouchez, D., A. Faure, G. Scherer, L.A. Tranie, and G. Antonini, "COM: The French Program, Preparation, Stabilization, and Handling of COM," Proceedings of Fourth International Symposium on Coal Slurry Combustion, May 1982.

55. Slepow, L.D., and A.S. Mendelssohn, "FPL's Sanford COM Demonstration Plant," Proceedings of Third International Symposium on Coal-Oil Mixture Combustion, April 1981.

56. Wagoner, C.L., C.F. Eckhart, and G.A. Clark, "At-Sea Test and Demonstration of Coal/Oil Mixtures as a Marine Boiler Fuel, Part I - Shoreside Testing," MARAD Report No. MA-RD-920-82016, April 1982.

57. Shang, S.D., T.Y. Fong, "Preliminary Tests on Coal-Oil Mixture and Its Mechanism Studies," Proceedings of Third International Symposium on Coal-Oil Mixture Combustion, April 1981.

58. Yin, C.F., and H. Borman, "Slurry Combustion Using an Ultrasonic Burner," Proceedings on International Symposium on Conversion to Solids Fuels, October 1982.

59. Schmidt, A.D., and J.L. Friedrich, "Full Scale Tests Firing Coal-Oil Mixtures in a 400 MW Steam Generator," Proceedings of Third International Symposium on Coal-Oil Mixture Combustion, May 1981.

60. Petersen, G.L., D.T. Buell, S.A. Scavuzzo, and J. Neidert, Jr., "Operation and Testing of the Paul L. Bartow Unit 1 on Coal-Oil Mixture," Proceedings of Fifth International Symposium on Coal Slurry Combustion and Technology, April 1983.

61. Bechtel Group, Inc., "Coal-Oil Mixture as a Utility Fuel," Volume 1, Conversion Guideline Handbook, CS-2309, EPRI Final Report, March 1982.

62. Lane, W.R., R.J. Burgess, and N.E. Takvoryzn, "Pilot Electrostatic Precipitator Test Results Sanford COM Project, Florida Power and Light Company," Proceedings of Third International Symposium on Coal-Oil Mixture Combustion, May 1981.

63. Shilling, N.Z., W.J. Morris, J.H. Brummer, and W.C. Love, "Performance Evaluation of an Intelligent Precipitator on a COM Fuel-Fired Boiler," Proceedings of Fifth International Symposium on Coal Slurry Combustion and Technology, April 1983.

64. Takahashi, Y., H. Hino, Y. Fujima, and A. Komori, "Basic Studies on COM Combustion," Proceedings of Second International Symposium on Coal-Oil Mixture Combustion, November 1979.

65. Nakabayashi, Y., and D.J.L. Lin, "Fine and Coarse Coal-Oil Mixture Combustion Tests of Air Atomized Burner in Japan," Proceedings of Second International Symposium on Coal-Oil Mixture Combustion, November 1979.

66. Foo, O.K., E.M. Jamgochian, J.C. Blake, and T.F. Skinner, "Market Assessment and Financial Analysis of COM Conversion," Proceedings of Second International Symposium on Coal-Oil Mixture Combustion, November 1979.

67. Burgess, L.E., Gulf & Western Industries, Inc., Personal Communication, 1982-1983.

68. Marnell, P., "Direct Firing of Coal-Water Suspensions State-of-the-Art Review," presented at Coal Technology 1980, Houston, November 1980.

69. Kelcec, G., P.O. Olivadoti, and A.F. Duzy, "Coal Slurry Firing at the Werner Station," ASME paper 62-PWR-3, 1962.

70. Schwarz, O., and H. Merten, "Preparation, Transportation, and Combustion of Coal-Water Suspensions," Brennst-Warme-Kraft, 18, No. 10, p. 474, 1966.

71. Schwarz, O., and H. Merten, "Direct Burning of Coal-Water Suspensions in Power Plants," Gluckauf, 2, p. 215, 1967.

72. Delyagin, G.N., "Regularities of the Combustion of Pulverized Coal-Water Suspension in an Air Stream," Inst. Goryuch, Iskop., p. 72, 1965.

73. Davydora, I.V., G.N. Delyagin, B.V. Kantorvich, and V.S. Levanesky, "Experimental Investigation of Combustion of Coal-Water Suspension," Inst. Goryuch. Iskop., p. 40, 1965.

74. Delyagin, G.N., and Z.N. Smirnova, "Interaction of Coal with Water in Combustion in a Coal-Water Suspension Layer," Inst. Goryuch Iskop., p. 126, 1965.

75. Scheffee, R.S. "Development and Evaluation of Highly Loaded Coal Slurries," Phase I Summary Report, FE-2667, May 1979.

76. Nakabayshi, Y., N. Sato, K. Fujii, S. Takao, and N. Suzuki, "Recent Study on Highly Loaded Coal Water Slurry," Proceedings of Fifth International Symposium on Coal Slurry Combustion and Technology, April 1983.

77. Kuroda, H., T. Masai, Y. Takahashi, and S. Watanabe, "Combustion of Coal-Water Slurry in a Multiple-Burner Furnace," Proceedings of Fifth International Symposium on Coal Slurry Combustion and Technology, April 1983.

78. Personal Communication from J.M.G. Tresaco, Hidroelectrica de Cataluña, S.A.

79. Allen, J.W., P.R. Beal and P.F. Hufton, "Small and Large Thermal Test Rigs for Coal and Coal Slurry Burner Development," Division of Fuel Chemistry, American Chemical Society, V. 28, N. 2, p. 62, March 1983.

80. Henderson, C.B., R.S. Scheffee, and E.T. McHale, "The Development of Coal-Water Slurries as Boiler Fuel," presented at the Fourth Annual Coal Utilization Conference, Houston, Texas, November 1981.

81. McHale, E.T., R.S. Scheffee and N.P. Rossmeissl, "Combustion of Coal-Water Slurry," Combustion and Flame, 42, 1982.

82. Ghassemzadeh, M.R., T.M. Sommer, G.A. Farthing, and S.J. Vecci, "Rheology and Combustion Characteristics of Coal-Water Slurries," presented at American Flame Research Committee, Chicago, Illinois, October 1981.

83. Manfred, R.K., R.W. Borio, D.A. Smith, M.J. Rini, R.C. LaFlesh, and J.L. Marion, "Current Progress in Coal-Water Slurry Burner Development," Division of Fuel Chemistry, American Chemical Society, V. 28, N. 2, p. 36, March 1983.

84. Hickman, R.H., and F.P. Buckingham, "Burner Development for Coal and Water Slurry Fuel," Proceedings of Fifth International Symposium on Coal Slurry Combustion and Technology, April 1983.

85. DeMichele, G., M. Graziadio, S. Ligasacchi, G. Saccenti, and G. Salvi, "Large Scale CWS Firing Tests at the S. Barbara Power Station," Proceedings of Fifth International Symposium on Coal Slurry Combustion and Technology, April 1983.

86. Rankin, D.M., R.P. Nicholson, H. Whaley, and I.D. Covill, "The Development of Coal-Water Mixture Technology for Utility Boilers in Eastern Canada," Proceedings of Fifth International Symposium on Coal Slurry Combustion and Technology, April 1983.

87. Basta, N., "Coal-Water Slurries: A Step Away from Success?" Chemical Engineering, p. 14, June 1983.

88. Kimel, E., and R. Kurtzrock, "Comparison of Coal Conversion Alternatives," Proceedings of Fifth International Symposium on Coal Slurry Combustion and Technology, April 1983.

89. Maier, G.A., R.E. Barrett, B.W. Rising, and R.D. Giammer, "Coal Slurry Fuels: Moving Toward Commercial Utilization," Proceedings of Fifth International Symposium on Coal Slurry Combustion and Technology, April 1983.

90. SYNFUELS, Fluidcarbon, Sweden, See Coal-Water Mixture at 60% of Oil Price in the U.S., November 5, 1982.

91. Foo, O.K., and E.M. Jamgochian, "Assessment of Coal-Liquid Mixtures in Cooperating IEA Countries," Technology Assessment of Coal-Liquid Mixtures in Cooperating IEA Countries, Volume III - Technology Assessment, MTR-83W87-04, The MITRE Corporation, June 1983.

92. Ford, F.W., A.M. Madgavkar, R.E. Moore, and R.J. McCormick, "Coal-Water Mixtures Commercial Development Evaluation," Proceedings of Fifth International Symposium on Coal Slurry Combustion and Technology, April 1983.

93. Malgarini, G., M. Giuli, L. Palumbo, and S. Palella, "Evaluation of CWM Use in Iron Making," Proceedings of Fifth International Symposium of Coal Slurry Combustion and Technology, April 1983.

94. Hoy, H.R., A.G. Roberts, and K.K. Pillai, "Combustion of Coal-Water Slurry in Pressurized Fluidized-Bed Combined-Cycle Plant," Proceedings of Fourth International Symposium of Coal Slurry Combustion, May 1982.

95. Clements, J.T., W.F. Fox, D.R. Gallagher, W.M. Poundstone, and L. Rheinfrank, Jr., "The Economics and Technical Feasibility of Utilizing Coal-Water Mixtures in Rotary Rock Dryer Applications and Other Small- to Medium-Size Industrial Combustion Applications," Proceedings of Fifth International Symposium of Coal Slurry Combustion and Technology, April 1983.

96. Mark, A., "Coal Aqueous Mixtures," Division of Fuel Chemistry, American Chemical Society, V. 28, N. 2, p. 23, March 1983.

97. L. Stigsson, "Combustion of Coal-Water Slurry in Sweden," Proceedings of Fifth International Symposium of Coal Slurry Combustion and Technology, April 1983.

98. Passman, R.A., "Coal Water Slurry Pilot Plant," Proceedings of Fifth International Symposium of Coal Slurry Combustion and Technology, April 1983.

99. Naka, A., S. Honjo, T. Imamura, Y. Mizuno, "Development of Additives for High-Content Coal-Water Slurry," Proceedings of Fourth International Symposium of Coal Slurry Combustion, May 1982.

100. Jones, J.F., R.A. Meyers, L.C. McClanathan, and J.L. Anastasi, "Front End Boiler Equipment Selection, Performance and Hydraulics for Using Coal-Water Mixtures," Proceedings of Fifth International Symposium of Coal Slurry Combustion and Technology, April 1983.

101. Barsin, J.A., "Commercialization of Coal-Water Slurries," presented at the 9th Energy Technology Conference, Washington, D.C., February 1982.

102. Sommer, T.M., and J.E. Funk, "Development of a High-Solids, Coal-Water Mixture for Application as a Boiler Fuel," ASME paper 81-JP6C-Fu-4, 1981.

103. Hargrove, M.J., A.A. Levasseur, and B.E. Davis, "Combustion and Fuel Characterization of Coal-Water Mixtures," Proceedings of Fifth International Symposium of Coal Slurry Combustion and Technology, April 1983.

104. Pan, Y.S., G.T. Bellas, R.B. Snedden, D.J. Wildman, and J.I. Joubert, "Effect of Coal Ash Content on Performance of a CWM-Fired Boiler," Proceedings of Fifth International Symposium of Coal Slurry Combustion and Technology, April 1983.

105. Barsin, J.A., "Commercialization of Coal-Water Slurries-II," Proceedings of International Symposium on Conversion to Solid Fuels, October 1982.

106. Cousin, S.B., and H.S. Schenkel, "Economic Evaluation of Fuel Conversion Options for Oil-Fired Power Plants," presented at the Joint Power Generation Conference, October 1981.

THE PREPARATION OF LOW ASH, LOW SULPHUR COAL   (Ref: G52/4)

Frederick Vickers* and Dr. Stephen Ivatt**

A majority of the published literature on Coal Liquid
Mixtures contains little reference to coal quality
requirements, ie. Coal Preparation (beneficiation) as
distinct from Slurry Preparation. A broad specification
has, however, been inferred and is discussed in relation
to the early stages of an investigation into the reduction
of the sulphur content of UK coals. Preliminary results
suggest that: the production of very low ash coals is
quite feasible; there is some scope for the reduction of
sulphur content but that it is extremely difficult; in terms
of solids concentration, there may be some advantage to
preparing coals for CLM rather than for conventional fuels.

INTRODUCTION

In the past decade there has been an increasing international interest
shown in coal both as a replacement for and source of fluid hydrocarbon
fuels. At the same time problems have been posed to the coal industry by
the increasing proportion of fine (below 500 μm) particles produced by
modern mining techniques, the development of hydraulic transport/mining of
coal and the environmental aspects involved both in the production and the
utilisation of coal. While this wide range of interests and problems stem
from largely independent origins it is interesting to observe that in terms
of coal beneficiation a common factor is emerging. This common factor is
the need to prepare high grade (low ash, low sulphur) coals of fine size.

* Head of Fine Coal Group, NCB,
Mining Research and Development Establishment

** Project Engineer, NCB,
Mining Research and Development Establishment

Coal preparation engineers have traditionally and with good reason studiously attempted to minimise the necessity for fine coal treatment on the grounds of both the high cost and the technical difficulty of the processes involved, particularly dewatering which has an important influence on the handleability of the products[1]. This minimisation of fines treatment has been achieved by two principal means, in Europe by the extraction of raw fines for subsequent blending with cleaned coal and in low coal production cost countries, notably the USA, by simply discarding the raw fine coal. The advent of Coal Liquid Mixtures (CLM) is therefore of great interest to coal processors in that, providing the required degree of beneficiation can be achieved, they offer a marketing outlet for fine coals of high moisture content (30 to 50% by weight).

The Mining Research and Development Establishment of the NCB have as yet no project specifically aimed at the preparation of CLM. There is however a priority project on the reduction of the sulphur content of coal by existing conventional processes. This project is part of a joint CEGB/NCB investigation into the reduction of SO_2 emissions and covers flue gas desulphurisation, new combustion techniques (eg. fluidised bed combustion) and new coal preparation techniques (eg. high gradient magnetic separation[2] and the Otisca dense liquid process[3]). Since, in order to remove the maximum amount of mineral impurity, it will be necessary to liberate that impurity by crushing/grinding, it follows that the techniques involved will be applicable to the preparation of CLM.

This article attempts to assess the degree of coal beneficiation required for CLM, discusses the types of conventional coal preparation process likely to achieve the degree of upgrading required and finally reviews the preliminary investigation being carried out at MRDE.

SPECIFYING THE COAL REQUIRED FOR CLM

A first requirement for the successful design of any type of process is a specification for the end product. Unfortunately, despite the abundance of current literature on CLM there is little information, other than particle size, on the coal phase of these mixtures. Given the variety of both the types of and potential markets for CLM it is appreciated that there can be no single ideal coal specification but some firm guidelines would greatly assist coal producers and preparation engineers in forward planning and research. It would appear, however, that there are four important basic parameters, particle size distribution, ash content, sulphur content and solids concentration. The first two are of relevance to combustion technology, the third concerns the environment while the fourth is important to the rheology of the mixtures. These parameters and their relevance to coal beneficiation are discussed below.

Particle Size Distribution

The particle size distribution of the coal phase of coal liquid mixtures is a compromise between the conflicting requirements of having a stable, easily pumpable fluid which is at the same time easily combustible. As a result a wide range of particle size distributions have been used in the transport and utilisation of CLM. A study of reports published to date does, however, indicate that the majority of utilisation processes would require a coal feedstock with a size distribution approximating to that of conventional dry pulverised fuel, ie. 80% below 75 µm.

In considering the coal beneficiation processes the precise particle size distribution required for CLM is unlikely to be a major factor. The economics and technical difficulties of such processes constrain the coal preparation engineer to prepare coals at the maximum particle size consistent with achieving the required product quality. Further, on the same grounds, the mean particle size at which it is possible to beneficiate coals by conventional means will probably be greater than that required by the mixtures.

Ash Content

The level of coal ash content that is required for coal liquid mixtures is, of course, dependent on their use. Since a prime motivation for the development of CLM is a reduction in the consumption of oil for power generation, the major market in the short term is as a direct replacement for oil in existing combustors. In this case the level of ash content that can be tolerated falls into two distinct categories dependent on boiler design. First, where the boiler was originally designed for oil burning the ash content must be low to minimise derating of the boiler due to ash deposition and coal ash contents of the order of 1 to 3% by weight appear to be required. Secondly, with boilers originally designed for coal firing and then converted to oil firing, there is a greater tolerance of ash and coal ash contents of up to about 10% would seem to be generally acceptable.

With respect to coal beneficiation the difference between these two levels of ash content, 2% and 10% by weight, is very great. Achieving a level of 10% ash content would present few problems, even with quite difficult coals. On the other hand, to consistently produce coals at 2% ash content it will be necessary to start with a good quality, easily washable raw coal, comparatively sophisticated process plant circuitry and closely controlled operation.

Sulphur Content

Internationally, increasing attention is being paid to the emission aspects of fuel combustion. Many countries now have strict emission control standards, particularly for SO_2 emissions. The definition of these standards varies throughout the world but there is a gradual movement towards definitions which in some way relate SO_2 emission to the heat value of the fuel; either directly as in the USA or in terms of power output as in Germany. Whatever definition is used, however, the various standards, in the case of burning hard coal, tend to correspond to a sulphur content of about 1% by weight.

Reducing the sulphur content of coal is probably the most difficult problem ever to face coal preparation engineers. There are two main forms of sulphur in coal, organic and pyritic and it is only the pyritic content which can be reduced by conventional coal preparation techniques. Problems of separation arise because the occurrence of iron pyrites in coal can vary considerably both in quantity and, more significantly, in its degree of dissemination within the coal. The main criteria pertaining to the reduction of coal sulphur content by conventional processing are; the degree to which the pyrites can be liberated from the coal and the particle size at which a specific degree of liberation occurs.

Solids Concentration

The fourth parameter which is important to the preparation of CLM is solids concentration. While this parameter principally concerns the rheology of the mixtures and their subsequent handling, transportation and storage it is of great importance, as previously stated, to coal beneficiation. To prepare coals with moisture contents as high as 30 to 50% by weight would give an unprecedented degree of freedom to coal preparation engineers to design for maximum beneficiation without the constraint of product handleability which currently restricts 25 mm to 0 product moisture contents to a maximum of about 10% by weight.

CONVENTIONAL COAL PREPARATION PROCESSES

Coal preparation techniques can be divided into two broad categories, coarse (above 500 μm) and fine (below 500 μm) coal processes. All of the coarse coal processes utilise differences in relative densities between mineral impurities and coal particles to effect a separation and, in principle, provided there is liberation of the impurities they can be readily removed by such processes. Fine coal processes on the other hand, particularly those which are capable of beneficiating very fine (below 150 μm) coals, rely on complex physico - chemical interactions to effect a separation. These are much more difficult to predict and control.

Although there are a large number of coal preparation techniques it is likely that it will be the fine coal processes which will predominate in any plant designed to produce low ash, low sulphur coals. The emphasis in the following notes is therefore placed on fine coal cleaning techniques and in particular on those which are included in the current MRDE investigation.

Dense Medium Cyclones

The dense medium cyclone is well known for both its flexibility and accuracy but only when operating at separating gravities in excess of a Relative Density of 1.3 R.D. on particle sizes greater than 500 μm. These limiting conditions do inhibit its ability to produce very low ash, low sulphur coals. With a majority of coals a separating density of 1.3 will be too high to yield coals with ash contents less than 2 to 3% weight while at particle sizes of greater than 500 μm there may be insufficient liberation of iron pyrites to allow much reduction in sulphur content. Attempts are being made to carry out dense medium cyclone separations at particle sizes down to about 100 μm with, it is claimed, some success. The most recent published work in this field comes from South Africa[4] and the USA[5].

The flexibility and accuracy of separation in dense medium cyclones will ensure that they have an essential role in the preparation of high grade coals, particularly if they can be adapted to treat sizes less than 500 μm.

Water Washing Cyclones

With the notable exception of the USA water washing cyclones have been largely ignored in coal preparation but there are signs of a revival of interest. Separating efficiency is rather poor for these devices although this can be improved by multiple staging. They have been mainly used for preliminary scalping, typically with the underflow feeding to concentrating tables and the overflow to froth flotation, but they are capable of producing very low ash, low pyritic sulphur products. There is a strong size classification aspect to their performance which can cause a quite sharp increase in separating density with decreasing particle size. In a study of single stage water washing cyclones Deurbrouck[6] reports evidence of quite substantial reductions in the ash and pyritic sulphur contents of several coals. Of the more recent work on these cyclones Mikhail and Butcher[7] report some interesting findings. Their work indicates that by operating water washing cyclones in the presence of up to 50% of coal slimes (below 150 μm) rather than clay slimes, much sharper separations at lower partition densities can be achieved.

Water washing cyclones can offer a relatively cheap, high throughput process and are worthy of consideration for coal cleaning circuits aimed at producing high grade coals.

Concentrating Tables

The concentrating table is another device which is used quite extensively in the USA but, until recently, comparatively little elsewhere. Its use has been limited mainly because it has a relatively low unit capacity and a poor separating efficiency in respect of ash content. It has long been known, however, that it is capable of effecting quite sharp separation of iron pyrites and this has brought about the revival of interest. Of the more recently reported work, that by Monostory[8] on tests carried out at pilot scale in Germany shows efficient separation of pyrites from a coal feed with a nominal top size of 2 mm.

There is clearly a good case for considering concentrating tables in cleaning circuits when low sulphur products are required.

Froth Flotation

This is the most widely used process in the world for the beneficiation of fine (usually below 500 μm) coal and is one of the few processes which can effect a separation of particles down to zero size. Separation is effected by utilising differences in the surface properties of particles and the process is capable of being very selective in respect of ash content, to the extent of separating individual coal macerals[9]. Efficiency of separation is to some extent dependent on particle size and reduces as particle size decreases. Although most widely used in coal preparation as a high capacity, single stage process, it can be used, as in the metallurgical industry, as a more sophisticated multiple stage operation. An interesting early example of the commercial application of multiple stage flotation for coal commenced operation in the UK in 1941[10]. In a three stage flotation process, the plant produced a 40% yield of coal at 0.7% ash content and 48% yield of coal at 2.6% ash content from a feed containing 7.8% of ash. The principal disadvantage of the process is that, unfortunately, iron pyrites appears to exhibit similar flotation characteristics to clean coal. As a result even well liberated pyrites will tend to float with the clean coal concentrate, particularly in single stage flotation. Manipulation of the process; multiple-stage flotation, inhibited flotation and depression of either coal or pyrites, can reduce the quantity of pyrites reporting to the clean coal product. Recent bench scale tests carried out by Carlson et al[11] show that some reduction of pyritic sulhpur content is possible using such process manipulation.

Given its ability to treat very fine sizes at high throughputs to produce low ash content products, froth flotation should be considered for any coal cleaning circuit which is required to produce such products.

Oil Agglomeration

Oil agglomeration is the only coal preparation process which can be claimed to provide efficient and selective separation, with respect to ash content, of very fine (below 100 μm) sizes. The principal disadvantage of the process is the high oil consumption required, 10 to 15% by weight of feed solids. This has led since the 1920's to intermittent commercial exploitation influenced mainly by cyclic fluctuations in the relative prices of coal and oil. The most recent commercial installations, based on the Olifloc[12] process, ceased operation in Germany in the late 1970's following the steep rise in oil prices. The advent of CLM has, however, led to a revival of interest, particularly for the preparation of Coal Oil Water (COW) and Coal Oil Mixtures (COM). A lot of research is taking place in both Canada[13] and Australia[14] where the use of oil agglomeration techniques are being considered as an integral step in the preparation of COM. A questionable aspect of the process is, as it is with froth flotation, the tendency to retain iron pyrites in the cleaned coal product. Investigations by a number of workers[15,16,17] have shown classical depressants to be relatively ineffective in depressing pyrites in oil agglomeration. On the other hand, there has been some apparent success when using mild chemical or bacteriological pretreatment[18,19,20].

An ability to efficiently separate very fine sizes makes oil agglomeration a potentially useful process for the preparation of the coal for COW and COM.

INVESTIGATIONS AT MRDE

As stated in the introduction, the primary objective of the current investigation at MRDE is to assess the potential of existing conventional coal cleaning processes for reducing product sulphur content. It is this aspect which is given prominence in this section.

Raw Coal Characterisation

Selection of type. It is obvious that given an unrestricted choice of raw coal from which to prepare a low ash, low sulphur product, coal preparation engineers would select a raw coal of low ash, low sulphur content. This type of raw coal is, however, in short supply on an international scale and most certainly in the UK. Taking a long term view, it was deemed unrealistic to carry out tests on such coals and the decision was taken to concentrate efforts on coals which are representative of both the present supply to the electricity generating industry and also the bulk of known, exploitable reserves. Such coals are in the lower range of the UK coal ranking system, ie. 500 to 900 on a scale of 100 to 900. This scale is based on volatile matter (less than 9% for 100 rank, greater than 32% for 900 rank) and caking properties.

Test Programme. A major problem in coal beneficiation is the fact that coal is heterogeneous and liable to show wide variations in character even within a particular seam at a particular colliery. As a result there is danger in over generalising when designing coal processing plant and each coal should be considered strictly on its own merits. The need to characterise a raw coal in some detail in order to assess its potential for upgrading presented an immediate difficulty to this investigation. Although there exist within the NCB comprehensive details on the washability of most British coals, the analyses comprise almost exclusively data on ash content and total sulphur content only and not on the forms of sulphur present. Since this investigation concerns the separation of iron pyrites, which can vary quite dramatically in quantity and degree of dissemination in some coals, it was necessary to initiate a programme of detailed testwork on selected raw coals. These examinations require a sizing analysis of the raw coal, float and sink tests on each size fraction over a range of relative densities and determinations of ash, moisture and forms of sulphur on each relative density fraction. In addition, in order to assess the potential for liberating pyrites, it is necessary to crush selected size fractions of the raw coal and repeat the detailed float and sink analyses.

Examinations of the type described above are expensive in both time and labour and at the time of writing only one study has been completed. From a major production unit in the Yorkshire Region of the NCB a sample of <31.5 mm raw coal feed to a Baum washbox was sized at 12.5 mm, 6.3 mm, 3.15 mm, 1 mm, 500 μm and 150 μm. Float and sink analyses were carried out on the sizes >150 μm at relative densities of 1.3 to 2.0 in steps of 0.1 relative density. Ash, moisture and forms of sulphur were determined on each float and sink fraction. The 150 μm to 0 fraction was analysed for ash, moisture and total sulphur only. In addition to the foregoing, a sub-sample of the 31.5 mm to 12.5 mm raw coal was gently crushed to below 3.15 mm and re-sized at 1 mm, 500 μm and 150 μm. Float and sink analyses and ash, moisture and forms of sulphur were determined as before.

Discussion of Results

Figure 1 shows the overall weight in pyrites in each sized relative density fraction of the 31.5 mm to 0.15 mm raw coal. An interesting feature of this distribution is the fact that more than a third of the total pyrites is contained in the relative density fractions greater than 1.8. Since analysis also showed the combined ash content of these fractions to be greater than 82%, there is a clear case for carrying out a primary high gravity separation on this coal before crushing and re-treating.

Figure 2 shows the fractional sulphur contents of the 31.5 mm to 0.15 mm raw coal. The two notable features here are: firstly, the highest concentration of pyrites is in the 'middlings' fractions with a peak of 2.5% weight in the 1.6 to 1.7 relative density fraction; secondly, the highest concentrations of organic sulphur occur in the lowest relative density fractions. Figure 3 shows this data plotted in cumulative form together with cumulative ash content. This indicates that as the coal reduces in ash content the organic sulphur content increases until, at the lowest ash content, it represents about 65% of the total sulphur content. Organic sulphur distributions of this type are common and pose a major problem to the beneficiation of coal by conventional processes.

The histogram, Figure 4, shows the effect on pyrites distribution of crushing the 31.5 mm to 12.5 mm raw coal size fraction to below 3.15 mm. All relative density fractions less than 2.0 show a reduction in pyritic sulphur content while the highest shows an increase, from 30% to 43% weight.

Figure 5 shows clearly the effect of crushing on the liberation of other mineral impurities. Washability data are plotted in the form of Mayer curves for both the original 31.5 mm to 12.5 mm raw coal and the same coal crushed to below 3.15 mm. The illustration clearly shows that whereas virtually none of the original large coal is available at less than 3% ash, there is almost 50% weight available in the crushed coal at a relative density below 1.4.

Future Tests. An important finding in these preliminary tests was the indication that it would be advantageous to 'de-shale' the raw coal before crushing and re-treating. This is now being followed up through a second series of tests. These involve collecting cleaned coal samples from the operating coal preparation plant on site and subjecting these samples to similar analysis, ie float and sink testing before and after crushing. Sub-samples of the crushed clean coal will undergo a coal processing investigation at both bench and pilot scale.

Raw Coal Beneficiation

The only fines treatment technique currently practised in the NCB is froth flotation. As a result most of the work carried out at MRDE in this field has been confined to studying that process and its associated operations. The principal equipment in this work is an 0.5 t/h pilot fines treatment plant which includes two banks of flotation cells, a thickener/clarifier and a vacuum filter. Water washing and classifying cyclones together with a concentrating table are to be included in the circuit to assist in the current investigation but they are not yet available. The following notes, therefore, comprise mainly background information on the flotation of naturally occurring fines from British raw coals with some limited data from laboratory scale tests on sulphur reduction.

Reduction of ash content. The natural fines from most British coals of higher than 600 rank (high volatile, medium caking) can be treated by single-stage flotation to produce cleaned products of less than 10% ash content. Since almost 20% by weight of the output from UK deep mines is below 500 μm in size, a considerable quantity of this grade of product is already available. For coals of lower rank, however, it would be necessary to introduce multiple-stage treatment. Figure 6 is a Mayer curve representation of the results of single and multiple-stage laboratory flotation of an 802 rank coal (high volatile, weakly caking). It can be seen that while the single-stage process produces a negligible quantity of coal below 10% ash content, three-stage treatment produces about 53% weight of coal at 9% ash.

The production of very low ash, 1 to 3%, coal is much more difficult. There are very limited quantities of coal of this quality available naturally although, as was shown in Figure 5, some can be liberated by crushing. In the single-stage flotation of natural fines this level of ash content may only be produced from the first or second cells of a bank of cells treating anthracite or prime coking coals but these are high rank (100 to 300) premium fuels which are already in short supply. Reasonable quantities of very low ash product can, however, be produced by multiple-stage flotation of lower rank coals, down to 500/600 rank. An example of this is given in Figure 7 which shows the results of laboratory scale flotation of a 500 rank coal. This indicates a potential for producing over 50% by weight of cleaned coal at 3% ash content from the third flotation stage.

Summarising: 6 to 10% ash fine coal is already available from higher rank (100 to 600) seams and can be produced from lower rank seams by multiple-stage treatment; very little 1 to 3% ash fine coal is available now but can be produced by multiple-stage flotation of coals of 100 to 600 rank, lower rank coals would require crushing and multiple-stage re-treatment.

Reduction of sulphur content by flotation. Single-stage flotation of British coals, as applied commercially, usually produces a clean coal concentrate with a total sulphur content which is higher than that of the original feed material. There appear to be two principal reasons for this: concentration of iron pyrites in the froth due to its natural flotability; concentration of organic sulphur in the cleanest coal fractions (ref. Figure 2). Re-treatment of the concentrate (ie multiple-stage flotation) can reduce the product sulphur content but would rarely be justified, economically, on the grounds of sulphur reduction alone. Typical examples of the results of multiple-stage flotation of a 600 rank coal are shown in Figure 8. Figure 8(a) shows the results of a pilot scale test and the increase in sulphur content from the first stage followed by a reduction in the second stage is clearly illustrated. Figure 8(b) shows the results of a bench scale four-stage test. The most notable feature here is the limited sulphur reduction at the successive flotation stages.

The depression of iron pyrites in coal flotation by chemical means has been the subject of much research but, as yet, has met with limited success. On the other hand, it has been demonstrated to be feasible to depress coal while floating pyrites, ie. reverse flotation[21]. The technique requires a two-stage flotation process wherein coal is floated in the first stage and then depressed in the second stage while floating pyrites. A variation of the method is to depress coal and float pyrites in the first stage and then float the coal in the second stage. An examination of these techniques was carried out at bench scale on a specially selected sample of high ash, very high sulphur fine coal of 902 rank. Results are presented diagrammatically in Figure 9. Flotation of coal in the first stage can be seen to be the better method.

In summary, multiple-stage flotation is an unlikely route for sulphur reduction. Reverse flotation, however, with flotation of pyrites in the second stage, is worthy of further study.

Reduction of sulphur by concentrating table. A preliminary assessment of this process has been made on a laboratory scale flowing film separator. A number of samples of natural low rank coal fines (below 500 μm) from operating cleaning plant circuits were tested. Examples of the results of these tests are shown in Figure 10. These results show that, in terms of sulphur reduction, the technique works well on British coals. Ash reduction was poor but predictable for material of such fine size.

In view of the encouraging sulphur reductions further tests were carried out using the technique in combination with froth flotation.

<u>Combining flotation with tabling.</u> A standard method of using concentrating tables in coal preparation is to pre-classify the raw feed by cyclone, treat the underflow by concentrating table and the overflow by froth flotation. The main disadvantages of this arrangement are: the cyclone overflow may contain a high proportion of fine pyrites which is unlikely to be removed in the flotation process; the low pyrites product from the concentrating table requires re-treatment to further reduce ash content. In an attempt to overcome these disadvantages and, in particular, to reduce the number of treatment stages, tests were carried out on simple combinations of froth flotation and tabling operations. The 500 μm to 0 raw fines for these tests were from a difficult to clean 900 rank coal and contained about 70% weight of solids below 63 μm.

The results shown in Figure 11 together with those of a single stage flotation test for comparison. These results are considered to be very encouraging with, as anticipated, those from the combination of flotation followed by tabling being the better. The major advantage of this particular combination is that it makes full use of the best features of both operations. Froth flotation is, in addition to being a good 'ash' separator, a very efficient desliming device and by using it as the primary separator the slimes loading for the concentrating table is considerably reduced. Further, the concentrating table is now only required to achieve a relatively simple separation between iron pyrites with a relative density of about 4.5 and clean coal with a relative density of about 1.4.

Further work will be carried out on this combined process at bench scale and subsequently pilot scale when the equipment becomes available.

CONCLUSIONS

In terms of ash content, test results show that it is technically possible to clean most British coals by conventional coal preparation processes to the standards dictated by the major markets proposed for Coal Liquid Mixtures.

Reducing coal sulphur content by conventional coal cleaning processes is undoubtedly difficult but test results indicate that worthwhile reductions can be achieved. It is necessary to liberate pyrites in order to separate it by these processes and test results suggest that secondary crushing of British coals can liberate at least some of the pyrites together with other mineral impurities. Test results also indicate, unfortunately, a tendency for organic sulphur content to be at a maximum in the cleanest coal fractions. This may mean that all of the pyritic sulphur has to be removed from some coals in order to meet combustion emission standards for sulphur compounds.

The development of Coal Liquid Mixture technology could bring an important advantage to coal cleaning operations. With the increasing quantity of fine coal produced by modern mining methods it is already difficult to reduce product moisture contents to a level compatible with

good product handleability for present markets. There is also increasing
pressure, on environmental grounds, to reduce coal sulphur content and it
appears that this can only be achieved by deliberately creating more fines.
The advantage of preparing coal for Coal Liquid Mixtures would, therefore,
stem from the much less stringent requirements for product moisture content.
This may more than offset any need for additional coal cleaning processes to
improve product grade.

ACKNOWLEDGEMENT

The authors wish to thank the NCB's Director of Mining Research and
Development for permission to publish this paper. Views expressed are those
of the authors and not necessarily those of the NCB.

REFERENCES

1. VICKERS, F — 'Problems in fine coal treatment'. Design 82. Instn. Chem. Eng. Symposium Series No 76. 1982.

2. HUCKO, R.D. — 'The application of HGMS in Coal Preparation' 9th International Coal Preparation Congress. New Delhi, 1982.

3. SMITH, C.D. — 'Otisca demonstration plant'. Coal Conference and Expo V. Louisville, 1979.

4. FOURIE, P.J.F. — 'Dense medium beneficiation of minus 0.5 mm coal in the Republic of South Africa'. 5th International Conference on Coal Research. Dusseldorf, 1980.

5. SEHGAL, R — 'Innovative Heavy Medium fine coal cleaning'. 9th International Coal Preparation Congress. New Delhi, 1982.

6. DEURBROUCK, A — 'Performance characteristics of coal washing equipment: Hydrocyclones'. US Bureau of Mines. RI7891, 1974.

7. MIKHAIL, M.W. — 'Novel techniques applied to the beneficiation of Western Canadian coals'. 9th International Preparation Congress. New Delhi, 1982.

8. MONOSTORY, F.P — 'Senkung des Schwefelgehalts von Kraftwerkskohle'. Gluckauf 115, 1979.

9. BROWN, D.J. — 'Coal flotation'. 50th Anniversary Volume Froth Flotation. AIME, 1962.

10. NELSON, H — 'The selective flotation of super low ash coal'. The Gas World. c.1946.

11. CARLSON, R — 'Effects of beneficiation of steam coal at the power plant'. 9th International Coal Preparation Congress. New Delhi, 1982.

12. BOGENSCHNEIDER, R.N 'The Olifloc process for the dewatering and cleaning of ultrafine slurries in Coal Preparation.' 15th Biennial Conference. Instn of Briquetting and Agglomeration. 1977.

13. IDLE, K 'The use of Selective Agglomeration in the preparation of COM'. 3rd Int. Symp. on COM Combustion, Orlando, 1981.

14. MESSER, L 'Coal beneficiation for oil admixture using the Spherical Agglomeration Process'. 4th Joint Meeting MMIJ-AIME. Tokyo, 1980.

15. GREGORY, F.W 'Oil agglomeration of coal fines' I.E.A. Mining Tech Cleaning House, R and D. Commentary. 1982.

16. ANON 'Oil agglomeration offers technical and economic advantages'. Mining Engineering. 32(8), 1980.

17. MEZEY, E.J 'Fuel contaminants: Vol 4, Application of oil agglomeration to coal wastes'. US EPA Report. EPA600/7-79-025b, 1979.

18. LEONARD, W.M 'Coal desulphurisation and de-ashing by oil agglomeration'. 2nd Symp. on Separation Science and Tech. for Energy Applications. Gathingburg, 1981.

19. McCREADY, R.L 'Removal of pyrite from coal by conditioning with Thiobacillus Ferroxidans followed by oil agglomeration' Hydrometallurgy, 5, 1980.

20. WHEELOCK, T.D 'Advanced development of fine coal desulphurisation and recovery technology'. Fossil Energy Ann. Report. Ames Lab. Iowa State University, 1980.

21. MILLER, K.J 'Coal-pyrite flotation in a concentrated pulp: a pilot plant study'. US Bureau of Mines. RI8239, 1977.

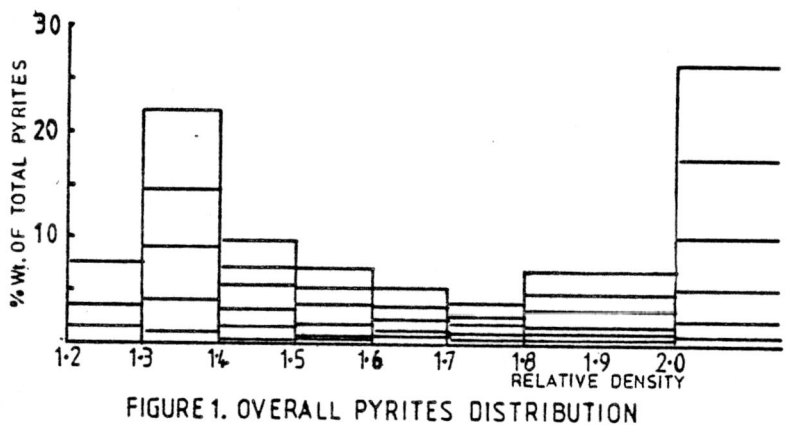

FIGURE 1. OVERALL PYRITES DISTRIBUTION

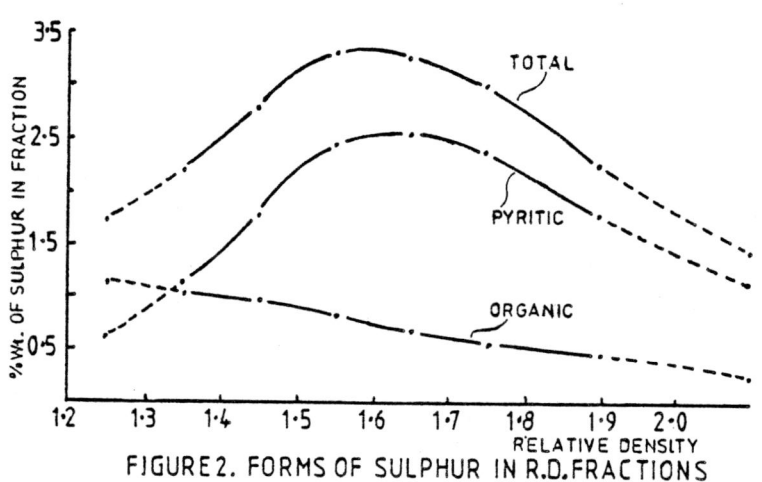

FIGURE 2. FORMS OF SULPHUR IN R.D.FRACTIONS

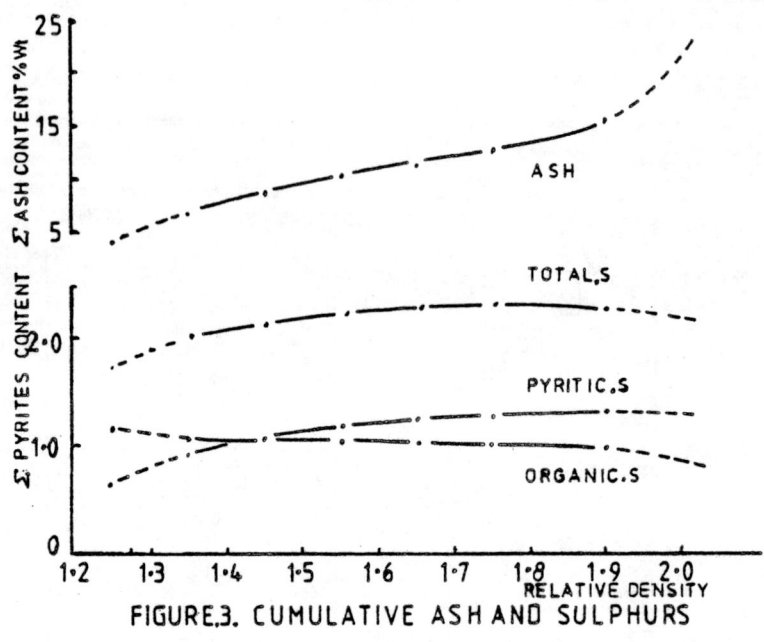

FIGURE.3. CUMULATIVE ASH AND SULPHURS

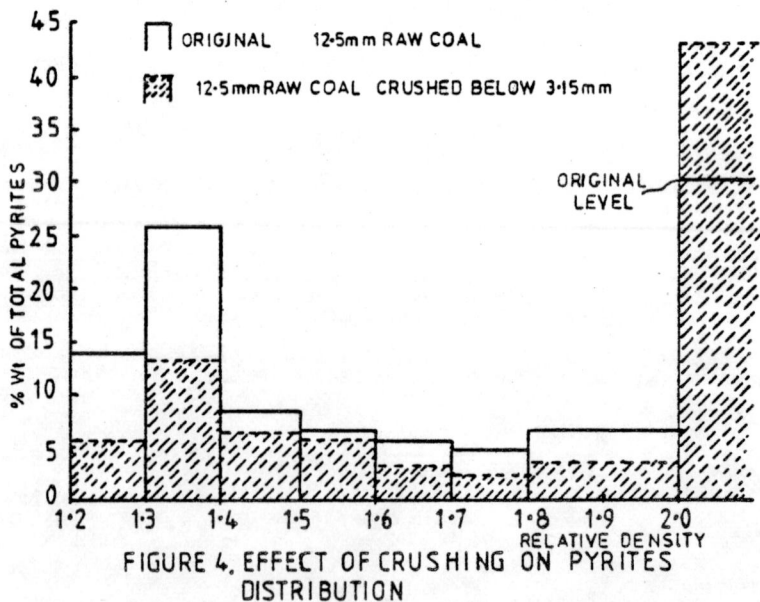

FIGURE 4. EFFECT OF CRUSHING ON PYRITES DISTRIBUTION

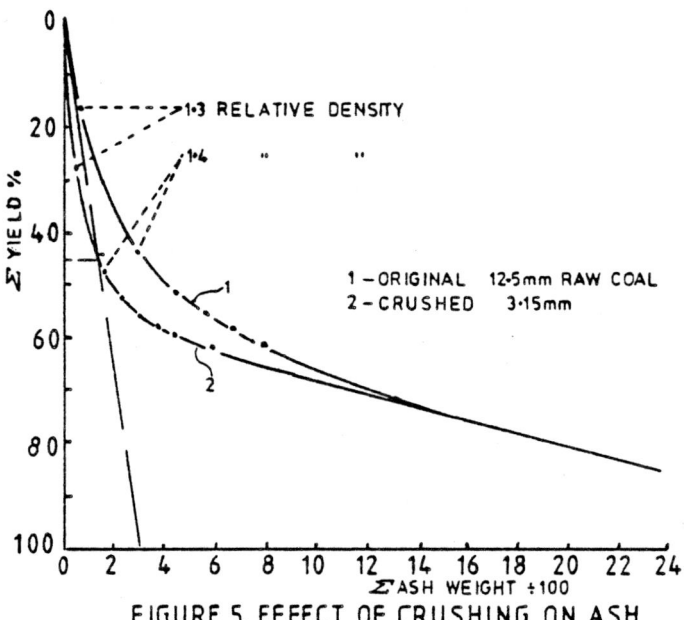

FIGURE 5. EFFECT OF CRUSHING ON ASH
RELEASE

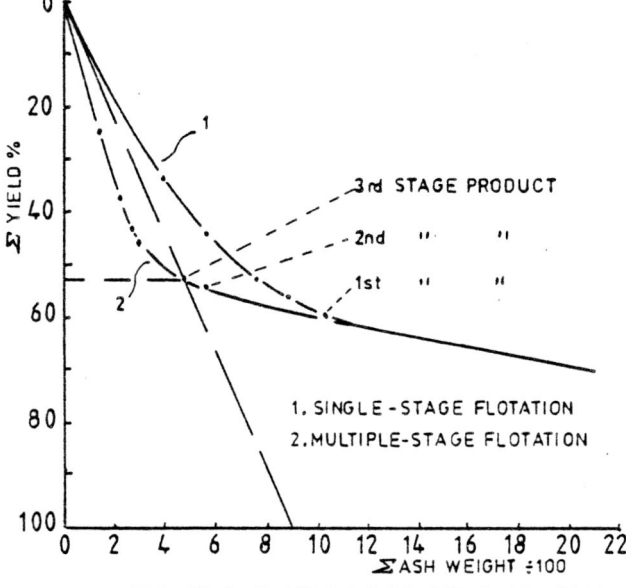

FIGURE 6. FLOTATION OF 802 RANK COAL

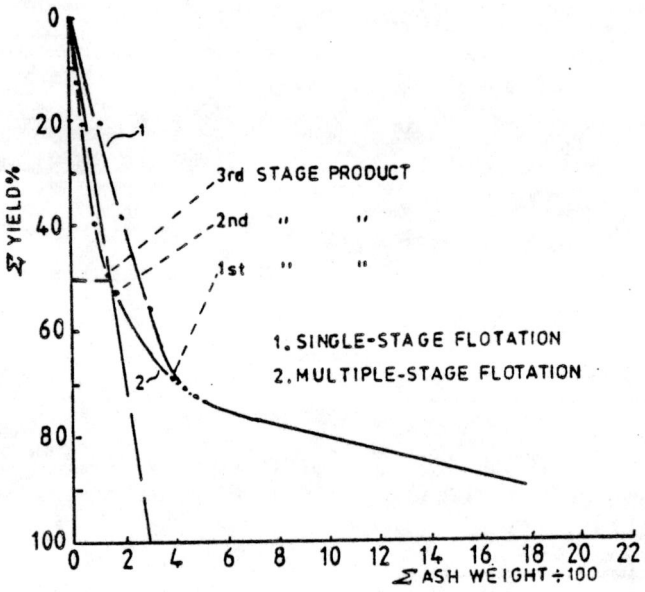

FIGURE 7. FLOTATION OF 502 RANK COAL

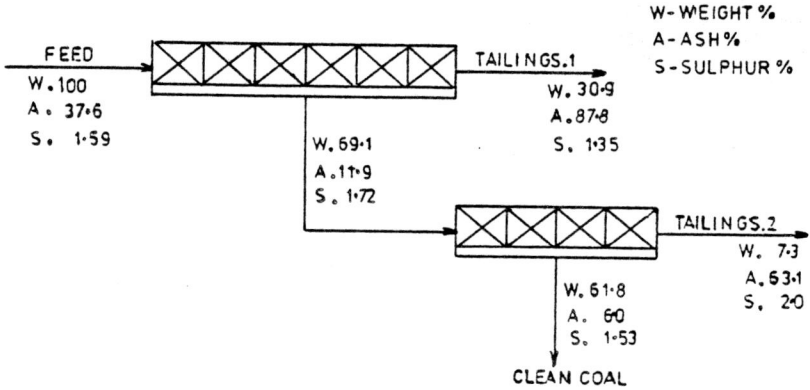

9(a) PILOT PLANT TWO-STAGE FLOTATION

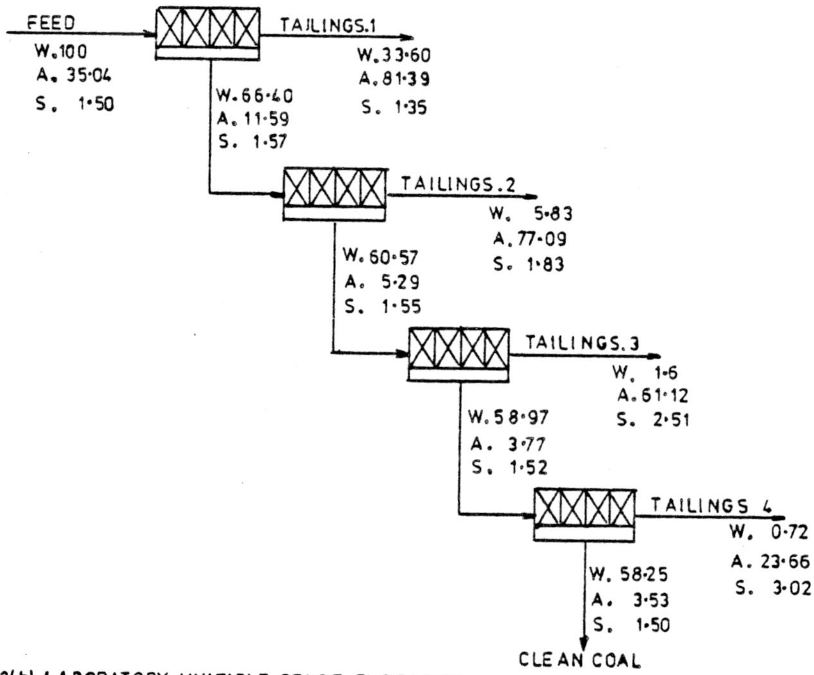

8(b) LABORATORY MULTIPLE STAGE FLOTATION

FIGURE 8. STAGED FLOTATION OF 600 RANK COAL

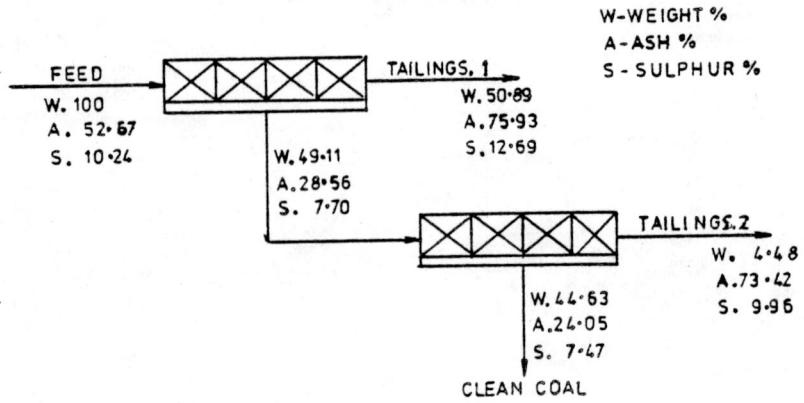

W-WEIGHT %
A-ASH %
S - SULPHUR %

9(a) TWO-STAGE FLOTATION

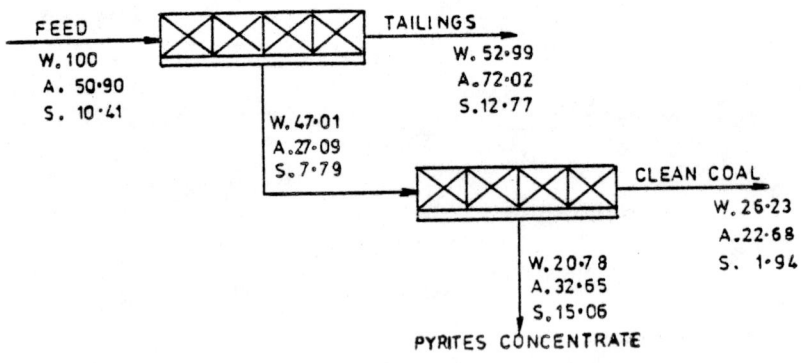

9(b) TWO-STAGE FLOTATION — REVERSE

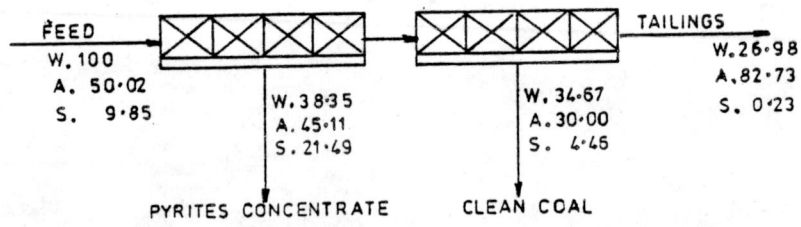

9(c) TWO-STAGE FLOTATION — MODIFIED REVERSE

FIGURE 9 ALTERNATIVE FLOTATION METHODS
(902 RANK COAL)

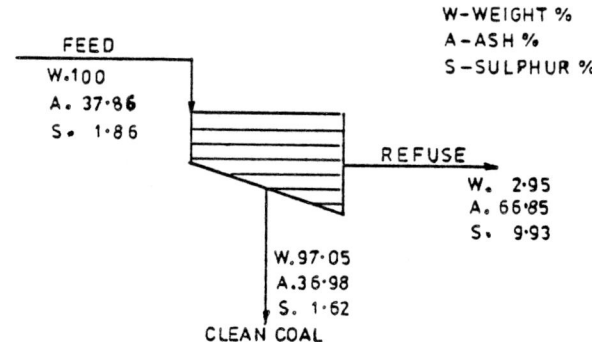

W−WEIGHT %
A−ASH %
S−SULPHUR %

FEED
W. 100
A. 37·86
S. 1·86

REFUSE
W. 2·95
A. 66·85
S. 9·93

W. 97·05
A. 36·98
S. 1·62
CLEAN COAL

10(a) 502 RANK COAL

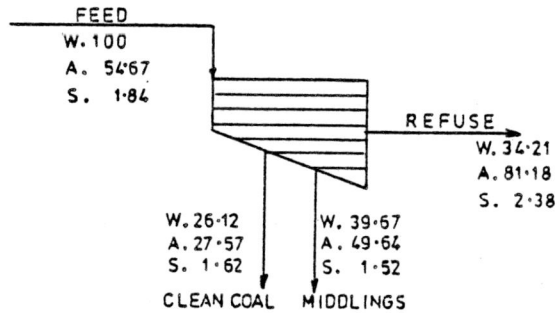

FEED
W. 100
A. 54·67
S. 1·84

REFUSE
W. 34·21
A. 81·18
S. 2·38

W. 26·12
A. 27·57
S. 1·62
CLEAN COAL

W. 39·67
A. 49·64
S. 1·52
MIDDLINGS

10(b) 602 RANK COAL

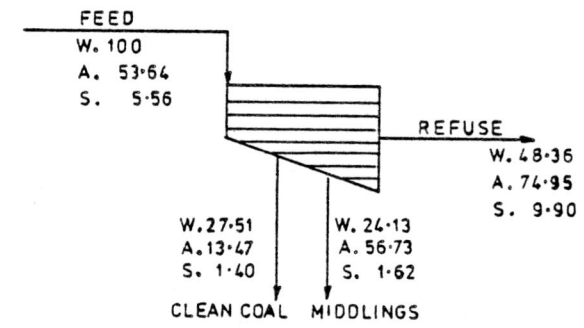

FEED
W. 100
A. 53·64
S. 5·56

REFUSE
W. 48·36
A. 74·95
S. 9·90

W. 27·51
A. 13·47
S. 1·40
CLEAN COAL

W. 24·13
A. 56·73
S. 1·62
MIDDLINGS

10(c) 902 RANK COAL

FIGURE 10. LABORATORY FLOWING FILM SEPARATIONS
ON <500μm COAL

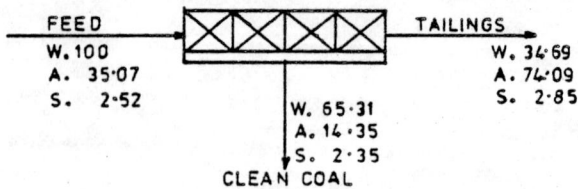

11(a) SINGLE-STAGE FLOTATION

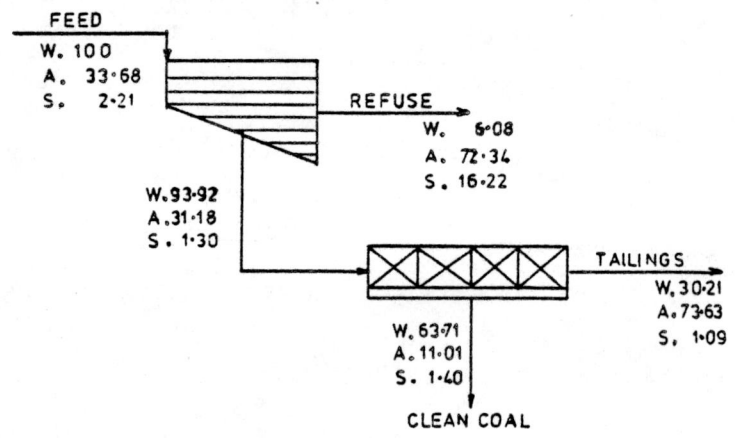

11(b) FLOWING FILM/FLOTATION

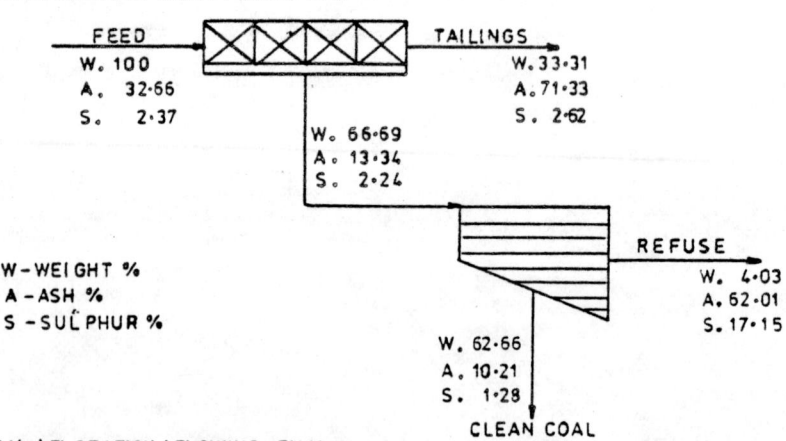

W – WEIGHT %
A – ASH %
S – SULPHUR %

11(c) FLOTATION/FLOWING FILM

FIGURE 11. COMBINATION PROCESSING OF <500 μm, 902 RANK COAL

SOME ASPECTS OF WET MILLING COAL-LIQUID SLURRIES

J.E. Jarvis

The commercial manufacture of coal slurries
for use as fuels or chemical feedstocks will
require machinery capable of continuously
processing large volumes of slurry.
The limitations of existing wet milling
processes and equipment are considered
for the production of coal slurries with
a majority of the solid particles less than
50 microns.

INTRODUCTION

A satisfactory coal-liquid slurry is a compromise which must combine low manufacturing cost and high coal content with trouble free pumping, storage and combustion.   Although coal-liquid mixtures can be manufactured by mixing dry crushed coal with a liquid without subsequent processing to reduce the size of the coal particles in the slurry, such mixtures are reputed to have poor storage characteristics, despite the use of suspension aids and to exhibit handling and combustion characteristics which limit their application to substitutes for heavy fuel oil.

It is possible that significant improvements in the characteristics and performance of coal-liquid mixtures can be obtained by wet milling the slurry to improve wetting and the homogenity of the mixture and reduce the size of the coal particles.   This will improve the handling and storage stability of the mixture, increase the rate of chemical reactions, including the combustion rate, and reduce erosion and wear both from the moving slurry and from solid combustion products.

THE MIXTURES

Although the coal-liquid mixtures of interest at this time are mixtures of coal with oil and coal with oil and water for use as substitutes for heavy fuel oils which are burned in furnaces and boilers , other mixtures and other applications are possible and may be of interest in the future. Some possible solid-liquid combinations and applications are listed in TABLE I.   It must be emphasised that some of the possible applications are highly speculative and may well be uneconomic in any circumstances.

Torrance and Sons Ltd., Bitton, Bristol.

CLM-E

A starting size for the coal in the order of 75 microns is assumed and a number of examples of wet milled slurries which have been produced, mainly on a laboratory scale for evaluation purposes, are listed in TABLE II together with an example of a "dry pulverize and mix" slurry. These slurries were produced by Union Process Inc. of Akron, Ohio as customer demonstrations of agitated ball mills or as a contract milling service, and are a selection of the forty or so trials of coal milling done by that company since 1979. Although the objectives of the laboratory trials were rarely revealed, it is understood that the main interest in slurries with an end requirement of 50% < 5 microns, is the investigation of the long term potential application of coal slurries as internal combustion engine and gas turbine fuels.

There appear to be many possibilities for wet milled coal/liquid slurries ranging from the six minutes residence time in a continuous tumbling ball mill needed to "refine" a coal-oil-water, heavy fuel oil substitute, where the action was probably only one of wetting, to the one hour residence time which resulted in a stable coal-oil mixture (95% less than 30 microns) without added water or surfactants. The trials detailed in TABLE II were carried out in agitated ball mills where the residence time is usually one eighth to one tenth that of a tumbling ball mill. This equates to residence times of up to ten hours in a tumbling mill to give a satisfactory internal combustion engine fuel.

WET MILLING

Coal slurries seem to be quite amenable to wet milling and from comparison of the laboratory trials detailed above with experience of wet milling a wide variety of slurries ranging from paint concentrates to biocides, it is reasonable to assume that coal at particle sizes below 75 microns behaves as a powder consisting of agglomerates which can be broken down by the usual process of dispersion. On this assumption, there are three possible physical mechanisms to be considered by which an agglomerated solid can be broken down by dispersion, neglecting the wetting effects for the time being. Firstly an agglomerate in a shear field will be exposed to forces which will depend upon the shear stress in the fluid and the orientation of the agglomerate to the shear field. The agglomerate will rupture if the stresses are sufficiently high and the orientation to the shear field is favourable. The particles resulting from this rupture are in themselves agglomerates which will suffer further reduction in size in favourable conditions. This is the most significant dispersion effect.

The second possible mechanism is disassociation of the fundamental particles. The agglomerate is considered to be a cluster of the individual particles with relatively weak bonds between them. Initially the hydrodynamically induced forces break the particle-particle bonds and then the individual particles move away from one another under the influence of local velocity gradients. (this mechanism is probably more true for the dispersion of surface treated synthetic pigments than coal particles).

The third mechanism is one of abrasion where agglomerates of the solid rub against one another or against other particles in the slurry and abrade away particles from the outer surface of the agglomerates.
The stresses causing particles to be detached from the outer surface in this way would be induced hydrodynamically. REF.1. This is a very doubtful effect. At small particle sizes the tendency would be for

touching particles to recombine.   It should be noted that the
mechanisms are ones of shear, induced hydrodynamically and that the
mechanisms of physical contact, of crushing and breaking, associated with
dry milling are not considered significant.

In most practical wet milling machines shear is induced in the fluid (the
coal slurry) by the relative motion between an agitator and the walls of
the container or between agitators or between free elements (grinding
media) moving in the fluid.   The wetting effects resulting from wet
milling are possibly two fold.   The usual physical effect of displacing
air or gas or liquid from the interparticulate spaces of the agglomerate
and also a chemical effect where the liquid reacts with the newly created
surfaces of the particles.   It is believed that this effect is
significant in maintaining the stability of some finely divided coal-oil
slurries and for the non-Newtonian rheologies of fine coal-water slurries.
Apparently chemical reactions of this type are not uncommon.   REF. 2.

There are a number of practical limitations which apply to wet milling
processes which affect the selection and application of machines.
These factors will help determine both the characteristics of the product
and the upper limits of proportionality when scaling-up to larger outputs.

A.   The shearing action must be transferred from the components of the
machine to the fluid (the coal/liquid slurry).   There will be some
limiting shear rate at which the velocity profile adjacent to the moving
surface will become discontinuous.   This will be influenced by the
rheology of the fluid.   This breakdown of the transfer of motion to the
fluid limits the maximum rotational speed and hence the output of disc
(colloid) mills, roller mills and "bead" mills.

B.   Where the shear field occurs in a thin film and the moving surfaces
are only held apart by the fluid film, film failure, because the film is
too weak or the forces on the moving surfaces are too great, results in
a loss of the dispersion action.   The contacting moving surfaces of the
machine are rapidly worn away.   Conversely the moving surfaces may be
so supported by heavy pastes and viscous fluids that their separation
prevents the creation of a shear field of sufficient intensity to effect
dispersion.
This sets the limits of the range of viscosities which can be processed in
any particular ball or bead mill and limits the roll pressures in roller
mills.

C.   For any (solid/liquid) mixture, the relationship between particle
size and particle separation remains constant, so the smaller the particles
become the closer they approach one another.   Ultimately the rates of
particle disruption and recombination will become the same and no further
effective particle size reduction will take place.   This effect is
dependent on the formulation of the mixture.

D.   In any given shear field the stresses imposed on a large particle with
a favourable orientation will be greater than those imposed on a small
particle and so large particles will be selectively disrupted at the start
of a dispersion process.   Ultimately all of the particles will have been
exposed, at their best orientation, to the maximum intensity shear field and
will have been disrupted to some minimum size at which the stresses imposed
by the shear field are inadequate to effect further disruption.   Particles
below this size will exist in the mixture as the products of earlier

disruptions of large particles.   This is the machine dependent limit
both of the ultimate particle size distribution and the rate of size
reduction.

E.   In practical wet milling machines only a small proportion of the fluid
being processed in the machine is in an area of high shear and is being
dispersed at any one time.   This limits the output of any machine and
accounts for the poor utilisation of space and energy in all wet milling
machines.

MACHINERY TO PROCESS COAL SLURRIES

From the characteristics of the coal slurries produced in laboratory trials
(TABLE II) and their behaviour during processing, it is possible to identify
the types of machines which are best able to make these slurries.

Although coal slurries are known to have been processed on a variety of wet
milling machines (including Colloid Mills and Vertical Bead Mills), there
are obvious advantages in preparing samples and small production quantities
in machines whose performance can be scaled-up to very large outputs if and
when a coal slurry becomes a commercially viable product.   Wet milling
practice is a well established and mature technology, despite the lack of
sound quantitative theory, and there have not been any significant
developments in processing equipment in the last twenty years.   It is
therefore reasonable to say that existing machines with maximum outputs
measured in tonnes per hour are not going to be suitable for applications
demanding outputs of hundreds of tonnes per hour, and that of the machines
listed in TABLE III only the two types of ball mill and the bead mills are
worth considering as production machines, and for a large scale commercial
process handling say 500,000 tonnes of coal per machine per annum (at a
one hour residence time) the only possible equipment seems to be the
continuous tumbling ball mill. FIG. I.
Large ball mills used for wet milling capable of giving this order of
output already exist and there is no evidence to suggest that they do not
perform as well as small ball mills.   There is also no indication that the
upper limit of size will result from a failure of milling performance.
The upper limit of size will probably be either a limit of the economy of
scale or some mechanical, structural or electrical engineering problem.

In the surface coatings industries and in some mineral dressing applications,
such as refining kaolin or whiting, tumbling ball mills used for wet milling
have been replaced by slow speed agitated ball mills (Attritors FIG. II)
or high speed small media mills (Beadmills FIG. III) REF.3.
These applications had used small capacity batch ball mills very often with
long residence times (24 to 72 hours) but at first sight there appears to be
no reason why both agitated ball mills and bead mills should not be equally
useful for short residence time high volume continuous applications such as
stabilizing coal-liquid mixtures.

For equivalent outputs both agitated ball mills and bead mills are smaller,
lighter machines than the tumbling ball mills they replace but for equivalent outputs the power consumed is little changed.   The size of the
machine, the initial capital cost of the machine, ball charge and
foundations and the continuous cost of wear are all very much reduced.

In order to give some indication of the possible limiting sizes of
processing equipment and the factors affecting both operation and scale-up,
a large ball mill and projections for both an agitated ball mill and a

bead mill are considered.   In each case the machine would be fed with a
mixture of dry milled coal with the liquid.   For the bead mill some form
of premixer would be required but both types of ball mill could be loaded
with either a premixed slurry or the separate components of the mixture.
The power consumed by large tumbling ball mills may be known but for both
agitated ball mills and bead mills, power consumption is not an important
criteria in most current applications and little data is available.
With the ball mills the power consumed by the machine is little affected by
the properties of the slurry and so the power consumed to achieve a
particular particle size distribution is solely a function of residence time
in the mill.

Again it must be emphasised that where the starting point is finely
divided coal powder ($<$ 75 microns) hydraulic shear is the major
mechanism of deagglomeration which allows the use of small grinding media
(spheres $<$ 12mm diameter).   The residence times considered may appear
short when compared with dry grinding practice.   A large tumbling ball
mill 7 metres inside diameter by 20 metres inside length, with a 50% ball
charge working at saturation with a mean residence time of 6 minutes would
process 1500 cubic metres of slurry per hour. (A six minute residence time
would be sufficient to "refine" a coal-oil-water, heavy fuel oil substitute)

With a one hour residence time (which would result in a stable coal-oil
mixture, 95% $<$ 30 microns, without added water or surfactants) the
output would be 150 cubic metres of slurry per hour.   This represents
70 tonnes of coal per hour or 500,000 tonnes of coal per annum at a
40:60 slurry.   Using this as a basis for comparison, to give some
indication of the reduction in size, the 7m by 20m ball mill projected
above would be replaced by a 50 cubic metre agitated ball mill to give the
same output (if such a machine were possible).   The ball charge weight
has been reduced by a factor of 8 and the filling ratio of the grinding
chamber has increased from .5 to .9.   REF.6

The size of a high speed continuous bead mill to replace the projected
7m x 20m ball mill would be 10 cubic metres.   The question is; are
either a 50 cubic metre Attritor or a 10 cubic metre continuous bead mill
practical, possible machines ?   With the continuous bead mill the answer
must be "no", the largest machines currently available have a chamber
capacity of approximately .5 cubic metres and although output is essentially
proportional to chamber volume for machines in the range 10 to 125 litres,
performance falls off markedly at the 250 litre size.

In this range of chamber sizes the agitator disc peripheral speed is in
the range 10 - 15 metres per second to effect dispersions over a wide
range of slurries from paint concentrates to biocides and that when
scaling-up for larger capacity machines in this range, the operating
peripheral speed is kept constant.   If the agitator discs are operated
at tip speeds greater than 20 metres per second, the agitator "disengages"
from the fluid and there is excessive wear of both the grinding media and
the agitator discs, heating of the fluid and decreased dispersion.
Efforts to use rotors shaped to act as impellors appear to have met with
little success.

Agitated ball mills (Attritors) appear more hopeful as a potential large
scale processing machine although the largest current production machine
again has a chamber volume of approximately .5 cubic metre.   With this
type of machine, performance scales with grinding charge weight across the
range from 5 litres upwards but under certain operation conditions "fall-

off" in performance has been observed which indicates the factors which will eventually limit the size of any projected machine. Unfortunately what that size is is not known.

The agitator rotates at a tip speed of approximately 3 metres per second and the grinding charge rotates in the chamber at a lower speed. Dispersion is effected by hydraulic shear in the nips between the elements of the grinding media when it is jostled and displaced by the agitator. The differential speed between the rotor and the grinding charge is maintained by the rolling friction of the layer of grinding media adjacent to the grinding vessel wall. The limit of proportionality of the process occurs when the layer of grinding media adjacent to the wall starts to slide, the rotational speed of the grinding charge increases, the differential speed between agitator and grinding media falls and the performance declines as the grinding charge "spins". Wear of the grinding media and vessel walls is increased.

Experience has shown that grinding media charge "spinning" can occur with ceramic media in batch machines where the vessel diameter is 1.3 metres and can be induced by pumping a process material upwards through a steel ball charge 6 millimetres in diameter at 12 cubic metres per hour in a vessel of .9 metres diameter. Although these values were not obtained milling coal slurries with steel media it would be incautious to speculate on the possible performance of a machine with a grinding vessel greater than 1.25 metres diameter or throughput rates greater than 20 cubic metres per hour. A machine within these limitations could be proposed to process coal-liquid slurries which require a one hour residence time in a tumbling ball mill with a residence time of 7.5 minutes. The gross volume of such a machine would be 7.8 cubic metres and the grinding chamber would be 6.4 metres deep. Eight such machines would theoretically produce the same output as the 7 metre by 20 metre ball mill proposed earlier. If operated as units in series the size of the grinding media would be reduced "down the line" to obtain the best relationship between grinding media size and the product particle size.

A machine of the dimensions considered above is feasible but it is very doubtful if such a machine will be built until the practice of wet milling cloal slurries in tumbling ball mills becomes well established.

The power consumption of agitated ball mills is at this time not known with any certainty. The .5 cubic metre machine is usually installed with a 50 kw electric motor. This motor is needed to start agitation because the grinding charge has a tendency to settle and "lock". Once agitation is established and process material is being pumped through the vessel, it is often very difficult to differentiate between the current being drawn under load and the no load current of the electric motor.

OBSERVATIONS

1. Coal-liquid slurries for burning can be produced by dry grinding and mixing but require surfactants or continuous agitation to give long term storage stability.

2. Wet milling coal-liquid slurries will improve the storage stability by improving wetting even where no significant change in particle size distribution of the coal is effected.

3. Where the coal-liquid slurry is intended as an internal combustion engine or gas turbine fuel the fineness of the product and the characteristics of the ash are critical for successful operation and wet milling is necessary to achieve the required particle size distribution. Various values are quoted from time to time. 99% less than 7 micron seems a reasonable target value.

CONCLUSIONS

1. Tumbling ball mills are the only large scale production machines available to wet mill fine coal slurries.

2. Agitated ball mills (Attritors) produce essentially the same results as tumbling ball mills and offer a convenient method of producing test samples and small pilot runs of coal slurries.

3. Agitated ball mills (Attritors) larger than current production models may be possible as dedicated production units making coal slurries for immediate use. (i.e. feed to a furnace or a ship's engine room).

4. High speed bead mills (small media mills) do not appear to offer any opportunities in coal slurry milling.

REFERENCES

1. SCHOFIELD, C and   "Pigment Deagglomeration in
    STEWART, I.W.   Paste Mixers"
                             Fourth European Conference on
                             Mixing.
                             Paper E.1. P.159

2. LOWRISON, G.C.   "Crushing and Grinding"
                             Butterworths London 1974
                             P.90

3. JARVIS, J.E.   "Dispersion Mixing in the
                             Paint Industry"
                             The Chemical Engineer July 1980.

4. U.P.I., Akron, Ohio.   Laboratory Reports
                             May 1981 to March 1983

5. COCHRAN, N.P.   "Oil and Gas from Coal"
                             Scientific American Vol. 234
                             No. 5 1976.

6. JONES, R.L.   "(Private) Report on Coal-Oil
                             Mixtures" as an alternative fuel.
                             Prepared for Stothert & Pitt Ltd.
                             1981.

ACKNOWLEDGEMENT

The author wishes to thank Torrance and Sons Ltd. for support in producing this paper.   The author's views on the problems of scale-up of existing machinery are his own.

TABLE I.   THE MIXTURES AND POSSIBLE APPLICATIONS

SOLID	LIQUID	POSSIBLE APPLICATIONS
Coal	Fuel Oil (s)	Substitute heavy fuel oil (for furnaces and boilers)
Gas Coke	Fuel Oil and Water	
Petroleum Coke	Water	Substitute diesel fuel for internal combustion engines
Wood Charcoal	Vegetable Oil (s) Ethanol Toluene Anthracene Oil Waste Organic Solvents	Chemical feedstocks for syn fuel processes as pumpable fluids in high pressure systems (Ref. 5) Source of purified carbon.

TABLE III   TYPES OF GRINDING MACHINES

TYPE OF MILL	LIMITATIONS ON PERFORMANCE *
Disc	Dry.
Ring	Dry.
Hammer	Dry.
Pan	Thick pastes non flowable.
Roller	Fine Milling at low throughputs.
Pin	Product greater than 50 micron.
Vibratory	Mechanical limitations on size.
Rod and Ball	Mechanical limitations only on very large machines.
Colloid	Large machines not possible.
Agitated Ball	Action probably scale dependant.
High Speed Bead	Action scale dependant.

* Limitations on performance relative to the production of large volumes of pumpable coal slurry.

TABLE II.   (a) COAL/LIQUID SLURRIES MILLED IN AGITATED BALL MILLS
REF. 4

Test No	Solid	Liquid(s)	% by Wt.	Media size mm.	Add (1)	Batch Volume	Test No
1	Coal	Diesel Oil	30 : 68	4	2%	360L	1
2	Coal	Anthracene Oil	30 : 70	6	-	15L	2
3	Charcoal	Diesel Oil	30 : 69	4	1%	5L	3
4	Coal	Water	44 : 53	4	3%	275L	4
5	Coal	Oil/Water	49 : 49 : 2				5

(1) Surfactants added to aid flow

TABLE II.   (b)

Test No	Particle Size			Note	Test No
	Start	Sample I	Sample II		
1	80% < 180 micron	50% 15 micron 1 hour	50% 3 micron 16 hours	Little change at 24 hours	1
2	70% > 300 micron	95% 50 micron 5 mins.	99% 50 micron 7 mins.	Continuous at (2) 55°C	2
3	50% > 180 micron	50% 3 micron 1 hour		Poor flow at higher solids	3
4	80% < 180 micron	50% 15 micron 2 hours	50% 5 micron 8 hours	(CF Trial 1.)	4
5	80% < 75 micron			Dry Mill and Mix. Settled.	5

(2) Size determined by wet sieving.

FIG. I   TUMBLING BALL MILL

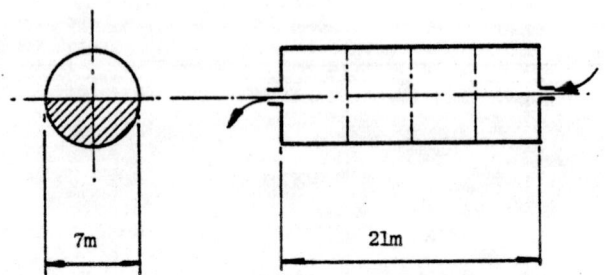

7m   21m

Charge weight (at 50% filling)   : 1900 tonnes
Volume of slurry (at saturation): 150 cubic metres

FIG. II
Agitated Ball Mill

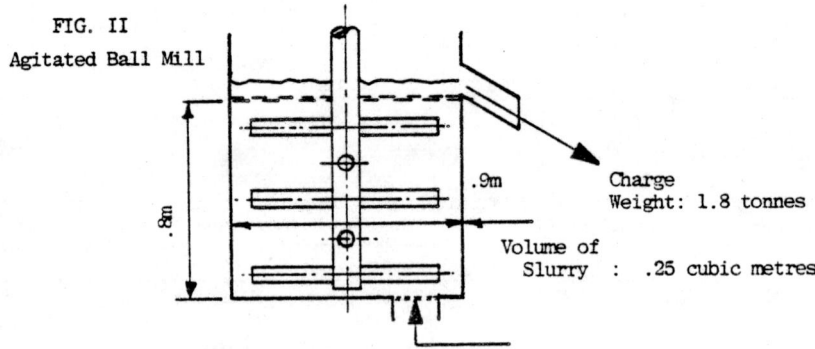

.9m

.8m

Charge
Weight: 1.8 tonnes

Volume of
Slurry : .25 cubic metres

FIG. III   High Speed Bead Mill

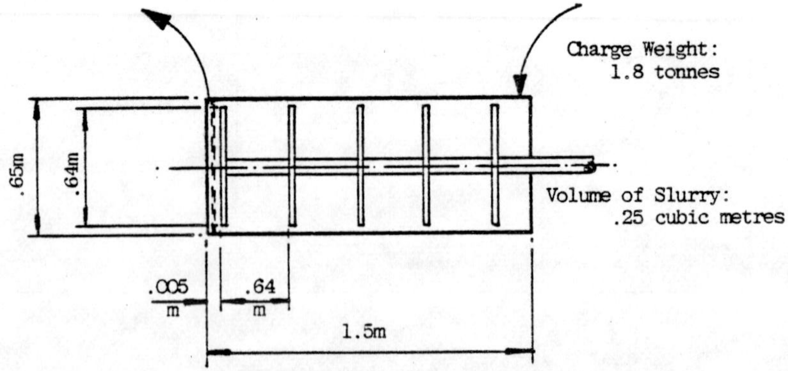

Charge Weight:
1.8 tonnes

.65m

.64m

Volume of Slurry:
.25 cubic metres

.005
m

.64
m

1.5m

84

PHYSICO-CHEMICAL FRACTURING AND CLEANING OF COAL

R. S. Sapienza, W. A. Slegeir, T. Butcher, and F. Healy

Mineral matter content and the costs of grinding are major
impediments to the use of coal-water slurries in existing
fluid fuel combustors. A process for the simultaneous
cleaning and fracturing of a variety of coals is being
explored at Brookhaven National Laboratory. This process
entails exposing coal to a carbon dioxide-water solvent
system under pressure. Substantial amounts of mineral matter are leached into the liquid phase, significantly lowering the concentrations of alkaline and alkaline earth
metals; some of the silica- and alumina-containing minerals
in the coal are also removed. The treated coal is more
easily crushed than untreated coal and grindability data is
reported.

INTRODUCTION

A great many R & D efforts are being aimed at future ways of burning
solid coal more effectively and at substituting coal for clean liquid fuels.
Increased demands for higher quality coal can be expected with the development
of new fuel forms, such as coal-liquid slurries. Coal-water slurry fuels
offer a practical, economic method for solid coal to replace the daily input
of 3 million barrels of oil per day used to fuel utility boilers, industrial
heaters and furnaces in the U.S.[1] However, the mineral matter in coal is a
major impediment to the direct use of this fuel in existing fluid fuel
combustors. Currently, only limited amounts of coal can be cleaned to acceptable levels and new cleaning methods for less tractable coals are needed.

The chemical composition, physical size, and mode of distribution of the
mineral matter in coal greatly affect the way in which it can be removed. For
any general cleaning method to be applicable to a variety of coals, it should
combine physical liberation and chemical separation techniques. Brookhaven
National Laboratory (BNL) has focused on this approach. With the proper
selection of reagents, swelling and selective solubilization of mineral matter
occurs with some weakening of the coal matrix. This work could be of value in
improving and extending the capability of current cleaning processes and providing a basis for the development of new coal cleaning methodologies.

This approach is flexible and is applicable to a variety of coals, since
the nature of maceral-maceral and maceral-mineral interactions are similar for
similar coals, and in a system that has been modified both chemically and
physically, particular mineral components may be attacked. It is believed
that improvements can be made which will be more effective in mineral matter
removal, in terms of degree and specificity, while potentially effecting a
route to organic sulfur removal.

*Brookhaven National Laboratory, Upton, NY 11973

BACKGROUND

We believe that a key element in chemical comminution is related to the swelling of the coal. Solvents that interact strongly with coal cause considerable swelling.[2-4] This interaction affords access to the internal structures and surfaces of the coal. Without this interaction, penetration of the coal matrix would be difficult and strongly dependent on the effective dimensions of the pores and the reagents.

A common axiom is that "like dissolves like" and this has been nicely quantified by Hildebrand and Scott.[5] The Hildebrand-Scott solubility parameter concept of cohesive energy densities has found use in a wide variety of applications. With smaller molecules, solubilization is easily measured. However, with molecules of large molecular weights, such as polymers, solvent interaction is conveniently measured by the degree of swelling. The aggregate nature of coal, with its macerals and minerals of variegated composition and generally intractable nature does not lend itself to simple solubility determination. However, coal does interact with a number of solvents and this interaction may be estimated by determining the degree of swelling associated with solvent systems of varying solubility parameters.

Solubility parameter spectra, in which the degree of swelling of a particular coal is plotted against the solubility parameters of a variety of solvent systems allows correlation of solvent interactions with particular components in the coal matrix.[6] Pyridine (δ=11 hildebrands), employed for extractability determinations, correlates with swelling of vitrinite macerals. Ammonia (δ=15 hildebrands) is known to cause spontaneous fracturing of coal[7] and this correlates with a pronounced peak corresponding to pseudovitrinite macerals.

Our interest was directed toward the swelling maximum of the fusinite maceral. Although a small percentage of the total coal, this high ash fraction[8] exemplifies mineral-organic interactions in coal. The high solubility parameters of fusinite (typically about 20) implicates the importance of polar bonds within the maceral and suggests the importance of hydrogen bond interactions between organic and mineral matter components. Hydrogen bonding in coal has been postulated by Orchin and Storch following their work on solvent extraction.[9] It seems reasonable that the factors involved in mineral-organic interactions in fusinite would apply for accessible mineral-organic interactions in other macerals.

Swelling strains the coal matrix, most probably leading to the rupture of hydrogen bonds. However, due to the lability of hydrogen bonds, the reorientation of the coal matrix allows reorientation of the hydrogen bonding structure. It is believed that an appropriate hydrogen bonding agent, if allowed to penetrate the swelled coal structure, would be capable of "tying up" hydrogen bonding sites in the coal. The combined effects of swelling and bond breaking would appear to allow included mineral matter to drop out of the coal structure while physically weakening the coal matrix.

The treated coal should be more amenable to crushing, and such breakage may occur along maceral boundaries, due to the combined effects of external hydrogen bonding and swelling, and this may promote the release greater amounts of mineral matter than achievable with coal as comparable particle-size distribution. This would reduce the cost of follow-on processing for separating impurities. This approach may also offer new ways of further coal clean-up since, during the comminution process, the coal structure is most susceptible to chemical attack.

TECHNICAL APPROACH

A wide variety of chemical reagents are capable of disrupting the internal hydrogen bonding of the coal. However, for a chemical coal comminution approach to be practical the hydrogen bonding reagent must be readily available, inexpensive, be of sufficiently small molecular size and not be expected to introduce unwanted chemical elements that will lead to corrosion or pollution problems with the processed, comminuted coal. These qualifications substantially limit the number of hydrogen bonding reagents, and our initial experiments were restricted to water combined with carbon dioxide to provide a reagent system which was tailored to the ideal cohesive energy densities of the coal samples being studied.

Hildebrand's rule states that the solubility parameter of a mixed solvent system is related to the volume fractions and solubility parameters of the respective components. With a CO_2/H_2O system, only the carbon dioxide dissolved would be effective in contributing to the solubility parameter of the mixed solvent, and this amount is strongly dependent on temperature and pressure. The strict application of Hildebrand's rule may be unwarranted because of complicating associations of CO_2 (carbonic acid formation and dissociation). Furthermore, any material extracted into the solvent system will alter the aggregate solubility parameter of the solvent system. The solubility parameter approach was chosen therefore, only as a guide to the component proportions. It is interesting to note that water (δ =23 hildebrands) retards the ammonia (δ =15) comminution process, no doubt because of unfavorable matching of the solubility parameter of the water-ammonia solvent system (δ between that of ammonia and water, depending on concentrations) with the coal.

This treatment is related to the unique effects carbon dioxide and moisture have on coal. Carbon dioxide readily and extensively penetrates the coal structure.[10] It has been proposed that CO_2 sorption may be used to measure the micropore structure of coal.[11] However, it has been recently shown that at higher CO_2 pressures, swelling of the coal takes place.[12] Liquid penetration through the solid coal may thus be promoted by CO_2 - enhanced swelling.

Several coals have been employed during the course of this study. The method can be applied to relatively unprepared coals and seems to be very effective for the removal of alkaline and alkaline earth metals which are related to boiler fouling characteristics; some removal of siliceous materials, which are most responsible for erosion is observed. With modification, other mineral groups could be removed. The system is flexible, may be modified both chemically and physically, and be integrated into or modify an overall coal preparation process. The method yields a more friable, "weaker" coal structure and could significantly reduce the energy requirements and cost of follow-on coal grinding and separation steps.

TREATMENT PROCEDURES

A variety of medium-to low-rank coals have been examined in the course of this work using CO_2/H_2O. These include Kentucky #9, Montana Rosebud, and Pittsburgh Seam coals and a variety of conditions have been employed. At present treatment is carried out in a 2L stainless steel autoclave, equipped with a gauge, liquid sampling and gas venting valves and a thermocouple. The thermocouple is connected to a proportioning band temperature controller, which in turn is connected to a heating mantle. This system affords precise and reproducible temperature control of the autoclave contents.

Generally, the coal sample (typically 500g), of appropriate mesh size, is added to the reactor followed by the appropriate amount of distilled water (typically 1L). After closure, the reactor is purged at least twice with CO and then brought to the appropriate CO_2 pressure (typically available tank pressure, about 750 psi); the weight of CO_2 can then being determined.
The reactor is then brought to the desired temperature (typically 80°C) and the pressure is again recorded. Once the temperature has stabilized, very little, if any, changes in pressure are observed. The temperature and pressure data may be used to calculate the amount of CO_2 dissolved in the water.[13]

At the end of the desired contact time, the reactor is removed from the heater, and the hot liquid phase is carefully transferred to a flask. After cooling, the coal is removed from the reactor and washed on a sintered glass funnel with distilled water. Final drying of the coal is carried out in a vacuum oven at 110°C. The cooled liquid phase, which frequently contains a small amount of powdered coal and precipitated mineral matter, is filtered. When appropriate, the water is evaporated to dryness to determine the quantity and nature of minerals leached from the coal.

The weight of the dried coal may be compared with that of the feed coal. When appropriate, the dried coal is ashed by ASTM method D-3174-73. Along with the treated coal, samples of untreated coal are concurrently ashed.

We have found that simply soaking the coal in water offers little change in mineral content or ease of grinding. Tests analogous to the CO_2/H_2O treatments were performed, using either nitrogen or helium in place of the CO_2, and only small changes in the coal were noted. These tests indicate that pressurized water alone is not important to the process. Rapid decompression tests using CO_2 in the absence of water were carried out to determine whether fracturing is due to pore-entrapped CO_2 causing stress on the coal structure during pressure release. Little change in the coal size or grinding time was observed. We believe these tests point to pronounced synergy of water with carbon dioxide.

The gas phase was examined for CO, H_2, CH_4, and SO_2 after some treatments, and in no cases were significant quantities of these detected. Within the liquid phase, a variety of inorganic materials are found (vide infra), but only traces of organics could be detected.

Grinding

Coal treated with CO_2/H_2O occasionally crumbles during the processing. It generally appears much more porous and often looks more like charcoal than the original coal. Most remarkably, the coal is very easily crushed. In fact, many samples are easily pulverized between the thumb and forefinger, in marked contrast to the starting coal. The resulting "crumbled" coal is generally quite granular, with little, if any, fines; the crumbled coal does not leave appreciable black stains on the fingers, unlike conventionally ground coal.

To assess more quantitatively the effects on grinding of this treatment, a laboratory-scale grinding system was needed and a batch ball mill approach was chosen. Comparison of the grindability of samples has been done using a method adapted from the Bond Work Index (BWI) concept.[14] The use of the BWI involves determining the energy input required to achieve a desired level of grinding with the Index being calculated by taking into account the energy input and the extent of size reduction. This approach is considered more flexible than other approaches, such as the Hardgrove, which are defined only for a given feed and product size. In addition the Hardgrove test requires only a small fraction of the coal to pass a given screen size and so may not provide reliable results if only surface fracturing is involved.

The actual grinding was carried out on a laboratory ball mill, using ceramic jars with dimensions of 5-13/16" internal diameter and 5-3/8" internal height. The overall volume is 1829cc. Initial experiments employed steel balls (seventeen of 3/4" diameter, eleven of 5/8" diameter and fifty of 1" diameter). The rate of jar rotation on the mill is nominally 80 rpm. The voltage and current feeding the ball mill are measurable and constant; therefore the grinding time is used as a measure of power input. Earliest experiments employed 200 g of coal, but to improve free space, 100 g samples were used in the remaining grinding runs.

In the BWI test undersized coal is removed at calculated times and fresh feed is added to simulate closed circuit grinding. This procedure is repeated until a constant mass of undersized product is produced per revolution. Simpler approaches, however, involving batch grinding have been shown to provide accurate results, particularly for comparing the grindability of two materials.[15] This later approach was used here for comparison of the grindability of treated and untreated coals. With the assumptions of constant power input to the mill, fixed feed size, and a similar product size distribution curve, the change in BWI and hence grindability can be approximated by the difference in time required to achieve a fixed percentage of the product coal passing a given screen.

Figure 1 compares dry grinding times required for Pittsburgh Seam coal (from Arkwrite Mine, NA1361) initially sized to 1-3/8" to 3/8". One curve depicts the weight of untreated coal remaining on a 18 mesh (US) screen after sieving finer material after specified grinding times. The second curve is for the same coal treated with CO_2/H_2O. For these samples, the grinding and sieving process was repeated to afford an indication of deviation in these processes. The figure indicates a pronounced improvement in grinding: two-thirds of the treated coal is below 18 mesh in about 5 minutes, while for the untreated coal about 70 minutes is required; seven-eights passage requires 10 minutes and well in excess of 150 minutes for the treated and untreated coals respectively.

For finer final grind sizes, the treated coal also appears to require much less grinding power. Figure 2 depicts the amount of coal remaining on a 200 mesh sieve as a function of grinding time. Again an order of magnitude reduction in grinding time is afforded for the treated coal, with 45% of the coal passing through the 200 mesh sieve after 35 minutes, while the untreated coal has reached that level only at about 400 minutes.

To determine effects of the treatment on still finer size (-400 mesh) modifications in the grinding and sieving procedures were required. It was found that significant reductions in grinding times and more realistic sievings could be achieved if both procedures were carried out wet with a small quantity of wetting agent present. Work in this area is currently underway.

Mineral Matter Removal

Even in early experiments, we noted that the liquid phase is frequently yellow to brown in color after separation from the coal. Many times, a brown precipitate or gold-colored mirror deposited on standing. After filtration and evaporation of water, a substantial amount of cream-colored residue is obtained. No melting or softening is observed upon heating this residue to temperatures exceeding 300°C, strongly indicating the material is inorganic.

Since the liquid phase is in contact with the coal and is responsible for mineral matter leaching, its composition would be expected to have a bearing on ash reduction in the coal. The solubility of CO_2 in the liquid phase increases as the CO_2 pressure increases, although not linearly. Table 1 summarizes a set of experiments directed toward determining the effect of aqueous phase concentration of CO_2 on the reduction of ash for Pittsburgh Seam coal. At low CO_2 concentrations, little ash reduction is observed, consistent with poor matching of the solubility parameter of water with coal. As the CO_2 concentration is increased, a significant reduction in ash is observed. However, increasing the concentration above 24 g CO_2/L does not result in significant ash reductions.

Table 1
Effect of CO_2 Concentration on Ash Content*

Concentration CO_2 Dissolved g/L	Reduction in Ash, %
8	5
11	0.5
24	16
37	15
43	15

* Amount of dissolved CO_2 determined from pressure at temperature (Ref. 13); conditions: 2L autoclave, 1L water, 500g Pittsburgh Seam Coal 4/8 mesh, 80°C, 20 h. Ashing performed in quartz crucibles as per ASTM. D-3174-73, with untreated coals, containing 6.6% ash, used as reference.

Emission spectrography was employed as a rapid and general method to characterize, in a semi-quantitative manner, the changes in the metal and metalloid contents of the coal due to the treatment, as well as to better characterize the composition of the leachate. Table 2 lists approximate concentrations of several elements frequently encountered in coal mineralmatter. The first two columns list concentrations found in the ash from untreated and treated samples of Kentucky #9 coal, and the third column lists relative concentrations in the solid formed from evaporating the liquid phase. The leached mineral matter contained very significant quantities of silicon, aluminium, calcium, magnesium, and sodium. The data indicate that the concentration of these last three were significantly lowered in the ash. Clearly silicon and aluminum were removed from the coal, but because of a reduction in the total amount of mineral matter, and their high concentration in the ash, their ash concentrations in the treated ash may not differ significantly from those in the untreated coal. Titanium was also leached. Iron is also removed from the coal, but the semi-quantitative nature of the analysis does not reveal the alteration in the treated and untreated samples. On standing the liquid phase often precipitates significant quantities of iron-containing compounds. Similar results have been obtained with Montana Rosebud coal. It is important to note that the more corrosive elements, the alkaline and alkaline earths, appear to be particularly responsive to this treatment. Additionally, these elements present particular problems in coal utilization since they lower ash fusion temperatures and may destabilize coal water slurries.

Table 2
Analysis of Leached Mineral Matter and Ash in Kentucky #9 Coal*

Element	In Ash, % Untreated	Treated	In Leached Mineral Matter, %
Si	>10	>10	>10
Al	>10	>10	>10
Ca	>10	0.03	>10
Mg	~ 1	0.05	10
Fe	~ 1	1	~ 1
Ti	0.5	0.08	~ 1
Na	0.1	0.01	>10
Mn	0.1	0.005	0.1
Ba	0.1	0.01	0.01
V	0.08	<0.01	0.05
Ni	0.03	0.005	0.1

*Semi-quantitative determinations by emission spectroscopy; treatment conditions for these early experiments were carried out in a 30 mL reactor, using 10.9 g coal, 20 mL water, 1170 psi at 80°C for 20 hours.

Some more precise data on the compositional differences in treated and untreated samples of Pittsburgh Seam coal are presented in Table 3. The standard 80°C, 1200 psi CO_2 treatment afforded some decrease in ash content and volatile matter. A significant decrease in sulfur was also observed. The change in heating value is indeed small and supports the belief that this process does not drastically alter the coal structure.

Table 3
Proximate Analysis, Sulfur, Calorific Value of Treated and Untreated Pittsburgh Seam Coal*

Parameter	Untreated	Treated	Change, %
Moisture, %	0.16	0.4	
Ash, %	6.5	5.9	- 8.5
Volatile Matter, %	38.3	37.7	- 1.6
Fixed Carbon, %	55.1	56.0	+ 1.7
Sulfur, %	2.5	2.0	- 21
Calorific Value, Btu/lb	14180	14200	+ 0.1

* On as received basis.

Ultimate ash analysis results for both treated and untreated Pittsburgh Seam coal are shown in Table 4. In accord with the results described above, alkaline and alkaline earths generally appear to be removed efficiently, although the results for sodium appear to be anomalous. Significant quantities of iron and titanium are also removed. The silicon concentration appears to remain constant while the aluminium concentration increases. We are planning to confirm these numbers, particularly those for sodium and aluminium.

Table 4

Ultimate Analysis (wt.%) of the Ash From Treated and Untreated Samples of Pittsburgh Seam Coal

Component	Untreated, %	Treated, %	Change*, %
SiO_2	40.5	42.5	+ 1
Al_2O_3	23.1	28.2	+ 11
Fe_2O_3	13.8	11.8	− 22
TiO_2	1.1	1.0	− 17
CaO	8.3	2.0	− 78
MgO	1.3	0.6	− 59
Na_2O	1.6	1.6	− 7
K_2O	1.4	.4	− 72

*% Change in component weight based on its weight in the starting coal.

We are continuing this work in an effort to assess better its industrial applicability. The effects on ash fusion temperature and sink-float cleanability determinations will be determined. Modifications employing small amounts of additives for specific mineral removal is an area that shows particular promise. Integrated grinding, a process in which the CO_2/H_2O treatment is combined with grinding, may lead to significant savings in grinding times and more effective leaching. Microscopical examination of the effects of the process may help confirm the ideas leading to this approach.

CONCLUSIONS

Although preliminary, these results suggest synergistic behavior in the interaction of carbon dioxide and water with coal. Significant removal (10-15%) of the mineral matter of the coal is observed, with the more corrosive alkaline and alkaline-earth metals removed preferentially. Along with mineral matter removal, significant improvements in the grindability of treated coal have also been observed in ball mill tests.

ACKNOWLEDGEMENT

This research partially supported by a grant through the Morgantown Energy Technology Center from the U. S. Department of Energy under contract no. DE-AC02-76CH00016.

REFERENCES

1. Marnell, P. and Krishna, C. R. "Coal-Water Fuel: A Technology Update" Presented at the 9th Energy Technology Conference, Washington, D.C., February 1982, BNL-30821.

2. Sanada, Y. and Honda, H. Fuel, 46, 451 (1967).

3. Aldrich, R. G., Fuel, 56, 345 (1977).

4. Keller, D. V., Jr. and Smith, C. D. Fuel, 55, 273 (1976).

5. Hildebrand, J. H. and Scott, R. L. "Solubility of Non-Electrolytes" Dover, New York 1964.

6. Weinberg, V. L. and Yen, T. F. Fuel, 59, 287 (1980).

7. Datta, R. S., Proc. Annu. Underground Coal Convers. Symp. 3rd (CONF-770652) 237 (1977).

8. McClung, J. D. and Geer, M. R. "Coal Preparation" 4th ed., J. W. Leonard, Ed., Am. Inst. Mining, Metal., Petr. Eng., New York 1979, Chapter 1.

9. Orchin, M. and Storch, H. H. Ind. Eng. Chem., 40, 1385 (1948).

10. Fuller, E. L., Jr. "Coal Structure" M. L. Gorbaty and K. Ouchi, Eds., Adv. Chem. Ser., 192, 293 (1981).

11. Walker, Jr., P. L. and Patel, R. L., Fuel, 49, 91 (1970).

12. Reucroft, P. J. and Patel, K. B., Fuel, 62, 279 (1983).

13. Wiebe, R. and Gaddy, V. L. J. Am. Chem. Soc., 61, 315 (1939); 62, 815 (1940).

14. Bond, F. C. Mining Eng., 4, 484, (1952).

15. Berry, T. F. and Bruce, R. W., Can. Mining J., p. 63, July 1966.

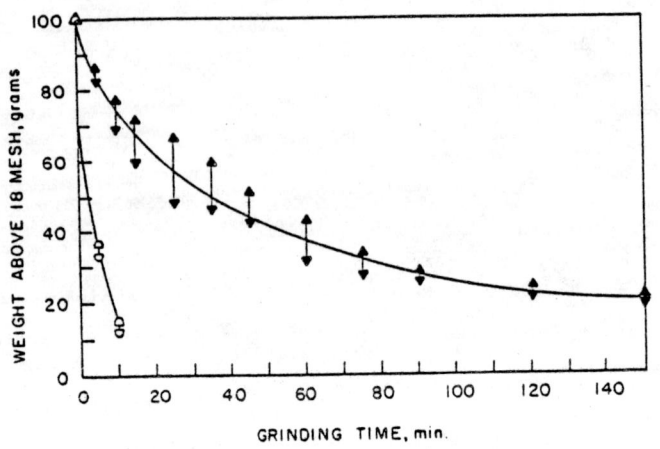

FIGURE 1

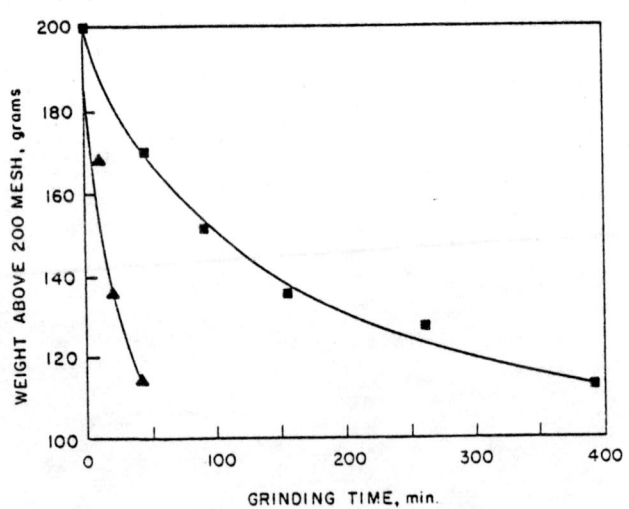

FIGURE 2

EQUIPMENT AVAILABLE FOR THE PREPARATION OF
PULVERIZED COAL FOR COAL LIQUID MIXTURES

DOUGLAS LEIVERS   C.Eng.   M.I.Mech.E.

SYNOPSIS

A selection of coal mill types available for fine grinding
applications is discussed together with a comparison of
the relative advantages and disadvantages of the main mill
types.

The impact on mill selection of coal types and slurry
preparation technology, coal beneficiation, partical size
distribution, etc., is also briefly discussed. Finally, a
high output CWM preparation plant is described.

INTRODUCTION

Production of COM and CWM relies primarily on preparing a suitable quality
of pulverized coal. Comminution of coal is not a new process but one that
has been practiced for well over 80 years. Early pioneers efforts were
applied to the firing of cement kilns with pulverized coal, and as time
progressed, coal in this form became known as PF (pulverized fuel). PF is
still extensively used throughout the world to fire boilers for the power
generating industry, industrial boilers for producing process steam and, of
course, the firing of cement kilns and other process furnaces.

To produce the PF for these units, coal grinding mills are used. The number
and size of these mills vary with the size of the unit, therefore, there is
a wide range of mill sizes and capacities available ranging from 3 to 300
tonnes per hour. Irrespective of the mill capacity, the fineness of the PF
produced is in the region of 70 per cent passing 75 microns (200 mesh), a
figure that has become basically standard throughout the world for PF
fineness.

Highlighted will be the main considerations in making a mill selection, they
are to give a continuous output with a suitable and consistent size grading,
the lowest capital investment cost of the plant to achieve these conditions,
overall power consumption plant reliability and an easily maintained plant.
Over the last three decades the use of PF in the fields of power generation,
kiln firing, etc., have declined and oil took over as the conventional fuel.
As oil reserves become more and more depleted, and cost increasing, new
interest is being shown not only in the re-application of PF but also in
coal liquid mixtures that could be used in place of oil.

The differences between the requirements for pulverizing coal for conventional PF and for the preparation of coal slurries are not clear. No standards have been laid down and various organisations throughout the world are preparing slurry fuels to their own specifications. Some require PF of a fineness very little different from that produced for the conventional direct fired PF systems, some require a fineness very much finer than the conventional, and yet others require a bimodal size grading structure. It is with the latter that current technology appears to be advancing.

It is the intention of this paper to discuss only the grinding systems and plant required to produce the PF fineness conditions currently required from the three localities of research. It is not intended to argue on these avenues of research, but only to outline the types of mills and the associated plant required to produce the liquid fuels currently being specified.

In this context it is intended to look at technologies of both the dry and wet grinding processes, to look at the wide range of coal that could be utilised, the effect this would have upon the plant and finally to place milling systems before you that will meet current demands.

MILL TYPES

There are three basic types of mill on the market today.

High speed impact mill.
Medium speed vertical spindle mill.
Slow speed tube ball mill.

HIGH SPEED IMPACT MILL

An example of this type of mill is shown in Fig. 1.

The impact mill is a high speed machine obtaining size reduction by means of attrition of the coal between fixed and moving pegs, or by impact from swinging hammers.

The latter type has a number of beater plates mounted on a horizontal shaft. The swinging hammers are loosely held between the plates by a series of rods inserted through holes drilled round the periphery of the plates. The shaft is extended to carry a whizzer type classifier and the induction fan. The body of the mill has three distinct sections, a heavily lined pulverizing chamber, followed by a classification zone and finally the fan casing. Air enters the pulverizing chamber with the raw coal and leaves via the classifier and fan, carrying with it the pulverized fuel. In certain cases the fan and classifier are mounted separately to the mill. The speed of the main shaft varies according to the type and size, but is of the order of 1500 RPM.

This type of machine is compact, reasonably quiet in operation and readily controllable, but its wear rate is high and power consumption substantial.

MEDIUM SPEED VERTICAL SPINDLE MILL

This type of mill, illustrated in Fig. 2, is used extensively throughout the world and there are several variations, one of which has a flat grinding surface and conical grinding rolls. The rolls are driven by friction contact between the coal on the rotating table, the action is crushing and attrition. The table is mounted on the vertical output shaft of a reduction gear unit, which is driven by a constant speed motor.

Pressure for crushing the coal is applied to the grinding rollers through the trunnion lever assembly from a hydro-pneumatic springing arrangement. Nitrogen filled accumulators take up the vibration in oil pressure in the system due to the vertical movement of the rollers when crushing the coal, thus giving a smooth and constant grinding action.

Coal enters the mill through a central feed tube where it is deposited onto the grinding table; centrifugal actions induced by the rotating table feeds the coal under the grinding rolls. Hot air or gas enters the mill below the grinding table and passes up through a stationary multi-vane ring situated at the periphery of the rotating table. The vane ring imparts a swirl pattern to the air, directing it over the table. The air entrains the ground coal leaving the table and, due to the rapid change in area local to the top of the vane ring, a fall in air velocity gives a primary separation of the ground produce. As the air fuel mixture passes through the mill a secondary separation takes place. The final separation is carried out in a classifier mounted on the top of the mill housing. The oversize fractions of the product return to the centre of the mill to join the raw coal feed for further size reduction.

This type of mill is exceedingly quiet when operation and runs with little or no vibration.

SLOW SPEED BALL TUBE MILLS

This type of mill (Fig. 3) is simple in construction, consisting of a lined steel tubular barrel containing a charge of steel balls. The shell is mounted on hollow trunnions which in turn are supported on either hydro-static or anti-friction bearings.

Coal feed and air is passed through one trunnion and the ground PF entrained in the air stream passes out of the mill through the other trunnion where it is conveyed by the air stream to a externally mounted classifier, where oversize particles are returned to the feed chute of the mill.

Due to the very heavy construction of this type of mill and the very large ball charge required, the total weight is high necessitating a far greater civil input in foundation design when compared to the other two types of mill.

The speed of rotation depends upon the size of the mill and can be from 10 RPM for the largest to 25 RPM for the smaller size. In order to obtain speed of this nature, a large speed reducing gear is required and is usually coupled to a girth gear mounted on the mill on one side and slow speed motor on the other.

This type of mill, due to its robust nature, is not vulnerable to damage by tramp iron, and wear of the grinding charge is made good by feeding new balls in with the coal feed.

COMPARISON OF MILL TYPES

Having described the three main types of mills available, to what effect can they be utilised for the production of PF for coal liquid mixtures? Impact mills, due to their small capacities, up to approximately 10 tonnes per hour cannot be seriously considered for commercial production other than the possibility of use on pilot plant application; whereas vertical spindle mills with capacities up to 500 tonnes per hour and tube ball mills with capacities up to 100 tonnes per hour when grinding to 70% passing 75 microns can be seriously considered.

Both vertical and tube mills can be again seriously considered for ultra-fine grinding, although in both cases the relative outputs will decrease significantly.

Apart from output, comparison of the three mill types can be made on a relative performance basis, when grinding coal of the same quality to a fineness of 70% passing 75 microns, against the power consumption per tonne of coal ground.

Impact mills	16.0 KWH/tonne
Vertical spindle mills	8.5 KWH/tonne
Ball tube mills	26.0 KWH/tonne

You will note from this the very high power consumption of the tube mill against the power consumption of a vertical spindle mill. This will have to be seriously considered when making a final plant evaluation.

So far we have discussed mills for producing dry PF. Part of known milling technology is that of wet milling. Although not greatly applied to coal grinding, it is extensively used in the processing of ores and slip (china clays) into slurries.

It is here that the tube ball mill comes into its own; of the mills described it is the only one suitable for wet grinding. Fig. 4 shows a wet tube mill. The primary difference is that the trunnions are usually smaller so allowing a larger ball charge than that of a dry mill. Water enters the mill along with the coal feed and the slurry produced overflows from the other trunnion. Wet grinding is normally carried out in a closed circuit system, the ground product being diluted to permit classification, the oversize material being returned to the mill. One great advantage to be noted when using a wet ball tube mill is the reduction in the power requirements over than required for dry grinding, up to 30% reduction is obtained on a fine grinding application.

DRY MILLING SYSTEM

The process of producing and bunkering PF has been known for a great number of years as central milling plant, independent of the final use of the PF. Due to the adaption of direct fired systems on generating and kiln installations, the use of the central milling declined. However, it has remained in constant use in the mineral dressing industry.

A typical central milling plant would comprise a raw coal bunker, coal feeder, pulverizing mill and classifier, cyclone, dust collector, exhauster fan and pulverized coal bunker. To produce the hot gas for drying, a hot gas generator would be included before the mill (Fig. 5).

Recent developments in high efficiency textile dust filters have allowed the system to be further simplified, dispensing with the cyclone and the interconnecting ducts, and the handling facility required below the cyclone (Fig. 6).

Where it is normal for the classifier to be mounted directly onto the top of a vertical spindle mill, when utilising a ball tube mill in this type of system, the classifier is mounted independent of the mill (Fig. 7).

These system types can be used for both normal grinding with product of 70% passing 75 microns (200 mesh) or fine grinding giving products of 99.5% passing 53 microns (300 mesh). Variation in the type of classifier used being the only modification. However, again it must be noted that the output of any mill will be reduced considerably when fine grinding.

It is normal practice when grinding in a central system to reduce the total moisture content of the pulverized coal to below 1.5 per cent. This reduces the potential fire hazard in the pulverized coal bunker. Inerting systems and explosion reliefs on cyclones and dust collector must also be provided to satisfy current safety regulations.

WET MILLING SYSTEM

The plant involved in the wet processing of raw coal to slurry is normally carried out in a closed circuit system comprising a raw coal bunker, feeder, tube mill, discharge sump and slurry pump, classification in this case being carried out in a hydrocyclone. To complete the loop and close the system, the oversize from the hydrocyclone is returned to the mill inlet to mix with the incoming feed (Fig. 8). With plant of this nature no inerting system is required.

So far we have covered systems to give a normal PF grading and a fine particle grading, now for the third type - Bimodal grinding.

Most, though apparently not all, highly loaded coal water slurries have what is generally referred to as a 'bimodal' coal particle size distribution.

The expression 'bimodal' in the context of slurry preparation means that the solid material held in suspension in the slurry has been prepared from two fractions - one of relatively coarse particle size distribution and the other of very much finer distribution. Typical ratios between the average particle sizes of the two fractions may be 5:1 or more with some 30% of the total solids weight being ground to the finer size distribution.

Such slurries are not truly bimodal of course, since both the coarse and fine fractions contain an extremely wide range of individual particle sizes with overlapping size distributions.

One fundamental - and purely mechanical - reason for using a bimodal size distribution can be seen in Fig. 9 which shows idealized monomodal and bimodal size distributions.

Spheres are used in these idealized representations. Coal particles are, of course, highly irregular in shape.

Nevertheless, it is clear from Fig. 9 that the theoretical solids loading in a bimodal slurry can be significantly higher than in the equivalent monomodal slurry.

Unfortunately, the theoretical mechanics of particle stacking form only one element in what is an extremely complex technology involving surface chemistry, electro-chemistry, colloid chemistry and a number of other more or less esoteric disciplines. It is not the purpose of this paper to examine any technology other than coal milling, but the bimodal nature of highly loaded (70% plus) slurries has a significant effect on apparent slurry viscosity.

Experimental evidence confirms that a single particle size (i.e. true monomodal) slurry has a peak solids loading of about 65% at which point the relative viscosity becomes infinite. In contrast, it can be seen (Fig. 10) that an idealized bimodal slurry offers the theoretical prospect of solids loadings in excess of 80% by weight with manageable viscosities. Such highly loaded coal slurries have, in fact, been reported by a number of slurry manufacturers.

Preparing the bimodal grading is carried out utilising two mills having carried out the primary grind in the first mill then the coarse fraction is removed and passed through a second mill. (Fig. 11) indicates a typical wet, two stage grinding application, the separation from the first stage being carried out in a "sieve bend" from which the coarse fraction passes to the second mill for further reduction.

USE OF CLASSIFIERS IN MILLING SYSTEMS

In each of the systems so far discussed, classifiers have been mentioned. There are several types.   (Fig. 12) indicates a static dry type classifier. Here the PF/gas mixture passes up through the annulus produced by an outer and inner cone to a section at the top of the inner cone where a series of adjustable vanes impart a swirl to the mixture as it passes through. This swirl effect imparts a centrifugal action to the particles, the heavier or larger particles being thrown out of the air streams to the periphery of the inner cone where they are rejected by gravity to return to the mill for further reduction, the finer particles passing out of the classifier to the next stage in the process. This is the normal type of classifier you will find these days for a standard classification to 70% passing 200 mesh.

The mechanical whizzer type classifier (Fig. 13) is normally used for fine grinding, the rotating element is driven by a variable speed motor. Classification is carried out by the centrifugal action imparted to the particles by the rotating whizzer blades. Variation of speed can maintain the required particle size and vary it, if found necessary.

Both the above types will be used in the dry process system. For the wet process system a hydrocyclone is used (Fig. 14). Fluid carrying the particles in suspension is introduced tangentially into cylindro-conical vessel, thus creating a spiral pattern of high centrifugal force which causes the larger particle to spiral down the cone wall to the outlet for return to the mill, whilst the remainder of the fluid and the finer particles move radially inwards to an inner spiral and to the central overflow outlet. To prevent short circuiting, the overflow outlet is provided with a tube which extends into the cyclone, this is known as the "vortex finder".

The sieve bend (Fig. 15) used as a product splitter has the screen made as a slotted deck by bars which are situated at right angles to the slurry flow. The feed is fed evenly across the width of the deck and as it passes each slot the depth decreases at a uniform rate. The result is a separation of the feed solids at a size far smaller than the openings in the sieve bend. The fines slurry is collected below the bend where it is piped for further processing whilst the coarse particles discharged from the bend are returned to the mill.

EFFECT OF COAL TYPE ON MILLING OPERATIONS

We have discussed mills, classifiers, and mill systems, but so far not coal or the effect coals have on performance of the plant. As you may be aware, coals vary in types over a wide range from anthracites to the low moisture lignites. In this range we will, at the moment, concern ourselves with the bituminous or mid range of coals. Each bituminous coal field will produce quite a large variation in coal qualities.

Some of these variations will affect the mill, classifier and the product grinding. Hardness of the coal gives by far the greater variation. Bituminous coals vary in hardness from 35 to 120 hardgrove index. Hardgrove index being one commonly used method of determining the coals grindability index. In the above range, 35 represents a very hard coal and 120 a very soft coal. You can well imagine the harder the coal then more power will be required to pulverize to at a specific grading, it will take longer to produce this specific grading, therefore, the product output from any mill will decrease accordingly. Finally, as the hardness varies so the natural breakdown of the particle varies. For a hard coal the slope of the grading analysis mill steepens (Fig. 16) producing less fines and less coarse, whereas for a soft coal the grading analysis flattens out giving more fines, but equally more coarse particles.

Ash in coal is one other variant, varying from as low as 2% to 25% by weight. As the low ash fuels are not common then beneficiation may be necessary before the final liquid mixture is processed.

FACTORS AFFECTING MILL SELECTION

In order to make a correct selection of milling plant, the mill design engineer must then have the following information placed before him:-

Product size analysis required.
Product quantity.
Hardgrove grindability.

Total moisture content of raw coal.
Full chemical analysis of the coal.
Full ash analysis of the coal.
The wear indices of the coal (if known).

If this information or part of it is unknown, then a substantial raw coal
sample should be provided.

This information will enable a mill section to be made. However, there is
always the possibility that the client may have some preference as to the
type of mill he requires.

Summarizing some general indications for the selection of pulverizers from
the three main types discussed, is as follows:-

Mill Type	High Speed Hammer	Medium Speed Roller	Slow Speed Tube Ball
Range of product fineness	Coarse-Medium	Medium-Fine	Medium-Fine
Power requirements	Medium	Low	High
Space/output	Medium	Low	High
Noise level	Medium	Low	High
Ease of maintenance	High	Medium	Low
Cost of maintenance	High	Medium	Low

Over a period of 2-3 years the maintenance cost of a tube mill becomes more
than that of a medium speed roller mill due to the barrel liners requiring
complete replacement.

From this table, and due to their low outputs, high speed mills can be
eliminated immediately. The medium speed mill will give the best all-round
performance. However, the ball tube mill, when used for wet grinding, has
considerable advantages over a tube mill for dry grinding, which are
summarized below:-

1. The tube ball mill power requirement is approximately 30% more for a
    dry grinding mill than that for a wet grinding mill when grinding the
    same coal to the same product fineness.

2. Unless an additional heating source is provided to dry the pulverized
    coal, then a dry grinding mill is limited to 2% maximum feed moisture
    entering the mill.

3. The wear rate on the grinding balls and liners per tonne of coal ground
    is lower than a dry grinding mill.

4. No dust collectors or inerting system are required.

5. A wet tube mill is normally smaller in diameter and has smaller
    trunnions than its dry grinding counterpart for the same product output
    and it is, therefore, cheaper to manufacture.

BENEFICIATION

In the context of the choice outlined above between wet and dry milling technologies one other factor may be of importance.

At least one potential market for coal water mixtures - as a substitute fuel for oil on industrial and power watertube boilers - will be more easily penetrated by slurries having the lowest possible ash content, i.e. 1% or less. This requirement implies that many slurry preparation plants are likely to have a coal beneficiation stage built into them. Approaches to the target of producing a low ash CWM vary, but such processes are in general wet based.

Thus, the concept of using a wet ball mill as one stage of the CWM preparation process not only has the obviously logical merit of preparing CWM in a wet mill, but also the additional merit of the wet mill product being used as finely divided wet coal feed (giving maximum ash release) for beneficiation by froth flotation or similar technologies prior to final slurry mixing and stabilisation.

CONCLUSIONS

Although a wide range of mill types is available for fine and ultra fine grinding of coal, it is clear that in the context of high output coal liquid mixture preparation plants there are relatively few types - perhaps only two (vertical spindle and tube ball) - which meet the output (100 tph and above) criterion and which are also fully proven in long term industrial service.

The following factors must also be considered:-

1. Bimodal slurries - probably requiring a two mill installation - will commonly be required

2. Coal beneficiation is likely to be widely used.

3. Slurry preparation plants are likely to handle a wide range of coal feed stocks of widely varying qualities. Mills must, therefore, be versatile and readily adjustable in order to maintain a reasonable grinding function.

With these points in mind it is possible to postulate that for large scale (100 tons/hr and above) CWM preparation plants for bimodal beneficiated slurries the optimum economic milling plant selection would be a medium speed vertical spindle mill, accompanied by a mechanical rotating classifier in order to maintain a reasonable range of size classification. The final milling system will combine both the dry and wet technologies of coal milling and take the form indicated in Fig. 17.

ACKNOWLEDGEMENTS

The author wishes to thank NEI International Combustion Limited for permission to present this paper and also for the help given by colleagues within NEI International Combustion Limited in the preparation of this paper which is gratefully acknowledged.

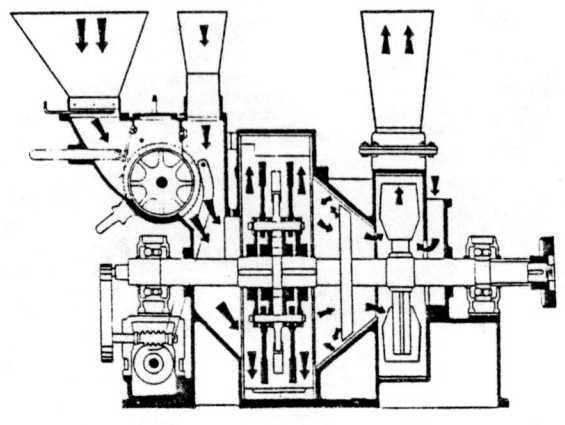

FIG. 1   HIGH SPEED SWING HAMMER MILL

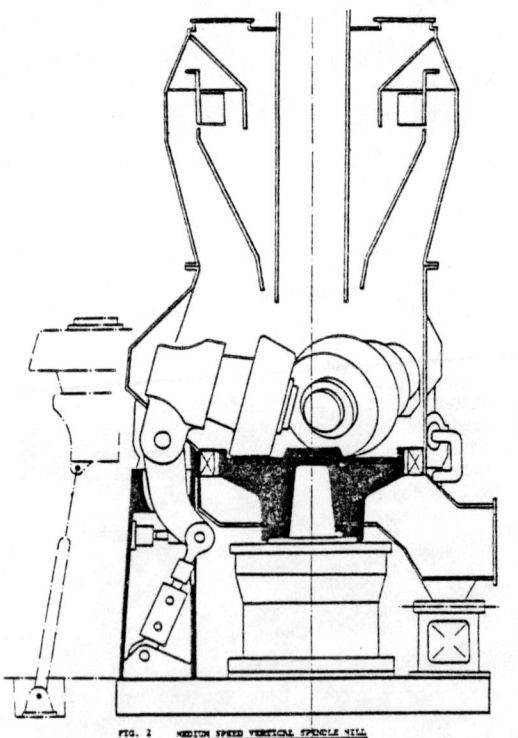

FIG. 2   MEDIUM SPEED VERTICAL SPINDLE MILL

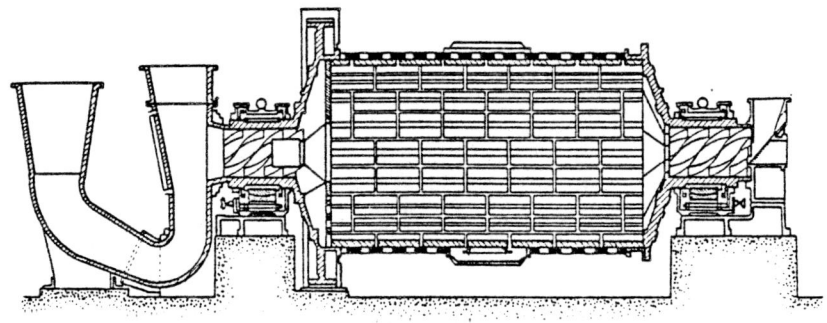

SLOW SPEED BALL TUBE MILL

FIG. 3

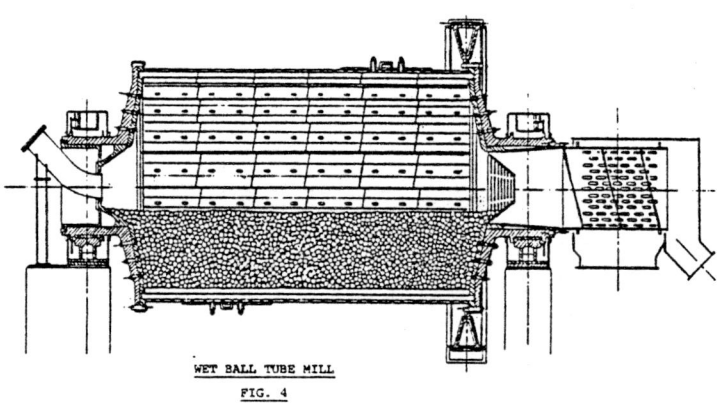

WET BALL TUBE MILL

FIG. 4

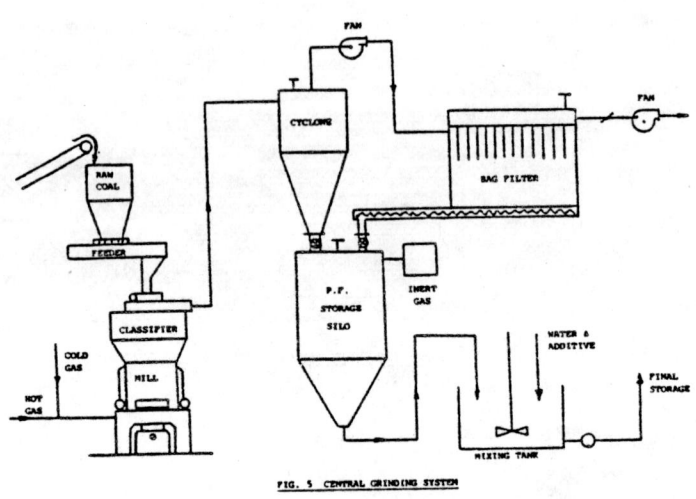

FIG. 5 CENTRAL GRINDING SYSTEM

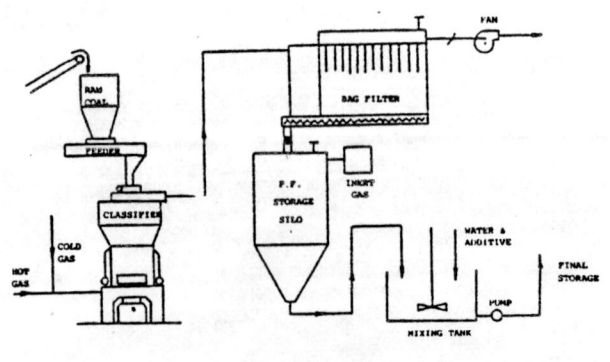

FIG. 6 CENTRAL GRINDING SYSTEM
WITH HIGH EFFICIENCY BAG FILTER

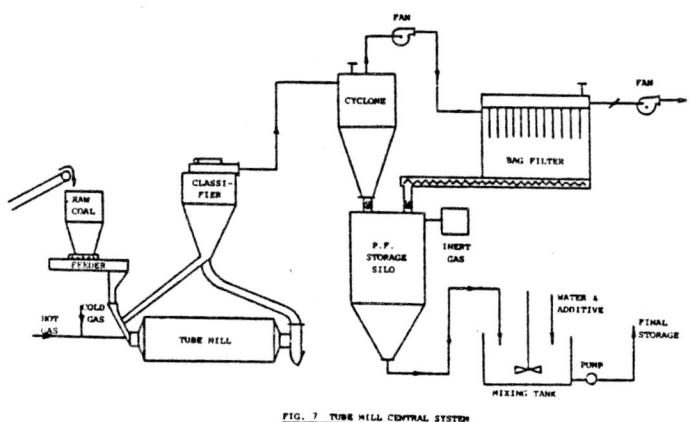

FIG. 7 TUBE MILL CENTRAL SYSTEM

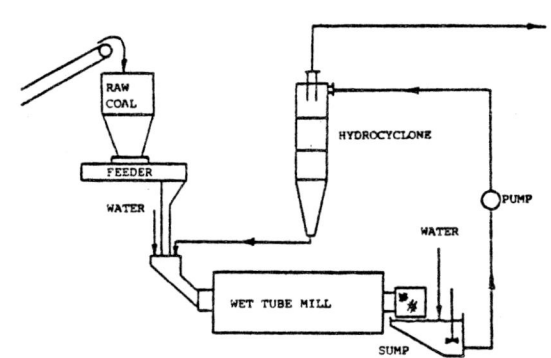

FIG. 8 CLOSED CIRCUIT WET GRINDING SYSTEM

CLM-H

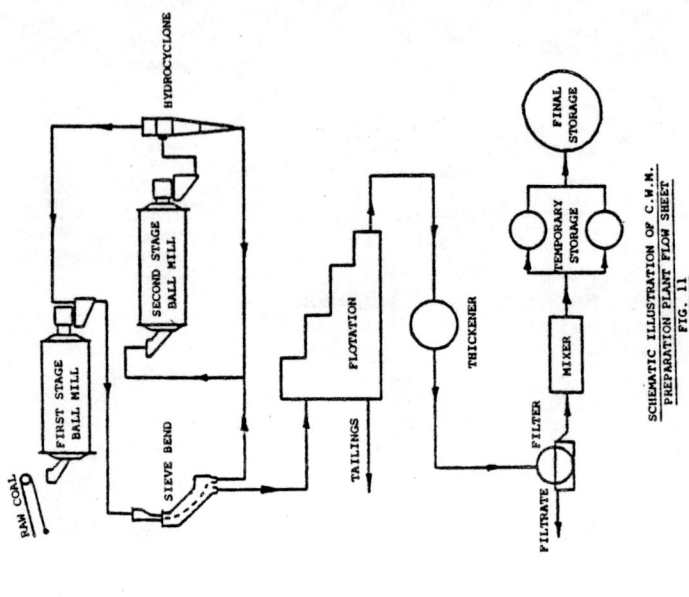

SCHEMATIC ILLUSTRATION OF C.W.M.
PREPARATION PLANT FLOW SHEET
FIG. 11

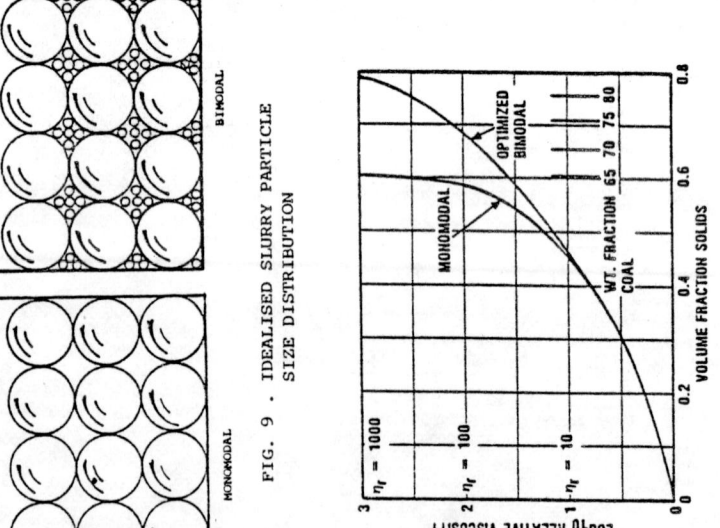

FIG. 9 . IDEALISED SLURRY PARTICLE
SIZE DISTRIBUTION

FIG. 10. EFFECT OF BIMODAL DISTRIBUTION
ON RELATIVE VISCOSITY

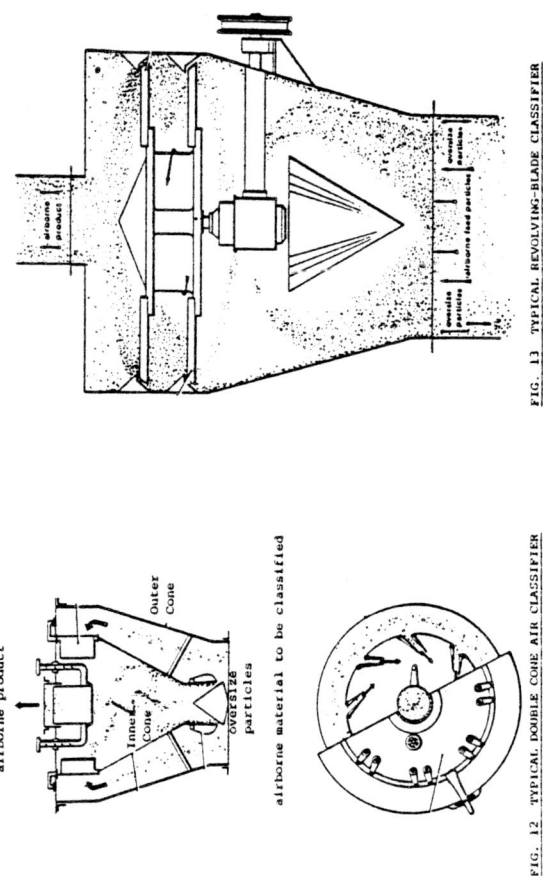

airborne product

Outer Cone

Inner Cone

oversize particles

airborne material to be classified

FIG. 12 TYPICAL DOUBLE CONE AIR CLASSIFIER

airborne product

oversize particles

airborne feed particles

airborne feed particles

oversize particles

FIG. 13 TYPICAL REVOLVING-BLADE CLASSIFIER

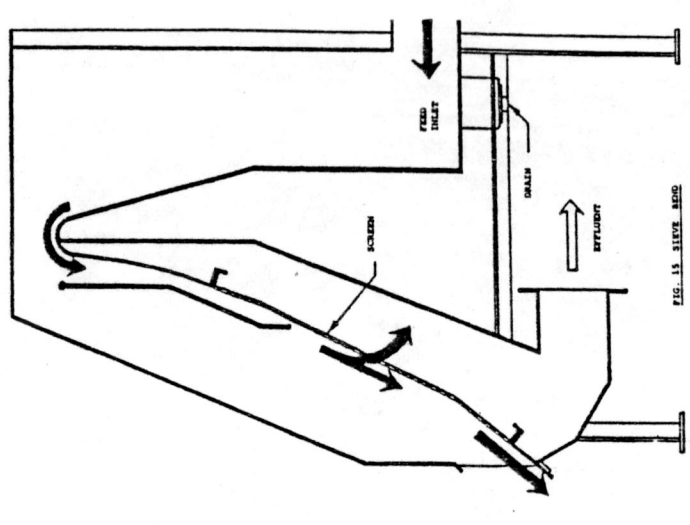

FIG. 15   DSM SIEVE

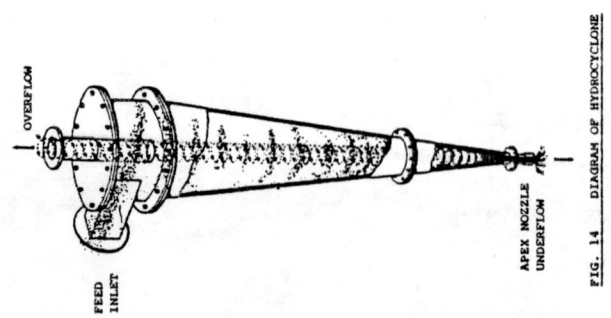

FIG. 14   DIAGRAM OF HYDROCYCLONE

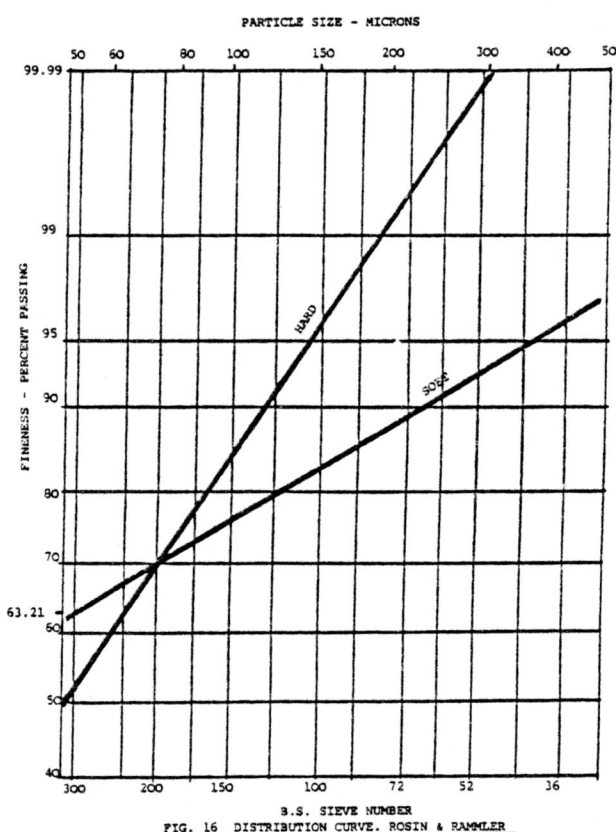

PARTICLE SIZE - MICRONS

FIG. 16   DISTRIBUTION CURVE. ROSIN & RAMMLER

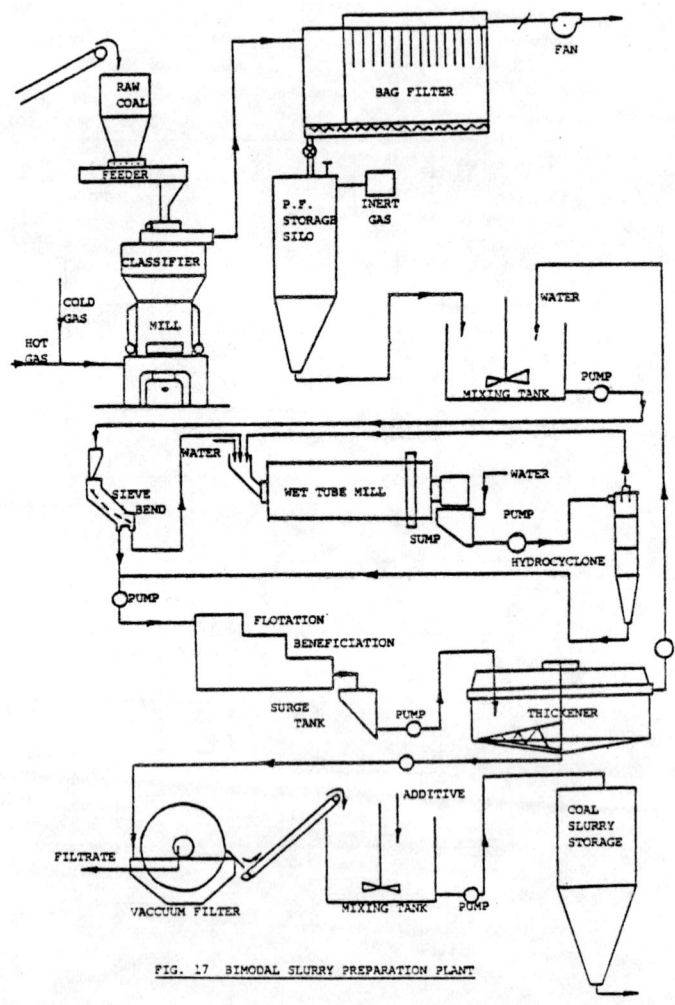

FIG. 17 BIMODAL SLURRY PREPARATION PLANT

PREPARATION OF COAL WATER SLURRIES
FOR THE TEXACO GASIFIER

Dr. B. Cornils, Dipl.-Ing. B. Richter, Dr. P. Ruprecht*
Dr. J. Langhoff, Dr. R. Duerrfeld**

The favourable characteristics of hydraulic conveyance of coal are utilized in the Texaco coal gasification process as realized in Oberhausen, West Germany by Ruhrchemie AG and Ruhrkohle AG. These advantages include the direct application of wet run of mine coal as well as the trouble free handling and transportation of coal. The inherent draw back of feeding large amounts of water to the gasifier have been minimized and even surmounted by Ruhrchemie/Ruhrkohle's efforts to reduce the water content of the slurry. By the use of suitable flow improving agents suspensions with a solids concentration up to 71 % can be prepared. Coal water slurries are non-Newtonian liquids. The fundamentals of these fluids are briefly summarized. Some results of the experimental work are given specifying the effects of some major parameters on the flow properties of the suspensions.

1. INTRODUCTION

After the first oil supply crisis in 1973 projects were started all over the world with the aim of replacing the fossil carbon in mineral oil by the carbon in coal. For the chemical industry, synthesis gas — a mixture of carbon monoxide and hydrogen — has proved to be an ideal link between the raw material coal and its chemical exploitation as a building block. The purpose of coal gasification is to produce synthesis gas from coal.

Of the various coal gasification processes of the "second generation", the Texaco process as used in the Ruhrchemie AG/Ruhrkohle AG plant in Oberhausen has proved to be the most successful to date. A characteristic of this process is the handling, conveyance and feeding-in of fine dust coal in the form of a coal-in-water suspension.

2. CHARACTERISTICS OF WET COAL FEED

The use of liquid feed has contributed much towards the success of the Texaco coal gasification process. The transportation of the fine coal as a slurry offers a large number of advantages if finely divided coal is to be converted under pressure with the gasification agents oxygen and water. Suspended in water, the fine coal resembles heavy fuel oil and can be easily fed into pressurised gasification plants by relatively simple pumps. This is a basic requirement for trouble-free automation of the process. An outline of the advantages of wet feed in coal gasification processes as compared with possibilities for dry feed is shown in table 1.

Table 1: Advantages of the RAG/RCH Texaco wet feed system

Of the advantages shown, the first two are evident and self-explanatory. How much drying energy can be saved, is illustrated in figure 1, which was taken from publications of competitors.

*   Ruhrchemie AG, D-4200 Oberhausen-Holten, W. Germany
**   Ruhrkohle Oel and Gas GmbH, D-4250 Bottrop, W. Germany

Fig. 1: Specific heat consumption for coal drying in
grinding/drying plants

The saving potential becomes obvious if one considers that a moisture content of under 3 % is
desirable for coal dust with good flow and conveyance properties.

It is also easy to appreciate the great advantage of trouble-free handling, measurement and conveyance of the coal-in-water suspension over the dry feed system when one considers the burning times of
finegrained coal dusts as illustrated in figure 2.

Fig. 2: Burning time of coal dusts as a function
of the particle size

These are fractions of a second with the grain sizes normally used in coal gasification processes. The
advantage of the wet feed system is also apparent considering the safety measures for coal dust of
table 2 recently published by H. Kreusing of Rheinbraun.

Table 2: Safety regulations for the storage and transportation
of coal dusts

These regulations were drawn up for the use of coal dusts in atmospheric and air-operated boiler
plants. Bearing in mind that with coal gasification it is necessary to feed partly pyrophorous coal
dust under pressure into a reactor containing oxygen and to convert it there, it soon becomes clear
that in the dry feed system good explosion protection design is of paramount importance and that
highly sophisticated instrumentation and control systems must be employed to ensure an exact
dosage of C/O during operation — certainly no easy task.

As comparisons of the capital expenditure involved clearly prove, the wet feed system is, in this
respect, far more simple and far less costly. In one single step, wet grinding of coarse-grained coal
produces the coal-in-water suspension required, which can be pumped without difficulty and whose
flow can be measured easily and with only a fraction of the instrumentation normally employed. In
the gasification reactor the steam acts as a moderator, thus making the system far less sensitive and
far safer than dry feed systems.

Over and above its effect as a moderator, the water fed in with the suspension has an advantageous
influence on the gasification process; the decomposition of the water due to the reaction with carbon
sets free H_2 and O_2. This has been shown by Nitschke et al. in figure 3.

Fig. 3: Specific synthesis gas production and oxygen
requirement as a function of the water/coal ratio ·

This leads to two extremely desirable effects: a partial saving of the gasification agent, oxygen, which
no longer has to be introduced from outside and an increase in the quantity of gas, or — in other
words — a rise in the specific yield of synthesis gas. Moreover, the synthesis gas obtained is rich in
hydrogen, which means less extensive conversion is required for downstream chemical syntheses and
that the resulting efficiency loss is reduced.

Above certain solid concentration levels — about 70 % — these two effects more than compensate for
the negative effect of water feed involved in slurry conveyance to the gasifier and safe control over
the coal feed becomes the outstanding advantage.

Thus wet feed is an integral part of the Texaco coal gasification process, which for many years has
proved its technical maturity and reliability in the Ruhrchemie/Ruhrkohle plant in Oberhausen.
Figure 4 shows its integration in the system as a whole.

Fig. 4: Flow sheet of the Texaco coal gasification
process in the Ruhrchemie/Ruhrkohle plant

3. PROCESS DESCRIPTION

The process contains the four main steps: preparation of the suspension, gasification, use of waste heat and slag treatment. Fine coal is stored in dried form or in suspension in a storage container and is fed in predetermined quantities together with water recycled from the process or, if necessary, fresh water into a wet grinding plant. The ground suspension is pumped out of the storage container into the gasification reactor and autothermally converted there with oxygen. The reaction gas containing liquid ash first enters a radiant cooler where it is cooled and the coarse ash removed. With a low content of largely solid fine ash, it then passes through a convection cooler. In both cooling stages high-pressure process steam is produced from the sensible heat of the raw gas.

The gas which is now largely cooled down is washed with water and then cooled even further by the removal of effective heat and is passed free of solids into downstream stages. The ash both from the bottom of the radiant cooler and the washing stage is suspended in water, withdrawn and classified. The fine ash which contains carbon can be recirculated together with the washing water to the grinding stage; the coarse ash which is largely free of carbon is either discarded or sent to a processor. Water-soluble inorganic impurities from the coal are removed from the system by a waste-water side stream.

Figure 5 gives an impression of the plant.

Fig. 5: Ruhrchemie/Ruhrkohle demonstration plant in Oberhausen-Holten

The plant is designed for a coal throughput of 6 t/h, a pure gas production of 10 000 m^3/h and an operating pressure of 40 bar. It has been in operation since January 1978. Until April 1982 the work was subsidised by the West German Federal Ministry of Research and Technology and from then by the Ministry of Economics, Small Business and Transport of the state of North-Rhine Westphalia. At this point we would like to take the opportunity of thanking both sponsors.

4. STATUS OF WORK

Table 3 outlines the results achieved so far.

Table 3: Current status of the test results of the Ruhrchemie/
Ruhrkohle demonstration plant for the Texaco
coal gasification process

In more than 13 000 h of real gasification time, over 70 000 t of coal have been converted to over 130 million Nm^3 of coal gas, which in turn has been converted to organic chemicals in the Oxo plants of Ruhrchemie AG on a regular basis over the past three years. The exclusive use of synthesis gas from coal as a feedstock for Oxo reactors has also been successfully demonstrated over periods of several weeks.

5. WET GRINDING SYSTEM

The rapid progress made in the development of the Texaco process is due to a large extent to the fact that a wet coal feed system is used. This, however, presupposes a suitable grinding system.

Basically it is perfectly possible to prepare a coal suspension by dry or wet grinding. Dry grinding would have had the advantage of being able to make use of the highly sophisticated and time-tested technical coal grinding installations already available. However, a process which operated with aqueous suspensions should be designed to use wet run-of-mine coal. This would make drying of the coal prior to dry grinding completely superfluous, which means a great energy saving and improved safety.

These undeniable advantages induced Ruhrchemie/Ruhrkohle to develop their own wet grinding system to fulfil the requirements of the Texaco coal gasification process. The main features of this system are:

- high wear resistance of the components
- easy control of grain size by adjustment of simple
  operating parameters

- it is unaffected by coal impurities in the form of
  gangue and tramp material

- high suspension concentration with good pumping
  properties and stability

- low energy consumption

- easy adaptation to commercial scale use as well as

- stationary trouble-free operation

The grinding capacity to be set depends on the grinding properties and the initial grain size of the feed coal as well as the grinding fineness of the suspension required.

The maximum size of the coal delivered should be 10 to 30 mm. A typical particle size distribution of the suspension required is shown in figure 6.

Fig. 6: Particle size distribution of coal suspension and feed coal

The residue on a 90 μm screen, which accounts for about 20 to 50 % of the grinding stock, has proved to be a reasonable criterion to characterise the grinding fineness, experience showing that the characteristic lines of the particle size in the Røsin/Rammler/Sperling grain network run more or less parallel.

Highly reactive coals are generally ground somewhat more coarsely than less reactive ones. This has a positive effect on the maximum attainable solid content of the suspension which generally decreases as the proportion of fine particles increases. The grinding fineness set in a certain case is usually a compromise between high suspension concentration and good gasification properties.

As would be expected, the selectivity of the gasification reaction towards carbon monoxide and hydrogen improves as the coal content of the suspension used increases. The aim of the coal dressing process is therefore to produce a free-flowing, easily convertible coal suspension with a maximum solid content.

Today suspension concentrations of up to 71 % can easily be attained with the wet grinding system developed by Ruhrchemie/Ruhrkohle. They contain additives which increase the solid content by 6 to 8 % whilst having no effect on the viscosity. By suitable selection of the kind and quantity of additives, the particle size distribution and viscosity, these additives not only improve the pumping properties but also increase the suspension stability, i. e. separation of the two-phase system is prevented.

The ground suspensions exhibit viscosities of up to 3 000 cP and with pressure drops within acceptable limits they can be transported through pipelines of quite considerable length.

6. FUNDAMENTALS OF COAL/WATER SUSPENSIONS

The viscosity of coal/water suspensions will now be discussed in more detail. Coal/water suspensions are non-Newtonian fluids which can be classified partly under dilatant and partly under pseudoplastic fluids. These terms are explained in figure 7.

Fig. 7: Rheological behaviour of Newtonian and
        non-Newtonian fluids

The flow curves can be described by the exponential law of Ostwald-De Waele, as shown in figure 8.

Fig. 8: The Ostwald-De Waele law on flow curves

K and n are constants for a given fluid. n is a measure of the deviation from Newtonian behaviour. From figure 9 it can be clearly seen that the parabolic velocity profile for the flow of Newtonian fluids (n = 1) has a flatter gradient for pseudoplastic fluids with flow exponential values of less than 1 and a steeper gradient for dilatant media where n is greater than 1.

Fig. 9: Velocity profiles of fluids with different values
        of the flow index n

Dilatant behaviour is sometimes observed in low-concentration coal suspensions. Pseudoplastic behaviour on the other hand predominates in suspensions with a higher solid content. Flow improvers lower the flow exponential value n of pseudoplastic fluids. Apart from lowering the viscosity, they also cause the fluid to approach Newtonian behaviour.

7. EXPERIMENTAL GRINDING RESULTS

The viscosity of a suspension is mainly determined by the sum of all shear forces which occur under stress between the particles. When the number of particles is increased, i. e. when the suspension concentration is increased, the total surface area of the particles is enlarged. At the same time the distance between them is reduced. Both factors lead to higher shear forces and thus higher viscosities. This is illustrated in figure 10 which shows the viscosity of the aqueous suspension of a typical Ruhr coal of a given fineness as a function of the suspension concentration.

Fig. 10: Viscosity as a function of the suspension concentration

The suspension has a 30 % proportion of particles of above 90 μm and contains a suitable concentration of flow improver. Viscosity was measured at a relatively low shear rate of 11.5 s^{-1}, as with pseudoplastic substances this is decisive for the high viscosities which occur during start-up of a pump for example. Pump tests have shown that the viscosity of a suspension in the plant in question should not exceed 3 000 cP at the given shear velocity, as otherwise the suspension could no longer be pumped satisfactorily.

A decrease in the fine grain proportion also leads to a sharp drop of viscosity. If the distribution of particle sizes is shifted so that there are more with larger diameters, the specific surface area of the particles is reduced and the distance between the particles increased. This relationship is illustrated in figure 11.

Fig. 11: Viscosity as a function of grinding fineness.

The coarseness of the particles attainable is limited by the combustion process and the technical installations, e. g. the geometry of the burner, the cross-sections of the pump valve openings etc.

One possibility of changing the coal particle size distribution whilst keeping the specific surface area constant is to increase the proportion of coarse and fine particles simultaneously in a certain way, i. e. by reducing the medium-size particle range. The application of this procedure finally leads to a bimodal distribution. The advantage over a Rosin/Rammler/Sperling distribution is that the packing density is higher for the same specific surface area. Ideally, the gaps in the particle size distribution are filled out by particles of smaller diameters. Our own tests have shown that bimodal distributions can increase the solid content by about 2 %. The commercial viability of this procedure seems, however, to be questionable in view of the equipment necessary.

The viscosity of a pure coal/water suspension generally depends on the grinding fineness, the solid content and the coal grades. Trials have shown that a grinding fineness of 30 to 50 % greater than 90 μm is optimal in both technical and economic terms. With this grain size the solid content of the suspension can amount to 59 to 64 %, depending on the kind of coal, without the critical viscosity of 3 000 cP being exceeded.

Whilst the viscosity is kept constant, the coal content of the suspension can be further increased by the addition of flow improvers. These additives are surface-active substances which adsorptively bind to the surface of the coal. After the lipophilic tenside molecule residue has attached itself to the surface of the coal particles, a uniform surface charge is built up which leads to mutual repulsion and thus better mobility of the particles and stabilisation of the suspension. The surface charge is the difference in potential between the surface of the particles and the fluid surrounding them, the so-called Zeta potential, and is responsible for the repellant effect. An additive to reduce the viscosity of suspensions therefore has the task of building up and maintaining as high a Zeta potential as possible.

As the additive concentration increases, the Zeta potential passes through a maximum value. This is explained by the attainment of a maximum packing density on the coal surface and by the start of micelle formation of the tenside molecules. With a constant solid content of the suspension an initial decrease in viscosity can therefore be observed as the additive content increases. At a critical additive concentration the viscosity attains a minimum value and then generally starts to rise again as the concentration increases further. This process is shown in figure 12.

Fig. 12: Relation between viscosity and additive content
of coal suspensions

If the influences of grinding fineness, coal concentration and additive content are combined, curves such as those in figure 13 result.

Fig. 13: Relation between grinding fineness and coal
and additive concentrations

These curves relate to a certain coal and are applicable to the critical viscosity of 3 000 cP. They therefore indicate the maximum solid content which can be pumped for a given grinding fineness and a given additive concentration.

These relationships were partly established in the laboratory and partly in plant trials. They formed the basis for the optimisation of the entire system and in fact were the foundation for the high efficiencies which are now possible with the wet feed system of the Texaco coal gasification process. However, the know-how now available is, in principle, not limited to the Texaco gasification process. Applications are conceivable everywhere where coal has to be conveyed hydraulically, whether over large distances as an alternative to rail transport or for use in normal power station boilers.

prevent potential sources of ignition	avoid spontaneous ignition	detect smouldering areas	minimize effects of explosion
• explosion protection in the bunker • avoidance of electrostatic charging (earthing) • no smoking	• avoidance of heat radiation • limiting of transportation air temperature • avoidance of uncontrolled air intake • inertisation during lengthy shut-downs	• temperature control • CO measurement • CH_4 measurement • sniffing test	• explosion-proof design • no admittance to bunkers during operation • diversion of the blast wave into a safe area • regular maintenance of explosion flaps • automatic interruption of unloading

RUHRCHEMIE AG
RUHRKOHLE AG

Safety regulations for the storage and transportation of coal dusts (to Kreusing) — **Table 2**

• Coal drying with its high energy requirements (reduction in efficiency) is unnecessary

• trouble-free handling, measuring and transportation of the coal feed and thus low capital expenditure

• water acts as a moderator in the gasification process, therefore costly safety precautions can be dispensed with

• At the high reaction temperatures of the Texaco coal gasification process part of the water undergoes decomposition

  → reduction of oxygen consumption

  → higher incidence of hydrogen (percentually and in absolute terms)

• possibility of trouble-free change over to other liquid raw materials

RUHRCHEMIE AG
RUHRKOHLE AG

Advantages of the RAG/RCH/Texaco wet feed system — **Table 1**

Total operation period	(h)	over 13 000
Total amount of coal	(t)	over 70 000
Total gas amount	(m_N^3)	over 130 MM
Coal throughput	(t/h)	up to 8,2
Quantity of gas	(m_N^3/h)	up to 15 200
Number of tested coals and residues	–	14
• ash content	(%m.f.)	6 – 28
• content on volatiles	(%m.f.)	16 – 44
• water content	(%)	1 – 55
• Hardgrove index	(^oH)	47 – 106
• ash melting point, reduc.	(^oC)	1280 – 1500
Solids content in slurry	(%)	up to 71
Temperature	(^oC)	1200 – 1600
Pressure	(bar)	up to 40
Gas composition	(Vol-%)	
CO		54
H_2		34
CO_2		11
CH_4		< 0,1
H_2S/COS		0,3
N_2		0,6
C-conversion	(%)	up to 99
Efficiencies (cold gas/thermal)	(%)	77/94

RUHRCHEMIE AG	Current status of the test results	Table 3
RUHRKOHLE AG	(optimum results of different test runs)	

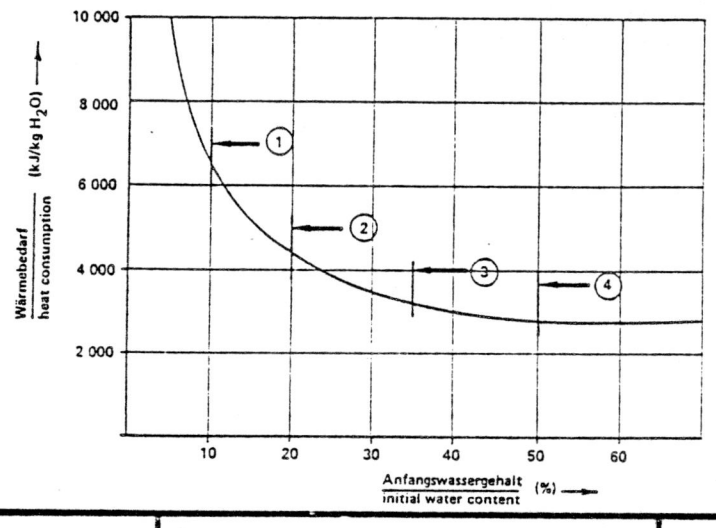

RUHRCHEMIE AG	Specific heat consumption for coal drying in	Fig. 1
RUHRKOHLE AG	grinding drying plants	

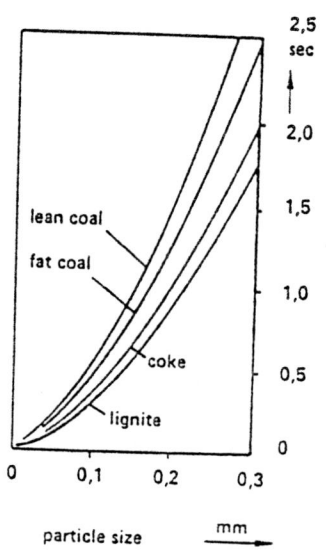

particle size   mm

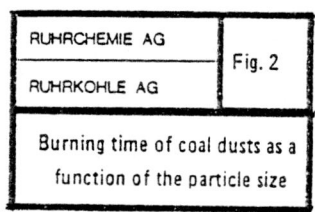

RUHRCHEMIE AG	Fig. 2
RUHRKOHLE AG	
Burning time of coal dusts as a function of the particle size	

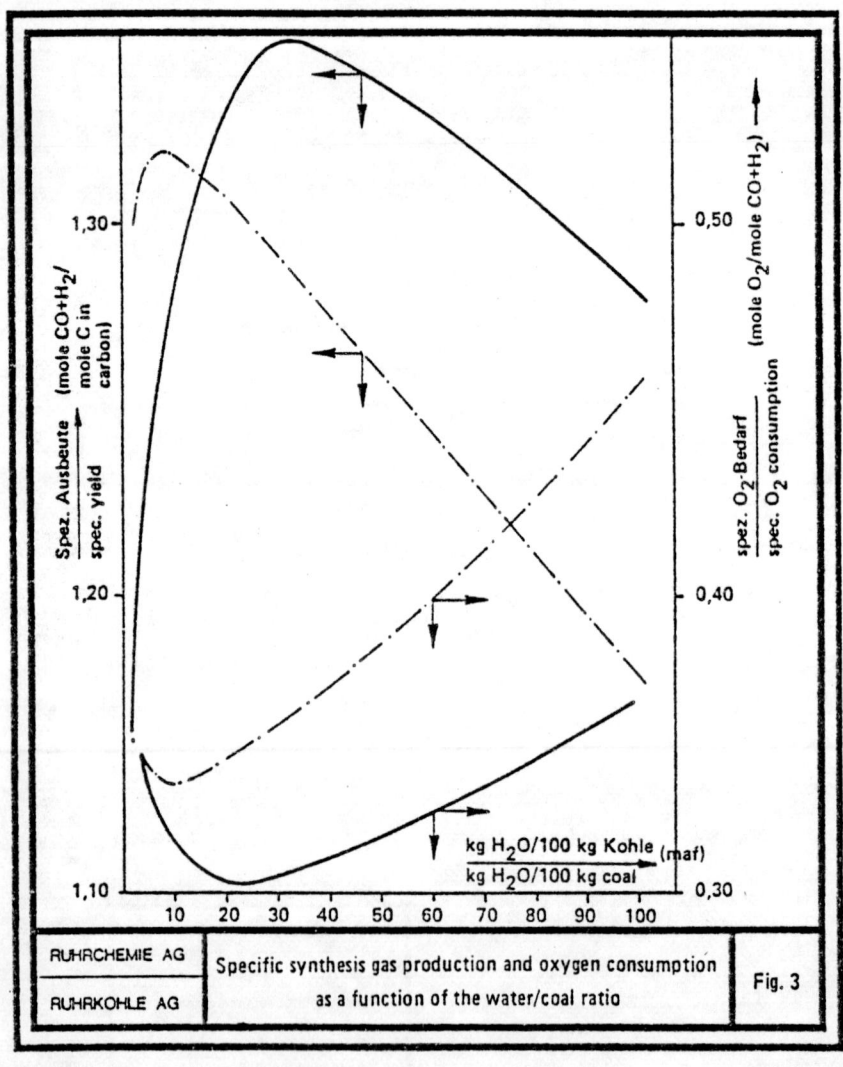

| RUHRCHEMIE AG | Specific synthesis gas production and oxygen consumption | Fig. 3 |
| RUHRKOHLE AG | as a function of the water/coal ratio | |

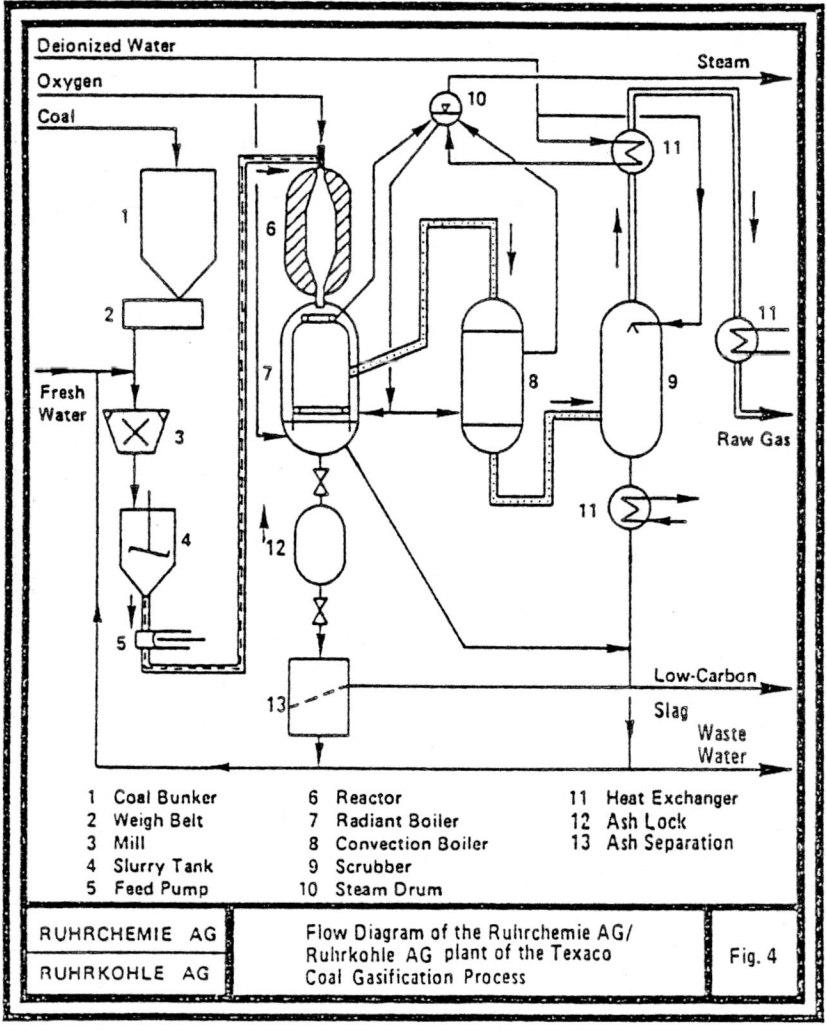

Deionized Water
Steam
Oxygen
Coal
Fresh Water
Raw Gas
Low-Carbon Slag
Waste Water

10
11
11
1
6
7
8
9
2
3
4
5
12
13

1	Coal Bunker	6	Reactor	11	Heat Exchanger
2	Weigh Belt	7	Radiant Boiler	12	Ash Lock
3	Mill	8	Convection Boiler	13	Ash Separation
4	Slurry Tank	9	Scrubber		
5	Feed Pump	10	Steam Drum		

| RUHRCHEMIE AG | Flow Diagram of the Ruhrchemie AG/ | Fig. 4 |
| RUHRKOHLE AG | Ruhrkohle AG plant of the Texaco Coal Gasification Process | |

CLM-I

FIGURE 5. RCH/RAG demonstration plant in Oberhausen-Holten

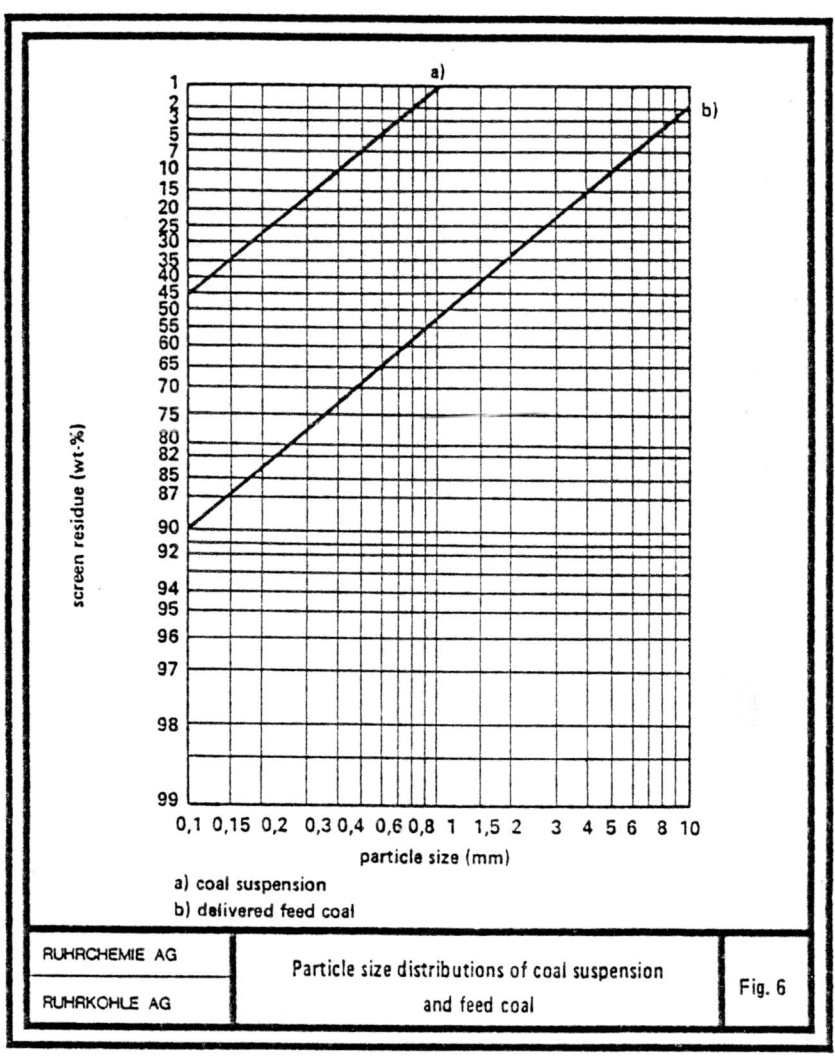

a) coal suspension
b) delivered feed coal

| RUHRCHEMIE AG | Particle size distributions of coal suspension | Fig. 6 |
| RUHRKOHLE AG | and feed coal | |

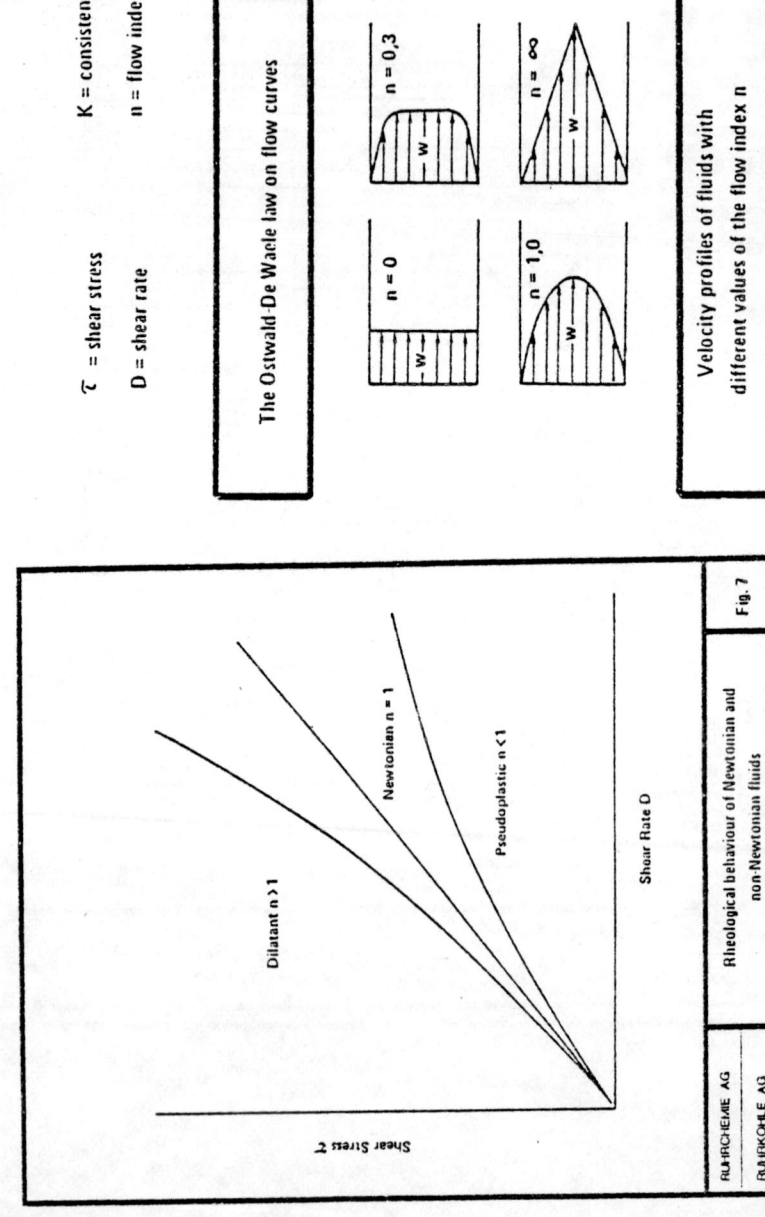

$$\tau = K \cdot D^n$$

τ = shear stress   K = consistency

D = shear rate   n = flow index

The Ostwald-De Waele law on flow curves   Fig. 8

n = 0   n = 0,3

n = 1,0   n = ∞

Velocity profiles of fluids with different values of the flow index n   Fig. 9

Dilatant n > 1

Newtonian n = 1

Pseudoplastic n < 1

Shear Rate D

Shear Stress τ

RUHRCHEMIE AG

RUHRKOHLE AG

Rheological behaviour of Newtonian and non-Newtonian fluids   Fig. 7

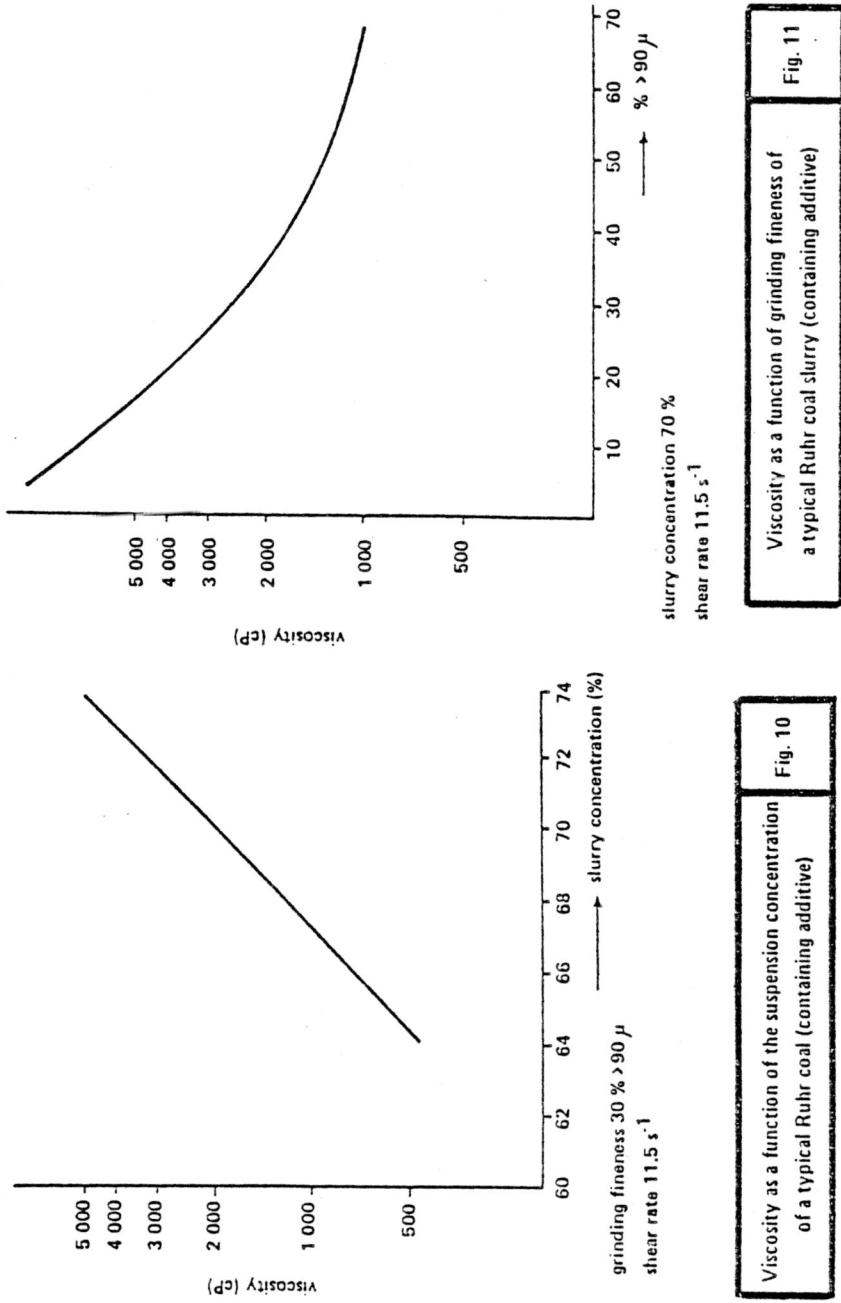

Viscosity as a function of grinding fineness of
a typical Ruhr coal slurry (containing additive)

Fig. 11

slurry concentration 70 %
shear rate 11.5 s⁻¹

Viscosity as a function of the suspension concentration
of a typical Ruhr coal (containing additive)

Fig. 10

grinding fineness 30 % > 90 μ
shear rate 11.5 s⁻¹

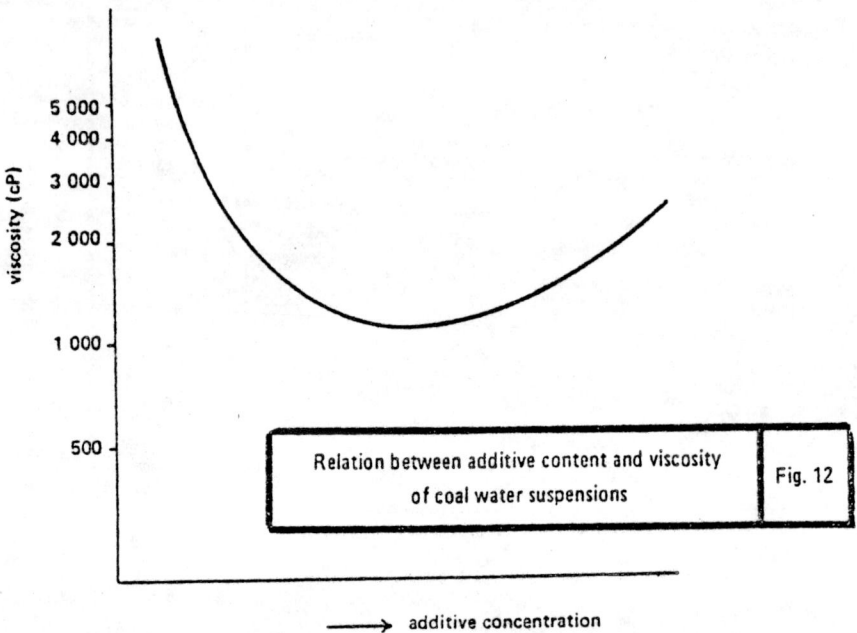

Relation between additive content and viscosity of coal water suspensions — Fig. 12

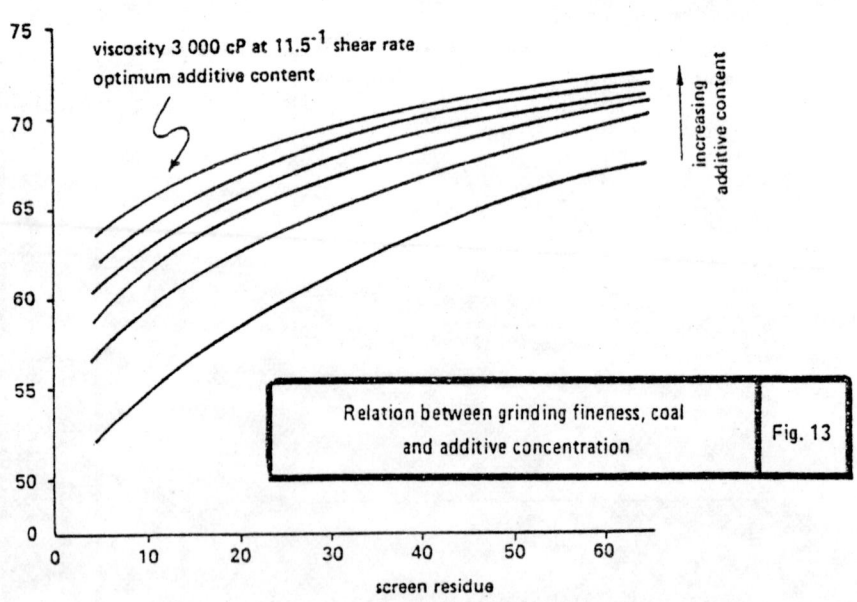

Relation between grinding fineness, coal and additive concentration — Fig. 13

SLAGGING/FOULING EVALUATION TESTS ON COAL/OIL DISPERSION FUELS

W.M.Urquhart*, W.I. Joyce* and A.G.Wootton**

Four COD fuels have been subjected to ranking trials in a test rig. This simulates furnace exit conditions in industrial boiler plant which is seen as the main market for COD retrofit conversions from HFO. From subjective/ qualitative and quantitative evaluations of slagging propensity two of the COD's were shown to be significantly superior to the other two. Inter-relationships have been established between deposition rates and categories, ash content and fusibility characteristics and the flue gas temperature at which deposition occurred. It is postulated that this could, after further validation with COD's covering a wider range of properties, be used for perform- ance prediction. It was found that the generally accepted Slagging and Fouling Indices for coal did not successfully predict COD slagging propensity.

INTRODUCTION

Discounting the current worldwide glut situation, which results largely from the 'short term' economic recession, the supply of crude oil in general and fuel oil in particular is expected in the longer term to become increasingly deficient. The eventual reduction in crude production resulting from the diminution of world crude resources will demand a deeper cut into the barrel to provide for upgrading the lighter fractions. Residual fuel oil supplies will thus become more scarce and expensive. This situation will eventually precipitate an almost complete replacement of existing oil fired plant by coal fired equipment. Such a conversion would clearly require massive capital expenditure and include the replacement of much oil fired plant which was still serviceable. The use of coal oil mixtures could, however, in the interim period facilitate a more gradual change over and extend the life of oil-designed plant.

Coal oil mixtures (COM's) are not new, (first patent 1879) but the low price differential between oil and coal has discouraged their commercial use. The oil crisis of 1973, however, changed the economic perspective and since then many R & D programmes have been initiated to establish the technical feasibility of firing COM in oil-designed plant.

Generally COM's have suffered from instability in storage and the consequent necessity for on-site preparation facilities has made them commercially less attractive. British Petroleum, however, have developed a method of preparing coal/fuel oil dispersions (COD's) which imparts excellent long term stability

* Research Centre, Babcock Power Ltd., High St., Renfrew, PA4 8UW
**BP Research Centre, Sunbury-on-Thames, Middlesex, TW16 7LN

characteristics. The fuels may contain up to 50% coal (35-40% normally) in finely divided form and no stabilising additives are required. Such a stable product offers the potential of central plant preparation and distribution through the normal fuel oil network.

In parallel with the manufacture and handling R & D programmes BP have undertaken an intensive combustion research study on the new fuel, covering all aspects from preheat through to gas clean-up. Following a feasibility study on a range of commercial combustion plant the research programme was orientated towards oil-designed industrial type steam raising boilers, which were identified as the potential U.K. market although the major market could be seen as large utility boilers in Japan and U.S.A. where conditions, in terms of furnace exit gas temperature, are typically less severe than their counterparts in the U.K.

Perhaps the most crucial question to be addressed by the study was that of the severity of ash deposition within the boiler. Oil-designed boilers have higher furnace heat release rates, higher furnace exit gas temperatures and closer pitched convection tubes than their coal-designed counterparts. For COD firing conversions, therefore, it was vital that fuel performance predictions could be made in order to minimise boiler derating. Feasibility studies on boilers had shown that the most critical deposition area was at the furnace exit. To generate the necessary fuel selection criteria experimental studies would be required to investigate the depositional tendencies of various COD's at a range of furnace exit conditions representative of commercial boilers. Realistically the work could only be undertaken using a suitable combustion rig and the one at the Babcock Power Ltd. Research Centre was used.

A pilot scale test programme was drawn up in which the slagging/fouling potential of a range of COD's with different coal ash properties was studied on a simulated superheater test section at the rig furnace exit. This paper charts the progress of these investigations and sets out some of the preliminary conclusions.

TEST PROGRAMME OBJECTIVES

The recognition of the furnace exit of a boiler as, potentially, the most critical area with respect to deposition defined the basic objective of the test programme. This was to rank a selection of COD's, in relation to Slagging/Fouling propensity, in the context of realistic boiler plant furnace exit gas temperatures (FEGT). By the same token the test rig format was also defined - that there should be deposit receiving test surface maintained at a realistic surface temperature in a COD flue gas environment capable of being controlled at temperature levels at and around the notional boiler FEGT's. In this position and at these temperatures the test surface would be exposed to both radiant and convective heat and these factors, according to a 1974 ASME Report [1], which classifies deposits and discusses Slagging and Fouling Indices, would categorise the deposits of the COD trials as Slagging.

The choice of "realistic" FEGT's was largely dictated by the temperatures prevailing in industrial type boiler plant of the type which, it was believed, would be the main U.K. market for oil to COD retrofit conversions. A study of design data for such boiler plant showed that, for three different boiler designs covering a range of 70-350 tonne h^{-1} steam output and operating conditions of 5-20% excess air the temperature at the furnace exit ranged from 1150-1300°C. By contrast large Utility type boiler plant in the U.K. can operate typically with 1450°C FEGT although there are potential overseas markets where less severe conditions prevail in Utility boilers.

The expectation, therefore, was that the major part of the test programme would be concentrated on FEGT's up to 1300°C but since there would be some interest in temperatures above this the rig design should allow the use of temperatures up to c.1500°C.

COMBUSTION TEST RIG

The test rig on which the evaluation trials were carried out was originally designed for HFO flue gas corrosion studies. As modified for the COD deposition programmes the rig is as shown in Figs. 1 and 2, diagrammatic and isometric representations respectively. The refractory duct which contains the 6 horizontal tube, steam-cooled test section, on which deposition occurs, is 340mm high x 230mm wide. A series of similar gas cooler sections further downstream and a spray-water cooler conditions the flue gas for discharge into the chimney. Adjustable, water-cooled heat absorbing surface in the furnace allows a FEGT range of 1100-1500°C to be achieved without recourse to wide variations in COD throughput which was approximately 35 kg.h⁻¹. An oblique camera/viewing port immediately upstream of the test section allows deposit build-up to be observed and recorded photographically.

COD firing into the double torroidal combustion chamber is by a steam-assisted pressure jet atomiser. The COD fuel supply, storage, pumping, heating and flow measurement system is shown in Fig. 3. It will be seen that an HFO firing system is also incorporated. This is used for rig preheating and test section preconditioning which is discussed later.

SLAGGING/FOULING EVALUATION

From the outset it was envisaged that the visual observation and assessment of the deposit build-up on the test section - a subjective qualitative evaluation - would provide the main basis for ranking purposes. It was hoped that supporting quantitative assessments could also be obtained. To this end arrangements were therefore made for the compilation of the following quantitative data during the course of a test run.

  a) - Heat transfer to the test section at intervals during the
       pre-conditioning operation on HFO and at intervals during
       a COD firing test run.

  b) - Pressure drop across the test section tube-bank at intervals
       during a test run.

  c) - Deposit surface temperature measurement by optical pyrometer.

In the event only the heat transfer measurements were found to be sufficiently reliable to warrant further attention.

The deposited material on the tubes at the conclusion of a test run was removed, collected, measured, photographed and weighed; this series of operations eventually became the main quantitative evaluation.

In the context of the considerations discussed earlier in the section on Test Programme Objectives the evaluations and ranking of the test COD's were thus based on

a) - the temperature (FEGT) at the test section

b) - the quantity of material deposited and rate of deposition

c) - the physical state of the deposit, its coherence and estimated ease (or otherwise) of removal.

As has already been indicated the first two of these items are measureable but the second cannot be regarded as a precise measurement depending, as it does, on post-test weighing when it was known that during some tests deposited material broke away from the surface before the conclusion of the test and was carried away in the flue gas stream. It also implied an assumption of a constant rate of build-up.

The third criterion listed above has to be wholly subjective and only when deposits from all 13 tests on 4 test COD's were available was an assessment sub-division made as follows.

a) - Category 1 - Very unlikely to cause any deposit problem - easily controllable soft deposit - safe

b) - Category 2 - Minor uncertainty but unwise to classify as wholly safe

c) - Category 3 - Major uncertainty - tending towards unacceptable categorisation

d) - Category 4 - Unacceptable - likely to give deposit removal and control problems - implies pronounced slagging/fusing of deposited material

During some early discussion of the test programme it was suggested that test duration defined by the appearance of molten slag or by bridging of the test section might also be an evaluation parameter but, in the event, although these concepts were used, on occasion, to decide the termination point of a test the variations in the physical nature of the deposit build-up were such that test duration per se was not a satisfactory evaluation parameter.

It was obvious that for strict comparability of deposit build-up the deposition surface had to be standard. Therefore, for all tests the upstream face of the test tube bank was maintained at a temperature of 450-550°C and it was subjected to a pre-conditioning period in HFO flue gas starting from a steam-cleaned condition. This pre-conditioning occurred during the heat-up to test conditions prior to the commencement of COD firing. The O_2 level during the tests was maintained at 1 - 2%.

TEST PROGRAMME/RESULTS

The analytical data for the four COD's on which the test programme was carried out are given in Table 1 which also includes Slagging and Fouling categorisations based on calculation methods described in Ref. 1.

It will be seen that although a Slagging range of Medium to Severe was encompassed the range of Initial Deformation Temperature, which, it was expected, could have some relationship with the test FEGT's and slagging propensity was more restricted than had been hoped when the test programme was originally planned.

An example of the test data compiled during a test run (on COD-B) at a target FEGT of $1330^{\circ}C$ is given in Fig. 4 which includes a record of the calorimetric determinations made during the course of the 6.5 hr test - test termination in this instance being based on the fact that molten deposit material was observed. Figs. 5a - f show a sequence of photographs taken at intervals during the test and showing the build-up of deposit material on the test tube-bank. Fig. 5g (colour original) shows the deposit removed from the tubes of the test bank and subjected to build-up rate measurement and the subjective category assessment described previously.

Table 2 includes the three parameters described in the previous section as the basis of the COD evaluations and rankings. It also includes in the last column a heat transfer criterion based on the type of information shown in graph form in Fig. 4. The quoted temperatures are the result of the extrapolation of the relationship between FEGT and the time taken to reach steady state reduction in heat transfer from the initial clean tube conditions to a nominal 8 hrs.

<u>COD RANKING ASSESSMENT</u>

For COD ranking assessments the following symbols are used. $\grave{>}$ implies "is a little better than", $>$ implies "is better than", $\gg$ implies "is much better than".

On this basis the visual/subjective assessment of deposit build-up in the context of FEGT level, the amount of build-up and its physical nature in relation to assessed ease or difficulty of removal by conventional soot-blowing techniques clearly established the following ranking order

$$COD\text{-}A \; \grave{>} \; COD\text{-}C \; > \; COD\text{-}B \; > \; COD\text{-}D.$$

To an extent the FEGT/Ash Fusibility relationship was also taken into consideration in that, for instance, with COD-C Category 2 type deposits were obtained at an FEGT level well above the Initial Deformation Temperature of the ash under oxidising conditions (IDT_O).

Turning now to the quantifiable assessments it can be seen from the final column of Table 2 that a ranking based on heat transfer considerations as described in the previous section gives

$$COD\text{-}A \; \grave{>} \; COD\text{-}C \; > \; COD\text{-}B \; \grave{>} \; COD\text{-}D.$$

The deposit build-up data, which is also included, has been subjected to fuller treatment including an examination of the inter-relationships between the build-up expressed both on a weight and rate basis and FEGT and FEGT/Ash Fusibility parameters. The only truly successful correlations were found to be those of the type shown in Figs. 6a and 6b, - i.e. the rate of deposit build-up (as determined from post-test weighing) and FEGT and its relationship with Initial Deformation Temperature. The Fig. 6b correlation using IDT_O rather than IDT_R is felt to be more applicable since, as Table 2 makes clear, all deposition build-up occurred in an atmosphere containing from 1% to 3% O_2. The ranking shown is

$$\text{Fig. 6a} - COD\text{-}A \; \grave{>} \; COD\text{-}C \; \gg \; \begin{array}{c} COD\text{-}B \\ COD\text{-}D \end{array}$$

$$\text{Fig. 6b} - COD\text{-}A \; > \; COD\text{-}C \; > \; COD\text{-}D \; \gg \; COD\text{-}B.$$

IMPLICATION OF ASSESSMENTS

In summary the assessments are as follows:

Basis	Assessment
Subjective	COD-A $>$ COD-C $>$ COD-B $>$ COD-D
Heat Transfer	COD-A $>$ COD-C $>$ COD-B $>$ COD-D
FEGT	COD-A $>$ COD-C $\gg$ COD-B / COD-D
FEGT - IDT_0	COD-A $>$ COD-C $>$ COD-D $\gg$ COD-B

It is obvious that there is a fair degree of consistency in these rankings
with COD-A invariably ranking higher, to a greater or lesser degree, than
COD-C and both ranking significantly higher than the other two COD's.

But since, as has already been discussed, the likely use of COD is in
industrial boiler plant the comparison of COD-A and COD-C should be in the
context of FEGT levels up to 1300°C when, as Fig. 6a shows there will be
little to choose between them.

It will also be seen that there is an inconsistency in the relative rankings
of COD-B and COD-D in that while the principal, subjective, assessment clearly
indicated COD-B as superior the rate/FEGT based correlations showed the
opposite, particularly Fig. 6.   This introduces the concept of COD ash
content which has not hitherto been considered since as an intrinsic property
of the COD it cannot be looked upon as an independent assessment parameter.
But it will be seen from Table 1 that the ash content of COD-B at 4.2% was
significantly greater than that of COD-D at 2.9% and this was obviously likely
to have an influence on a rate-related assessment.

Fig. 6b shows a cross-curve which separates the Category 2 and 3 types of
deposit - i.e. those which, it was considered, would not be troublesome and
those which, it was fairly certain would be.   It appeared to be worthwhile
to investigate the inter-relationship of the Fig. 6 data with COD ash content.

The result is shown in Fig. 7; it should be noted that for this correlation
the deposit rate ordinate data have not been normalised with respect to COD
firing rate.   It will be seen that there is a reasonably coherent relationship between the deposit build-up rate and the amount of ash in the COD for a
given FEGT in relation to the IDT_0 value of the COD ash and the deposit
category.   It would, however, be unwise at this stage to use such a graph for
predictive purposes, since it has been compiled on the basis of only 4 COD
trial series.

Table 1 includes calculated Slagging and Fouling Indices which, are generally
accepted as indicating no more than a four-part categorisation -

Low/Medium/High/Severe

On this basis we would have a ranking order for Slagging and Fouling of

COD-B $>$ COD-A $>$ COD-D $>$ COD-C

The considerable difference between this calculated ranking order and the experimentally determined ranking is further emphasised by Fig. 8 which relates the calculated rankings to deposit categorisation and the $FEGT/IDT_0$ correlation of Figs. 6b and 7. It will be seen that the experimental trend is the opposite of that which would be predicted by the Indices.

However, it must be pointed out that the calculated Indices purport to predict the behaviour of <u>coal</u> ash and are based on analytical data from <u>coal ash</u> from Eastern USA bituminous coals and it may be unwise to expect COD ashes to conform because

a) - the combustion environment of the COD ash is different, both physically and chemically

b) - COD ash contains both coal and oil ash, the latter a small part of the total but inevitably on the outside of the ash particle and possibly having a disproportionate influence

c) - the COD grinding process may have resulted in the pick-up of iron which, on combustion, will be oxidised to Fe_2O_3; it is significant that all COD ash analyses showed between 1.8% and 4.6% higher Fe_2O_3 contents than the corresponding analyses of the parent coal.

It is therefore not necessarily the case that COD deposition behaviour should conform to derived Indices whose provenance is based on USA coals; the experimental results certainly show that they do not.

COD COMBUSTION IN BOILER FURNACES

Although the primary objective of the work was the ranking of the COD's in relation to FEGT described in the previous section the implication was, as mentioned in the Introduction, that there should be consideration, in the light of the findings, of the possible de-rating of boiler furnaces designed for oil firing when firing COD.

The assessment of ash deposition and slagging in coal-fired boiler furnaces is complex and includes three major aspects which have to be considered

a) - Previous experience

b) - Detailed furnace arrangement

c) - Fuel (and ash) properties

Such assessment can only be carried out in full in the context of design data specific to designated candidate boiler plant. It has also to be realised that the type of deposit categorised as unacceptable in this report is not necessarily unacceptable in all boiler plant even in circumstances where deposit fusing had occurred. The determining factors are adequate tube spacing and ash removal equipment sufficient to keep the build-up situation under control. The narrow tube pitching on the rig allows tube bridging to occur with relative ease and it is not possible to determine the effect of typical tube spacing on the candidate boilers referred to above which, although they will have comparatively narrow tube spacing in the convective surface at the furnace exit, the pitching will not necessarily be as close as that of the rig.

However, the outcome and experience of the COD deposition trials can still be discussed in general terms in relation to boiler use, in particular by consideration of determined "acceptable" FEGT levels - i.e. temperature levels giving deposits which, subjectively, were considered unlikely to be troublesome in a boiler as described previously.

It is considered that the limits of FEGT in relation to heat transfer as discussed in this paper are probably directly applicable but it is believed that some "correction" factors must be incorporated if the data is to be applied to furnace conditions. It is suggested that in the context of a real boiler the acceptable FEGT will be lower because

a) - Furnace gas temperature will be higher than the FEGT in a boiler

b) - Local reducing atmosphere conditions occur in boiler furnaces.

Using sub-scripts B and R to designate boiler and rig and where TF_B is boiler furnace temperature and IDT_{Ox} and IDT_{Red} are the Initial Deformation Temperatures of the ash under oxidising and reducing conditions we thus might postulate

$$FEGT_B = FEGT_R - C_1(TF_B - FEGT_R) - C_2(IDT_{Ox} - IDT_{Red})$$

There is no data to assist the quantification of the correction factors C_1 and C_2 but it can be assumed that $C_1(TF_B - FEGT_R)$ will be small, possibly negligible since although TF_B may be high there is, in most boiler furnaces, a thin layer of cooler downward flowing gases next to the furnace wall and it is this last temperature which is finally significant in terms of deposition. But correction $C_2(IDT_{Ox} - IDT_{Red})$ is likely to be more significant and while, ideally, it might be proportioned relative to the degree of reducing atmosphere close to the furnace walls it is suggested that, until such time as there is evidence that the factor for this aspect is less than unity, the full $IDT_{Ox} - IDT_{Red}$ correction be made.

On this basis the maximum acceptable $FEGT_R$ levels - corresponding to the upper limits of Category 2 deposits as indicated in Fig. 8 have been modified to allow the full correction for the difference between IDT determined under oxidising and reducing conditions. These "corrected" values, as shown in Table 3, have then been interpolated in the established relationships between gas temperature and boiler rating for three different sizes of industrial boiler plant at 5% and 20% excess air levels. These relationships are included in the Babcock Power Design Data Manual, and, as such, are confidential information. Table 3 also includes the calculated attainable boiler ratings which are quoted as a percentage of the design Maximum Continuous Rating.

With COD-A and COD-C some derating of the small industrial type boiler would have to be accepted - even with 5% excess air but otherwise no de-rating is required for COD operation. With COD-B and COD-D a greater or lesser degree of de-rating to between 61% and 94% is required depending on size and operating circumstances - the large boiler requiring less de-rating.

The overall rating is essentially as listed in Table 3.

$$\text{COD-A} \atop \text{COD-C} \gg \text{COD-D} > \text{COD-B}$$

Although some test $FEGT_R$ levels were in the Utility boiler range no attempt has been made to assess de-rating requirements for COD operation but in view of the much higher $FEGT_B$ levels of such plant it is likely to be substantial.

The foregoing is, of course, speculative and is based on the premise that, while the furnace exit heat transfer surface is the most critical area, gas temperature conditions there cannot be divorced from what is happening upstream in the furnace and the conditions there. Further work will be required to determine whether or not the adjusted FEGT's and consequent de-ratings quoted in Table 3 are valid.

CONCLUSIONS

The work described was a comparative evaluation study to rank four COD fuels in terms of their anticipated performance in industrial type boiler plant which is foreseen as the main U.K. market for retrofit conversions from increasingly expensive heavy fuel oil. However, the wider potential in some overseas markets, already referred to, should be noted.

A basis for a predictive technique to correlate the various evaluation parameters was also developed. From this, with a knowledge of the COD ash content and fusibility and the proposed furnace exit gas operating temperature a rate of deposit build-up and estimation of its ease of removal can be obtained. It is, of course, accepted that the prediction has been compiled on the basis of test data from only four COD samples whose properties did not cover as wide a range as was intended when the test programme was originally conceived and that it will require further validation.

It was clearly established, however, that Slagging and Fouling Indices calculated on the basis of ash analyses (and developed in the context of coal ash studies) were not applicable to COD performance as evaluated by test rig operation despite the fact that considerable effort had been expended to ensure that this operation was a realistic representation of boiler plant conditions.

The authors thank the directors of Babcock Power Ltd. and BP for permission to publish the results of this collaborative work.

REFERENCE

1. Winegartner, E.C. (Ed.), ASME Research Committee Report on Coal Slagging and Fouling Parameters. (1974)

TEST FUEL		COD-A		COD-B		COD-C		COD-D	
FUEL ANALYSIS									
Ash	%	2.0		4.25		2.1		2.9	
Sulphur	%	2.9		2.6		2.8		2.8	
Moisture	%	0.45		1.15		0.9		0.65	
Gross Cal. Val. MJ.kg^{-1}		38.95		37.43		38.27		38.38	
ASH ANALYSIS									
SiO_2	%	44.95		71.85		28.9		46.75	
Al_2O_3	%	29.55		15.05		21.2		24.25	
CaO	%	1.5		0.86		8.7		4.45	
MgO	%	0.55		0.13		0.8		0.48	
Fe_2O_3	%	14.95		8.3		18.0		10.6	
TiO_2	%	1.0		0.7		0.75		0.99	
Na_2O	%	0.94		0.41		2.7		2.55	
K_2O	%	1.25		0.7		1.16		1.6	
Mn_3O_4	%	0.03		0.04		0.17		0.07	
P_2O_5	%	0.04		0.3		0.82		0.18	
SO_3	%	1.30		1.15		15.1		6.55	
BaO	%	0.10		0.07		0.2		0.26	
V_2O_5	%	0.68		0.26		0.7		0.45	
NiO	%	0.14		0.05				0.12	
Cr_2O_3	%	0.35				0.3			
ASH FUSIBILITY		REDUC	OXID	REDUC	OXID	REDUC	OXID	REDUC	OXID
Initial Def. Temp.	°C	1200	1320	1160	1330	1150	1240	1150	1220
Hemisphere Temp.	°C	1500	>1500	1400	1470	1230	1330	1260	1350
Fluid Temp.	°C	>1500	>1500	>1500	>1500	1380	1400	1370	1420
Fouling Index		0.24 Medium		0.05 Low		1.66 Severe		0.7 High	
Slagging Index		0.92 Medium		0.54 Medium		2.37 Severe		1.26 High	

TABLE 1 - ANALYSES OF TEST COAL/OIL DISPERSIONS

TEST COD	IDT_0 °C	TEST NO	TEST DUR-ATION hr	MEAN FEGT °C	O_2 LEVEL RANGE %	MAX. DEP. RATE (1) $g.cm^{-1}.h^{-1}$	DEP. EVAL. CAT. (2)	FEGT-IDT_0	HEAT TRANS. CRIT. °C
A	1320	A1	12.0	1348	1.0-2.5	0.057	2	+28	1260
		A2	9.9	1482	1.5-3.1	0.042	3	+162	
B	1330	B1	7.7	1230	2.0-2.3	0.070	1	-97	1080
		B2	6.5	1320	1.6-2.2	0.140	2	-10	
		B3	6.5	1337	1.0-2.6	0.248	3	+7	
		B4	6.3	1341	1.0-2.6	0.197	2	+11	
		B5	3.6	1446	1.1-2.7	0.464	4	+116	
C	1240	C1	12.3	1277	1.7-1.9	0.096	1	+37	1250
		C2	12.0	1364	1.6-2.2	0.048	2	-124	
		C3	5.9	1430	1.8-2.1	0.090	3	+190	
D	1220	D1	5.7	1155	1.4-3.0	0.023	1	-65	1070
		D2	9.0	1204	1.5-2.2	0.040	2	-16	
		D3	1.5	1315	1.4-3.0	0.162	3	+95	

(1) - Normalised to mean COD throughput
(2) - See Section on Slagging/Fouling for Deposit Categorisation

TABLE 2 - SUMMARISED TEST DATA AND RESULTS

COD	MAX. ACCEPT-ABLE $FEGT_R$ °C	IDT_0-IDT_R °C	EST. MAX. ACCEPT-ABLE $FEGT_B$ °C	TYPE A SMALL		TYPE B MEDIUM		TYPE C LARGE	
				EXCESS 5%	AIR 20%	EXCESS 5%	AIR 20%	EXCESS 5%	AIR 20%
A	1415	120	1295	89	100	100	100	100	100
C	1390	90	1300	90	100	100	100	100	100
D	1260	70	1190	67	75	77	87	83	94
B	1320	170	1150	61	67	68	77	74	83

TABLE 3 - INDUSTRIAL BOILER DERATING ASSESSMENT

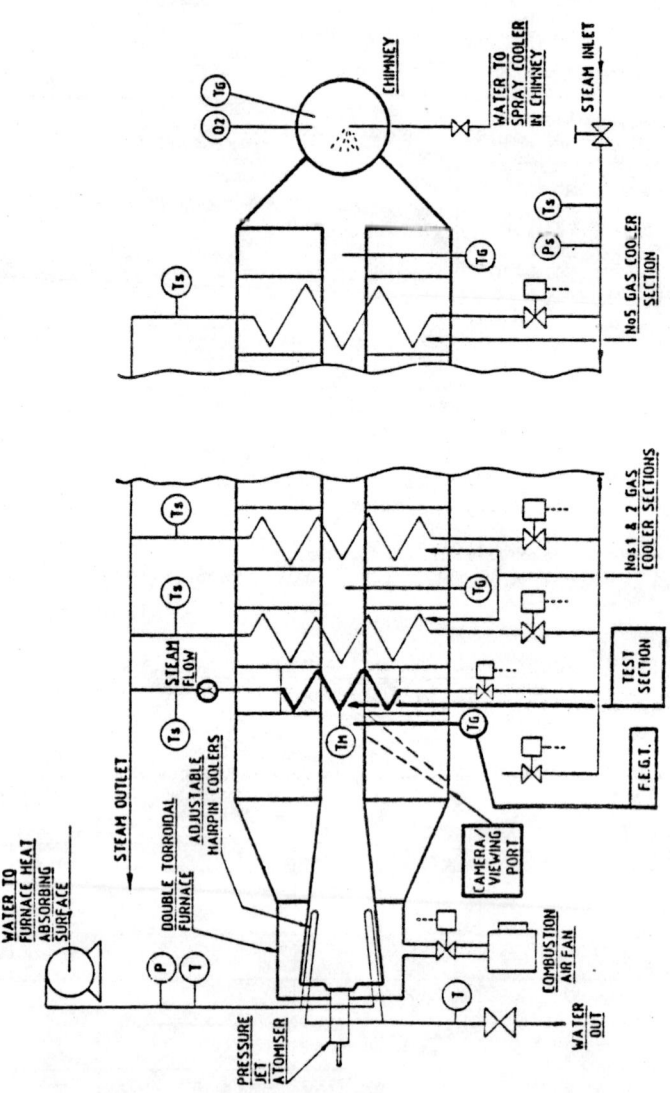

Fig.1 Diagramatic Arrangement of Test Rig

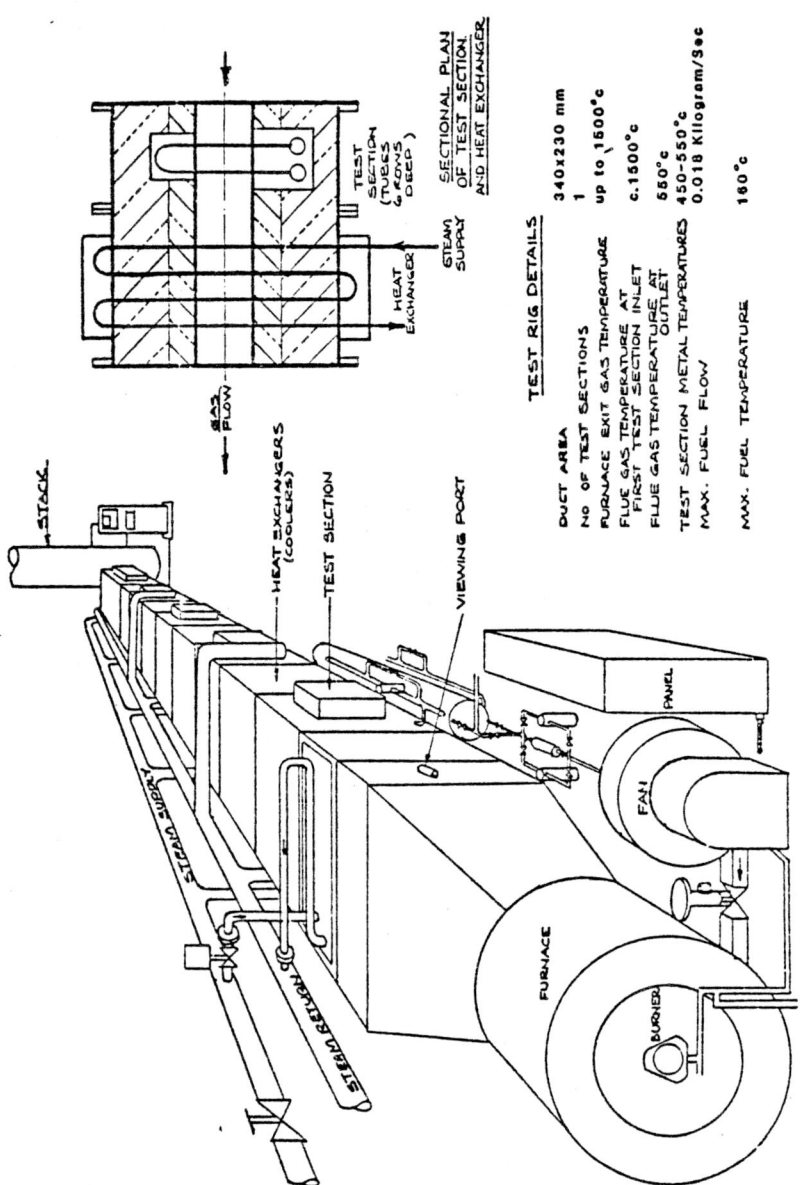

TEST RIG DETAILS

DUCT AREA	340 x 230 mm
NO OF TEST SECTIONS	1
FURNACE EXIT GAS TEMPERATURE	up to 1500°c
FLUE GAS TEMPERATURE AT FIRST TEST SECTION INLET	c. 1500°c
FLUE GAS TEMPERATURE AT OUTLET	550°c
TEST SECTION METAL TEMPERATURES	450-550°c
MAX. FUEL FLOW	0.018 Kilogram/Sec
MAX. FUEL TEMPERATURE	160°c

Fig.2. COD Slagging/Fouling Evaluation Test Rig

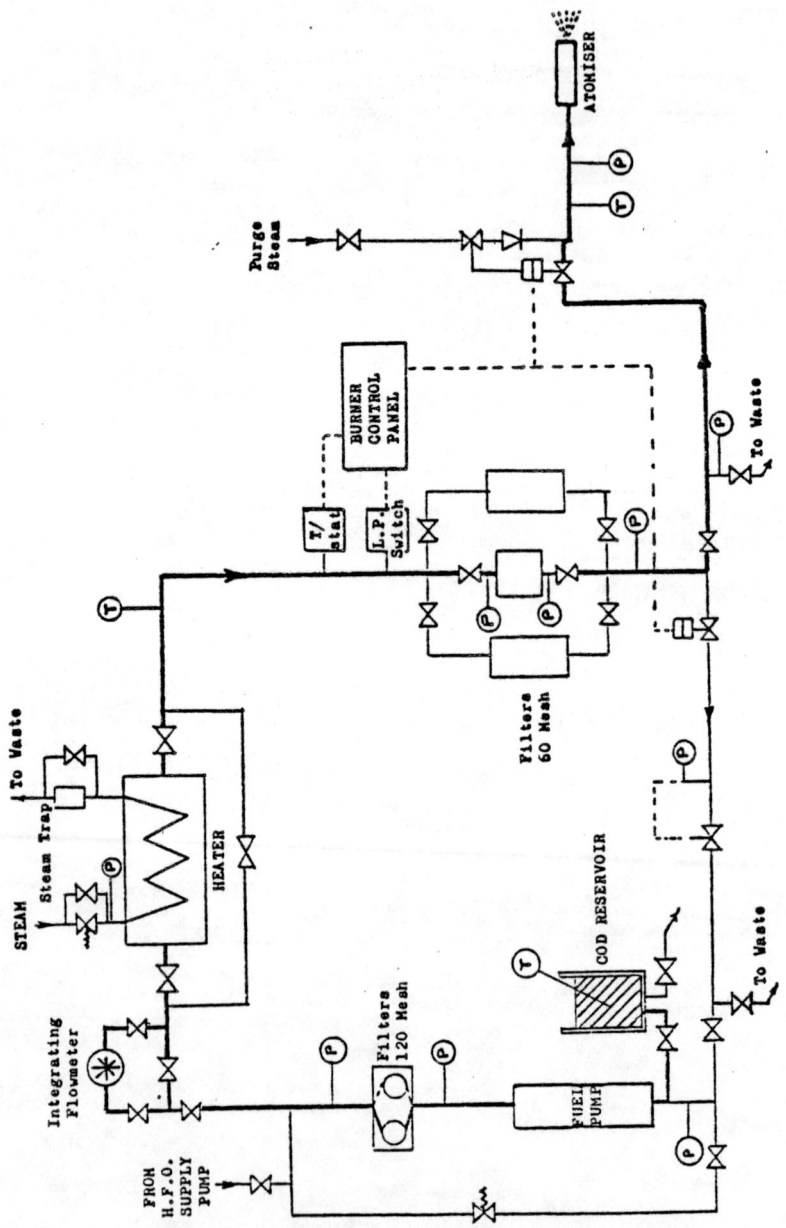

Fig.3. C.O.D. Fuel Supply System

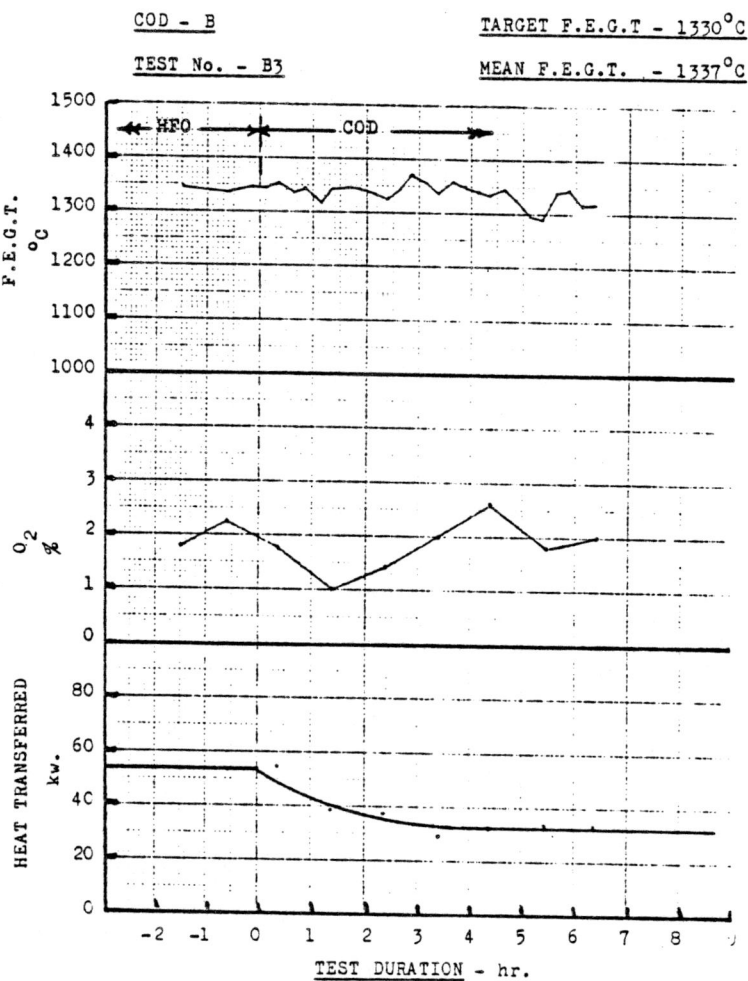

Fig.4. Summarised Test Data

Test Fuel - COD-B Test B3   F.E.G.T. Nominal - 1330°C

                                             Actual - 1337°C

Fig. 5a - Since Start 0.7h

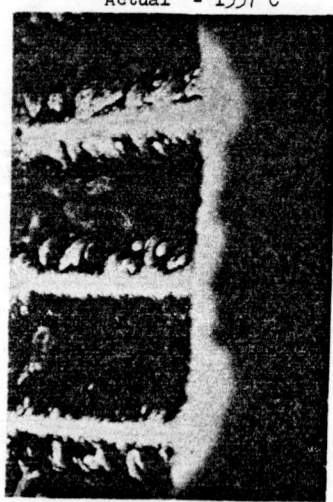

Fig. 5b - Since Start 1.0h

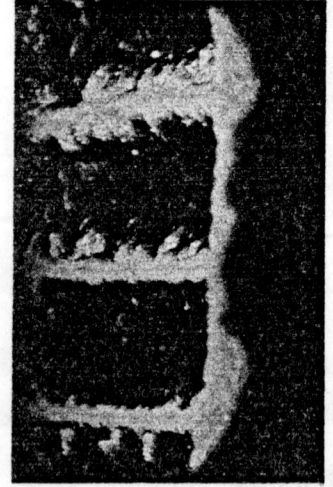

Fig. 5c - Since Start 1.6h

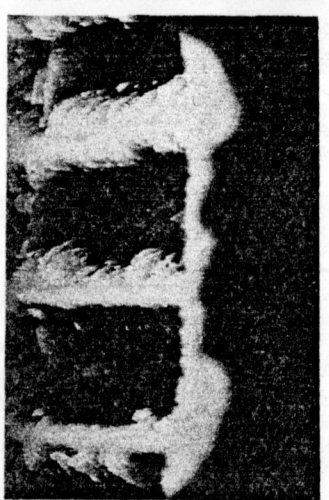

Fig. 5d - Since Start 3.7h

Fig.5 Deposit Build-up

Test Fuel - COD - B   Test B3

Fig. 5e - Since Start   4.7h   Fig. 5f - Since Start   6.4h

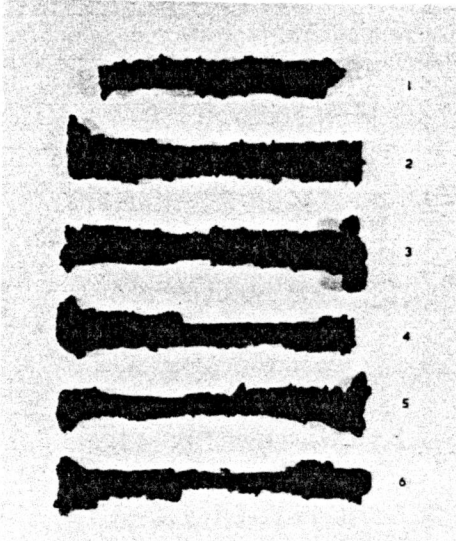

Fig. 5g -
Deposit Build-up
on Test Bank at
end of Test
(Colour Original)

Fig.5. Deposit Build-up

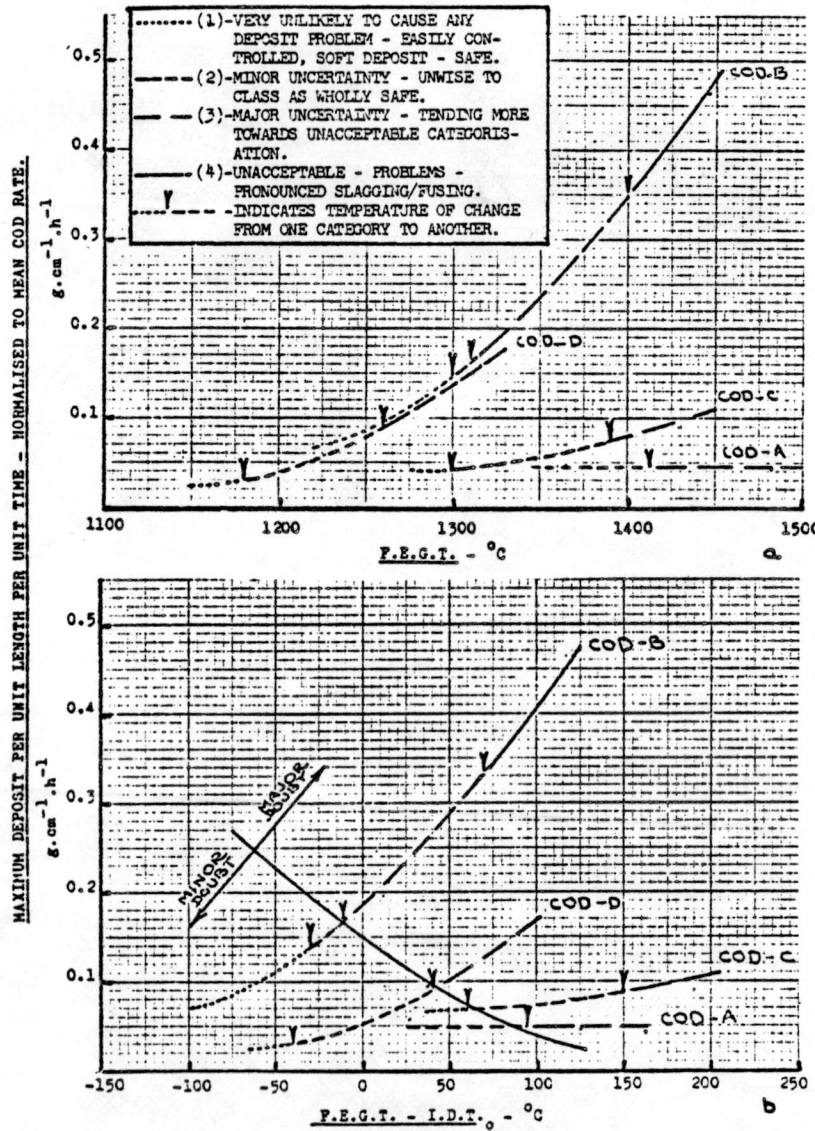

Fig.6. Relationships Between Deposition Rate, F.E.G.T. And Ash Fusibility

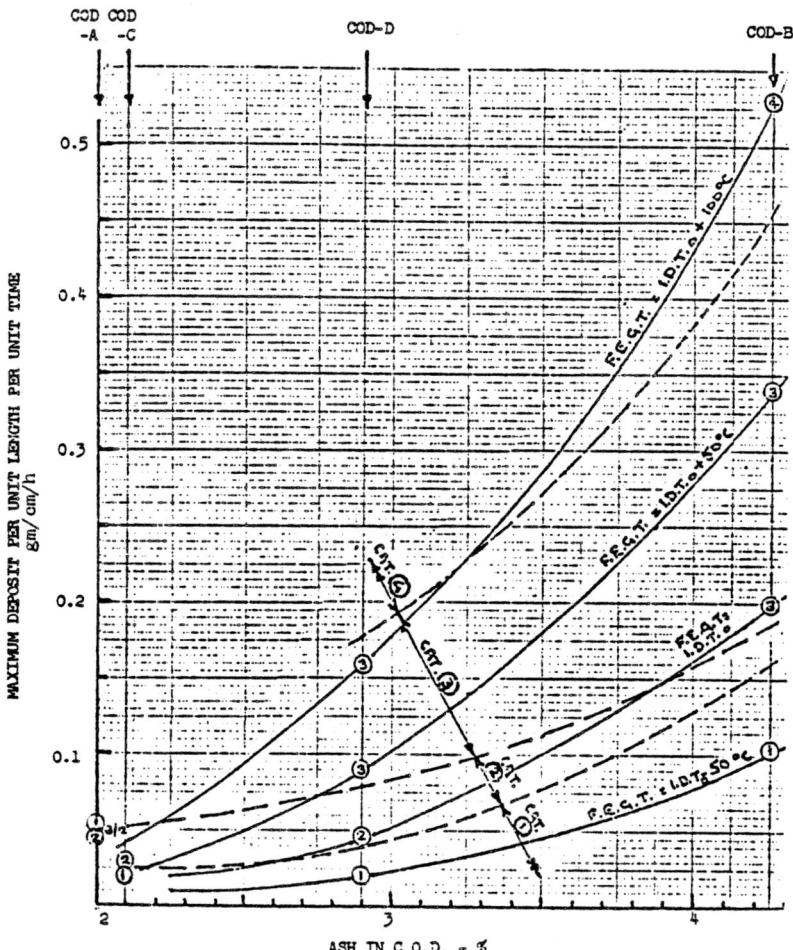

ASH IN C.O.D. - %

Fig.7. Correlation Between Rate Of Build-up, Deposit Category, Ash Content And F.E.G.T. In Relation To I.D.T. Of C.O.D. Ash Under Oxidising Conditions

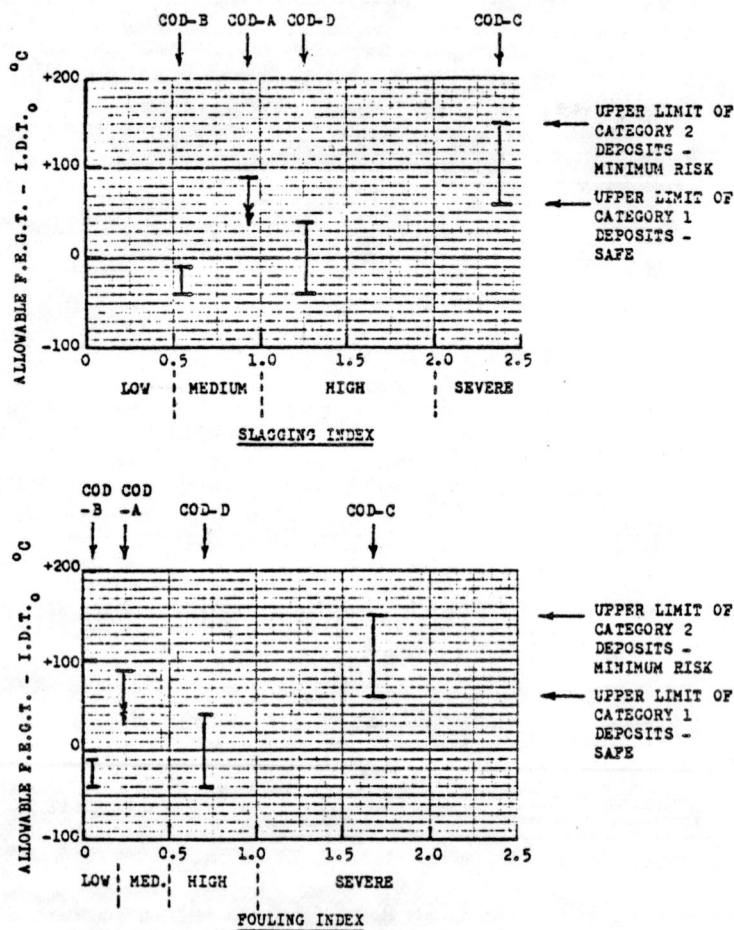

Fig.8. Relationship Between Allowable F.E.G.T., Deposit Category And Slagging/Fouling Indices

THE COMBUSTION OF COAL-WATER SLURRIES

M. Pourkashanian* and A. Williams*

The combustion of a coal-water slurry is a complex
process and little is known about the detailed mechanism.
In the present paper the combustion characteristics of
single suspended and freely-falling droplets of a coal-
water slurry in heated air were investigated. The
temporal variation of droplet mass, temperature and size
were measured. The results are compared with those for
heavy fuel oil. This work shows that extensive disruption
and microexplosion of the droplet occurs during the flame
life-time and this is governed by the coal type. A
largely carbonaceous residue is formed after the volatile
matter has been rapidly released and burned which take a
relatively long time to burn.

INTRODUCTION

Because of dwindling global resources of natural gas and oil, steps have to
be taken to assure the most efficient and economic use of existing energy
resources and the development of new alternative sources. For reasons related
to both availability and economic factors, major R & D programmes are
currently under way aimed at converting coal into alternative liquid and
gaseous fuels. To minimise processing costs and energy requirements, it is
possible to produce a slurry fuel by mixing finely pulverized coal with water
together with a small amount of a stabilising agent. The major attractions
for the combustion of coal-water slurries include (a) the fuel can be burned
in a similar way to heavy fuel oil with concomitant minimal design
modifications to existing oil-fired appliances, (b) flexibility of slurry
composition to achieve the required rheological, thermal and combustion
properties, (c) the feasibility of the pipeline transport of mixtures of coal
in water with a reduction in transportation costs (1). However the combustion
of a coal-water slurry in conventional oil-fired appliances can pose some
technical problems, such as an increase in wear and blockage of mechanical
components, flame instability, changes in heat transfer, stability and flow
problems during storage and pumping. Particulate emission and other pollution
problems such as changes in NO_x and SO_x yields may also occur. Coal-water
slurries are also a very useful feedstock for gasification plant.

Whilst in the last five years there has been considerable interest
(2) in the combustion of coal-water slurries there still remains a lack of
technical data on the mechanism and rate of combustion of droplets and sprays
of slurry fuels. The techniques developed to study the combustion of single
suspended and freely falling droplets has provided much valuable information
in the past on the behaviour of liquid fuels (3-5). Thus it seems reasonable

* Department of Fuel and Energy, The University of Leeds, Leeds LS2 9JT.

to extend these proven techniques to a study of the single droplet combustion
of a coal-water slurry to investigate the similarities and differences
between heavy fuel oil and coal-water slurries.

The work described in this paper is concerned with the burning of a
single droplet of a coal-water slurry suspended at the end of a fine quartz
fibre or thermocouple bead. Also some experiments have been undertaken
involving freely-falling droplets in a vertical furnace simulating real
combustion situations. In this way it is possible to determine the mass
losses of a fuel droplet without the uncertainties resulting from the
presence of the suspending fibre.

EXPERIMENTAL TECHNIQUES

In the present investigation three techniques were used to study the
combustion of a single droplet of coal-water slurry, and these are outlined
in the following sections.

Fuel Sample Preparation

The coal-water slurries were prepared gravimetrically with the coal
being dispersed by stirring in water and continual agitation was maintained
throughout the experimental period to maintain the uniformity of the mixture.
The composition of the coal-water slurry used in the present experiments was
60% wt dry coal, 39.5% wt water and 0.5% wt of a starch stabilizer. Table 1
gives the main characteristics (to British Standards Specification) of the
coal used. Each of these coals was crushed and well mixed to prevent
separation of its components. It was sieved and the -37 μm portion was used
to prepare the mixtures, the arithmetic mean diameter (based on number) of
the coal particles was determined by a Quantimet 720 Image Analyser and was
found to be 29 ± 3 μm.

The fuel oils used were (a) a medium fuel oil (1000 sec Red I,
density 0.92 kg/m^3) with a low (0.4% wt) asphaltene content and (b) a heavy
visbroken fuel oil (3500 sec Red I, density 1 kg/m^3) with a 6.6% wt
asphaltene content.

Thermometric Technique

In these experiments the thermometric suspended droplet technique
was used to investigate the temperature change of droplets of heavy fuel oil
and coal-water slurries during combustion (6). An electrically heated
horizontal tubular furnace was used. A droplet of about 1.3 mm diameter was
suspended on the bead of a Pt-Pt/13% Rh thermocouple made with 0.05 mm
diameter wires. A water-cooled tube protected the droplet while the furnace
was positioned around the droplet. On withdrawing the water-cooled tube the
droplet was exposed to the hot air in the furnace. The droplet was burned in
air at a known furnace temperature and the thermocouple output was recorded
on a U.V. oscillograph together with the output of a photodiode which
indicated the instant of droplet ignition. The droplet behaviour was
simultaneously recorded by a cine camera.

Gravimetric Technique

The initial studies showed that the combustion of suspended droplets
of coal-water slurries behaved in an analogous way to suspended droplets of
heavy fuel oil combustion, and that considerable droplet swelling,

micro-explosions, and disruption occurs as the droplet burns (7).
Consequently it is not possible to relate the measured diameter of the
droplet during combustion to its mass because the varying density of the
liquid is unknown. A gravimetric technique was previously developed by us (8)
to overcome the above problem. In this method, a small quantity of fuel
sample, typically 0.56 mg was attached to the end of a quartz fibre which was
hanging from the arm of a continuously recording microbalance. As described
in the section on Fuel Sample Preparation a water-cooled tube protected the
droplet whilst the furnace kept at 1173 K was raised into position. The
water-cooled shield was then withdrawn and the change in droplet mass was
recorded on a U.V. oscillograph. Any vertical convection currents generated
by the hot furnace were minimised by arranging for the water-cooled tube,
once withdrawn to seal the lower end of the furnace tube.

Moving Droplet Furnace

In this technique coal-water slurry droplets were ignited by a pilot
flame and allowed to fall through a vertical electrically heated furnace. The
furnace consisted of an alumina tube about 5.4 cm in diameter and 2 m long.
In order to provide viewing and sampling ports 1.75 cm diameter holes were
drilled at 10 cm points along the tube. At the top of the furnace a drop
generator was fitted. This simply consisted of a fibre from which a droplet
was initially suspended, the droplet being injected into the furnace by
agitation with a 100 Hz vibrator. The temperature within the furnace was
reasonably uniform despite its height and was kept at 1123 ± 50 K. Direct
gravimetric methods were used to measure the mass loss of drops passing
through the furnace. A water-cooled probe was placed in the furnace so that a
ceramic boat could be positioned in the furnace to catch the falling drops.
These could be collected on weighed strips of glass fibre filter papers
placed inside the ceramic boat. The sampling procedure was standardised as
was the storage and weighing of the filter paper strips. In this way it was
possible to examine and calculate mass losses from a single drop of
coal-water slurry by catching it on a filter strip. The mass loss as a
function of time was determined from measurements of the initial droplet mass
and the mass of the droplet taken from successive viewing ports down the
vertical furnace.

Scanning Electron Microscopy

Residues were taken from the suspended droplet fibre and mounted on
aluminium stubs, splutter-coated with Au/Pt and examined with an electron
beam energy of 10 keV with a magnification of between 300-1500 times.

DATA ACQUISITION AND MANIPULATION

The data obtained by the combustion of coal-water slurries may be
interpreted in a number of ways and is not intrinsically as simple to handle
as data from a single phase fuel such as a liquid fuel. However, to
facilitate comparison, data has been interpreted in a way which is as similar
as possible to that previously used by us for heavy fuel oils as follows
(9,10):

Size and Mass Measurements: From the photographic records taken in both the
gravimetric and thermometric experiments, curves of relative droplet diameter
(d/d_o) against time were plotted, where d is the transient equivalent
diameter and d_o the initial equivalent droplet diameter (mm). By assuming
that the filmed silhouette of the droplet is elliptical, the equivalent

diameter, d, of the sphere of equal volume, is given by $d = (d_1 d_2^2)^{1/3}$ where d_1 is the length of the major axis and d_2 that of the minor axis. The micro-balance produced a voltage output directly proportional to the measured droplet mass. For each droplet burned a mass-time history was obtained from the U.V. trace. A curve of relative mass, m/m_0, where m is the transient droplet mass and m_0 its initial mass, was then plotted against time. The combustion rates of the various fuels can then be characterised (6) in terms of burning rate constants based on (a) the overall mass burning rate between the instant of ignition and flame extinction, (K_S), this being calculated from the best fit through the data points, and (b) a burning rate constant (K_p) defined by Peterson (11) as $K_p = (2/3)(d_0^2/t_e)$ where t_e is the time at which the droplet mass has fallen to 1/e times the initial mass. Likewise in the falling droplet experiments the variation of the mass of free falling droplets of both the heavy fuel oil and the coal-water slurry were plotted against residence time. The average rate of mass loss (dm/dt) is given by the slope of the regression line (K_f).

Temperature Measurements: In an analogous way the U.V. oscillograph traces of the droplet temperature were obtained and from these measurements droplet temperatures were plotted against time for each droplet.

Ignition Delay: The time to the onset of ignition (t_i) and the size of the droplet at that moment were also measured. Reduced ignition delay times defined (4) by t_i/d_0 (s/mm) were thus obtained. In addition, data on the point of the extinction of the diffusion flame were obtained because this marks the end of 'droplet' burning and the commencement of the heterogeneous combustion of the carbonaceous residue or char.

RESULTS AND DISCUSSION

The General Features of Combustion

Typical temperature-time and relative diameter-time histories for droplets of a heavy fuel oil and coal-water slurries with initial diameters of 1.2 and 1.3 mm respectively are shown in Figures 1-4. From these and similar traces the combustion of each droplet can be divided into three regimes as follows:

The Induction Period: Experimental data on the ignition delay periods for droplets of each type of fuel burned are shown in Figure 5. The coal-water slurry droplets exhibit a significantly longer ignition delay compared with the heavy fuel oil. This increase in ignition delay is due to the time needed to evaporate the water before any combustible volatiles can be released from the coal. The addition of water to the coal thus represents an energy penalty (or loss) to the fuel. In the slurry containing coal type B with a gross calorific value of 33.75 MJ/kg, a 60% loading of coal will result in an energy penalty of about 4.2% because of the heat required to vaporise the water. Clearly the presence of the water reduces the overall rate of droplet temperature rise, thus reducing the rate of fuel devolatilisation. In addition the evaporation of water from the surface of the slurry droplet reduces the partial pressure of combustible fuel volatile matter within the gaseous boundary layer surrounding each slurry droplet. This process thus acts also to impede the rate of formation of a flammable fuel volatiles-oxidant mixture in gas phase environment of slurry droplet. It is therefore expected that long ignition delays and burning time are found compared with heavy fuel oil and which can lead to flame instability and reduced overall combustion intensity. The ignition delay of similar sized droplets showed

Arrhenius dependence with the furnace mean temperature, i.e. an equation of the form

$$t_i = A \exp (E/RT)$$

From the data, values of A and E for different fuels were calculated and these are given in Table 2.

It is also clear that the ignition delays associated with the coals A, B and C follow the expected pattern, i.e. as the % volatile matter of the coal increases the activation energy E decreases, but overall the activation energy for heavy fuel oil is smaller than for the coal-water slurries.

Disruptive Combustion (T_f): This is the period in which the volatile matter of the fuel burned visibly around the droplet. The flame life-time is dependent on the type of coal present in the slurry. Experimental results indicated that slurry containing anthracite coal with a volatile matter of 5.7% did not produce visible flames and only surface combustion was observed. For the slurries containing coals type A, B and C flame extinction occurred at temperatures within the range 1549 to 1663 K .

Residue Combustion (T_R): When the visible flame due to the volatiles disappeared the entire coal particle had a bright orange glow and burned heterogeneously. The initial combustion of the residue took place with the particle remaining approximately constant or slowly decreasing in diameter. Oxidative attack occurs on the outside of the particle particularly around the pore mouth, and to an increasing extent internally until, at an advanced stage of combustion, the coal particles broke into several pieces.

For coal-water slurry droplets peak temperatures of 1864-1926 K were attained soon after final flame extinction, and the temperature remained fairly constant during surface combustion before decreasing rapidly when fragmentation started. However in the case of heavy oil peak temperatures of 1721-1776 K were observed soon after final coke formation; these decreased rapidly when coke combustion was complete and just before fragmentation started. The time required for combustion of a char particle is quite long compared to the time required for the coke combustion of heavy fuel oil. Because of the relationship between the rate of combustion of char and the rate of heat transfer to the walls and other surfaces the design of coalwater slurry burners is likely to be dependent on accurate data on the rate of particle combustion.

Mass Loss Data

Figures 6 to 8 illustrate typical results obtained for the loss of mass of the fuel droplet in a furnace at a temperature of 1163 K. The examples given are for heavy fuel oil, and coal-water slurries containing coal types A and B. For the first and second stages of coal-water slurry droplet combustion just before the surface combustion starts, the average relative ignition mass loss ($(m_o - m_i)/m_o$, Peterson (11) rate constant K_p (mm^2/s) and mass burning rate (K_s s^{-1}) were calculated and shown in Table 3.

For coal-water slurry droplets the rate of pre-ignition loss of mass was higher than that for heavy fuel oils due to vaporisation of the water initially present before ignition of the droplet occurs (e.g. typically 15% wt water remains at ignition). Finally it was noted, as expected, that both the size and mass of the residue remaining after surface combustion of these samples depends on the coal ash content as illustrated in Figures 7 and 8 for coal types A and B respectively.

The variation of the mass of free falling single droplets of heavy fuel oil and coal-water slurry (containing coal type A) with residence time are shown in Figure 9 for a furnace of $1123 \pm 50K$. The fuel droplets all have the same mean initial mass and from the fitted regression line the rate of mass loss (dm/dt) was calculated. In addition the overall burning rate coefficient (K_0) was calculated by plotting (mass of droplet)$^{2/3}$ against residence time (3). These values and the rate of mass loss for a freely falling droplet are given in Table 3. In the case of the coal-water slurries the plot mass against time is not linear and approximately can be divided into two regimes representing (i) vaporisation of water combustion of coal volatiles, and (ii) slow char combustion. Pairs of values of K_f and K_0 are therefore given in Table 3 for these combustion regimes respectively. The results indicate that the value of the overall burning rate coefficient will depend on the ignition delay time which in turn depends on the rate of surface water vaporisation from the droplet. The mass burning rates of the free falling droplets are greater than suspended droplet mass loss rates because of the differences in the air velocity past the droplet. These differences can be calculated by standard techniques but will not be discussed further here.

Char Reactivity

The time required for consumption of a char particle is a significant part of the coal-water slurry droplet reaction process. The rate of combustion of char residue from coal-water slurries is assumed to be governed both by physical and chemical control as in the case of char produced by coal particles alone. There are a number of chemical reactions which may occur on the surface of the droplet or in the gas surrounding the droplet. Perhaps the most important surface reactions are the heterogeneous reactions:

$$2C_{(s)} + O_2 \rightarrow 2CO \qquad (i)$$

$$C_{(s)} + CO_2 \rightarrow 2CO \qquad (ii)$$

followed by the gas phase reactions

$$CO + OH = CO_2 + H \qquad (iii)$$

However the heterogeneous reaction process is complicated by diffusion of the reactants, reaction by H_2O, O_2, etc. reactants, particle size effect and variation with temperature. In the case of coal-water slurry the vaporisation of water in the first and second stages of droplet combustion will increase the concentration of H_2O vapour present in the gas phase to a higher level than that associated with char combustion. The water vapour presence will serve as a catalyst to the CO oxidation by generating hydroxyl radicals thus:

$$H + H_2O = H_2 + OH$$

$$O + H_2O = 2OH$$

The OH radicals not only react with the gas phase CO thereby reducing its concentration, but also react with any particulate soot present and the char thus

$$C_{(s)} + OH \rightarrow CO + H$$

Unlike coal particle combustion, for coal-water slurry the reaction of both H_2O vapour and CO_2 with carbon is large and cannot be ignored compared with reaction with oxygen (12), clearly the problem of calculating gas concentration around a droplet in such a case is very complicated. However making assumptions about the gas concentration at the surface of the droplet Field (13) proposed a method of calculating the rate of combustion of carbon.

It is assumed that the overall rate of reaction is controlled by mass transfer and the rate of reaction may be written:

$$R_C = [3.42(T_m d_p)]^{-1}(2D_{O_2}P_{O_2} + D_{H_2O}P_{H_2O} + D_{CO_2}P_{CO_2})_g$$

where R_C indicates the rate of burning of carbon particles (g/cm²s), T_m is the mean of gas and surface temperature (K), d_p is the particle diameter (cm), D is the diffusion coefficient (cm²/s), P is the partial pressure (atmos) and g indicates that the gas partial pressure refers to free stream.

The values of P_{O_2}, P_{H_2O} and P_{CO_2} were estimated by considering the volume of furnace, and calculating the concentration of major gases. From the above equation the estimated values of the burning rate at the surface of the droplet were calculated and are shown in Table 3. However a more accurate rate of consumption of carbon can be obtained by chemical analysis of the gases around the droplet. At present we are attempting to produce a theoretical model for the combustion of coal-water slurry based on the work carried out by Matalon (14).

Scanning Electron Microscopy Studies

Scanning electron microscopy has revealed fundamental structural differences between the particulates residues from heavy fuel oils and coal-water slurry fuels. Plates 1 and 2 show the surface and edge of residues from heavy fuel oil, taken from the suspended droplet fibre. Plate 2 shows the tiny "blow holes" where trapped volatiles escaped from the interior of the droplet through the expanding skin of the residue. These "blow holes" vary from 1 μm to 10 μm in diameter. In some samples for heavy fuel oil ripples were observed running around the surface holes and giving the appearance of a solidified melt. However this effect was not observed for the coal-water slurry residues.

Plates 3 and 4 show that residues from coal-water slurry contain partially burned out cenospheres. Plate 3 shows that there is a marked difference in structure of the particulate residue from that of heavy fuel oil. Plates 3 and 4 show a large number of holes, tens of microns across of irregular shape on the surface of residue where the trapped volatile matter and water vapour (or droplets) escaped. The devolatilisation and vaporisation take place rapidly and therefore a large number of holes are caused in the outer wall of the residue. The small coal ash particles can be identified on the surface of the rounded, pitted solid, which is the incompletely burned out original slurry droplet. Almost all the residues were hollow and roughly spherical. These differences between coal-water slurries and fuel oil are not surprising but the different structures clearly lead to different burn-out behaviour of their respective carbonaceous residues as shown in Table 3.

CONCLUSIONS

The combustion of droplets of heavy fuel oil and three different types of coal-water slurries has been studied, and the following conclusions can be drawn. Suspended droplets of coal-water slurries burn in the same general way as very heavy fuel oil droplets; they did not behave in the same way as a pulverised fuel which they would if the coal particles were released from the droplets by explosive vaporisation. A pre-ignition delay period is followed by ignition and combustion of volatiles. A carbonaceous residue is formed by agglomeration of the individual coal particles, which then oxidises, undergoes fragmentation and then burns out. Differences in combustion rates and in combustion mechanism do however exist, and in particular there is an increase in the ignition delay time, and the temperature of droplets of coal-water slurries during surface combustion is about 6% higher than heavy fuel oil. The present results indicate that single droplet experiments can yield a sensible comparative estimate of combustion performance and produce data on the mechanism of combustion of fuels. However the interpretation of the results of suspended drop experiments relative to the behaviour of droplets in practical spray flames must be undertaken with care. Heating by radiation heat transfer in real systems must be taken into account when examining combustion systems involving coal-water slurries.

ACKNOWLEDGEMENTS

We are indebted to the British Gas Corporation for support for part of this work.

NOMENCLATURE

A	Pre-exponential constant
D	Diffusion coefficient (cm^2/s)
d	Transient droplet diameter
d_o	Initial equivalent droplet diameter (mm)
d_p	Particle diameter (cm)
g	The gas partial pressure
K_f	The rate of mass loss for freely falling droplets
K_O	Overall burning rate coefficient for freely falling droplets
K_p	Peterson rate constant (mm^2/s)
K_S	Mass Burning rate (s^{-1})
m	Transient droplet mass
m_o	Initial mass of droplet (mg)
m_i	Mass of droplet at time of ignition (mg)
P_{O_2}, H_2, CO_2	Partial pressure of O_2, H_2 or CO_2 (atmos)
t_e	Time at which the instantaneous droplet mass has fallen to (1/exp) times its m_o
T_f	Flame life time (s)
t_f	Furnace temperature (K)
T_i	Pre-ignition time (s)
t_i	Ignition delay time (s)
T_m	Mean of gas and surface temperatures (K)
T_R	Coke combustion time (s)
t_{vap}	Water-vaporisation time (s)

References

1.   Scheffee, R.S., Rossmeissl, N.P., Boyd, T.J., Henderson, C.B. and McHale, E.T. 'Development and burning of coal-water slurries', Proceedings of the Third International Symposium on Coal-oil Mixture Combustion, Orlando, Noyes Data Corporation, April 1981., p. 46-56.

2.   Essenhigh, R.H. and Bailey, E.G.. 'Comparative combustion mechanisms in dry and slurry coal firing', Proceedings of the Engineering Foundation Conference, Combustion of Tomorrows Fuels, April 1983.

3.   Hottel, H.C., Williams, G.C. and Simpson, H.C. 'Combustion of droplets of heavy liquid fuels', 5th Symposium (International) on Combustion, Reinhold, 1955, p. 101.

4.   Miyasaka, K. and Law, C.K. 'Combustion and agglomeration of coal-oil
        mixtures in furnace environments', Combustion Science and
        Technology, 1980, 21, p. 1-14.

5.   Law, C.K. 'Recent advances in droplet vaporization and combustion',
        Progress in Energy and Combustion Science, 1982 vol.8, No.3, p. 171.

6.   Jordan, J.B., Kimber, S.M. and Williams, A. 'Ignition of droplets of
        liquid fuels solvent extracted from coal', Fuel, 1977, 56, p.417.

7.   Jacques, M.T., Jordan, J.B., Williams, A. and Hadley-Coates, L.
        'Combustion of water-in oil emulsions and the influence of
        asphaltene content', 16th Symposium (International) on Combustion,
        The Combustion Institute, 1977, p. 307.

8.   Jordan, J.B. and Williams, A. 'Combustion of droplets and sprays of
        some alternative fuels', Progress in Astronautics and Aeronautics,
        1978, 62, p. 180.

9.   Braide, K.M., Pourkashanian, M. and Williams, A. 'Combustion of
        single droplets of coal-oil mixtures', Proceedings of the
        Engineering Foundation Conference 'Combustion of Tomorrow's Fuels',
        April 1983.

10.   Braide, K.M., Isles, G.L., Jordan, J.B. and Williams, A. 'The
        Combustion of single droplets of some fuel oils and alternative
        liquid fuel combinations', J. Inst. E., 1979, 52, p. 115.

11.   Peterson, F. 'The mass of an oil-droplet during combustion',
        Combustion Institute European Symposium 1973, Academic Press, 1973,
        Paper No. 62, p. 366.

12.   Khitrin, L.N. and Grolovna, E.S. 'High temperature technology',
        1964, p. 485.

13.   Field, M.A., Gill, D.W., Morgan, B.B. and Hawksley, P.G.W.
        'Combustion of pulverized coal', Leatherhead, 1967, p. 186-209.

14.   Matalon, M. 'Complete burning and extinction of a carbon particle in
        an oxidising atmosphere', Combustion Science and Technology, 1980,
        24, 3, 115.

Table 1. Properties of Coals Used

Coal Type/Source	NCB Rank No	Moisture % wt	Ash % wt	Volatile Matter % wt	Swelling Number	Calorific value MJ/kg
A Bituminous Langwith, Midlands	802	6.46	5.96	37.6	1	29.85
B Bitumonous Blackhall, Durham	501	1.12	4.30	34.4	7	33.75
C Bituminous Selby, Yorks	-	-	8.24	37.1	-	-
D Anthracite South Wales	101	1.50	3.35	5.6	Non-Swell	34.88

Table 2. Ignition Delay Parameters

Fuels	A	E KJ mol^{-1}
Heavy fuel oil	6.09×10^{-2}	25.02
C-W slurry (coal type A)	5.35×10^{-2}	34.82
C-W slurry (coal type B)	1.52×10^{-2}	42.42
C-W slurry (coal type C)	2.5×10^{-2}	37.80

Table 3. Pre-ignition mass loss, Peterson rate constant K_p, mass burning rate K_S, rate of mass loss of falling droplet K_f, overall burning rate coefficient of freely falling droplet K_O (standard deviation for individual results shown in brackets)

Sample	$(m_o-m_i)/m_o$	K_p	mass burning rate(K_S)	$K_f \times 10^3$	$K_O \times 10^3$	R_C
		(mm^2/s)	(s^{-1})	gm s^{-1}	cm^2 s^{-1}	mg/m^2 s^{-1}
Medium fuel oil	0.168 (0.009)	0.425 (0.004)	1.075 (0.011)	4.85	17.14	-
Heavy fuel oil	0.186 (0.044)	0.456 (0.002)	1.199 (0.041)	3.66	15.89	-
C-W slurry (coal type A)	0.191 (0.064)	0.194 (0.018)	0.56 (0.037)	2.83 1.16	15.25 4.5	0.44
C-W slurry (coal type B)	0.227 (0.017)	0.166 (0.074)	0.394 (0.031)	2.71 1.71	15.01 4.69	0.447
C-W slurry (coal type C)	0.215 (0.007)	0.188 (0.016)	0.380 (0.056)	-	-	0.431

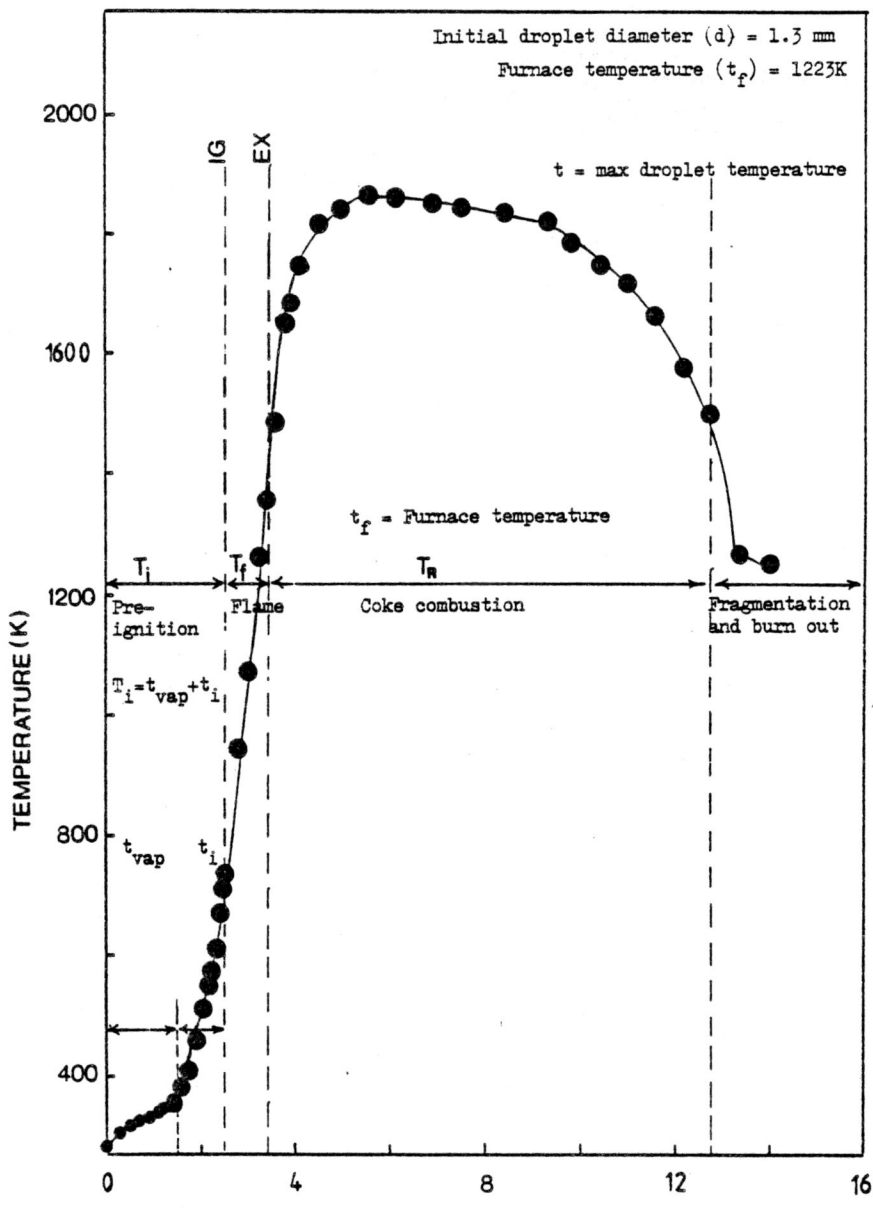

Fig. 1. Typical temperature - time history of coal-water slurry droplet

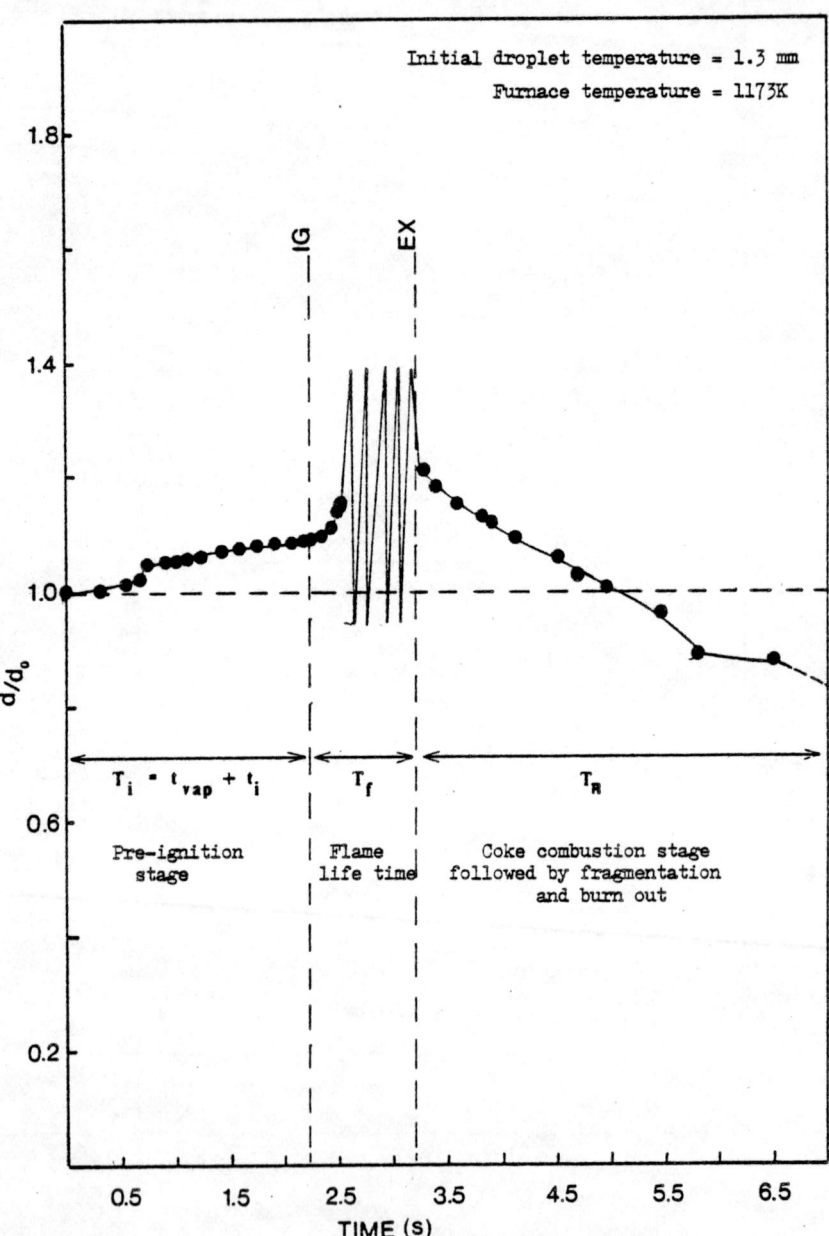

Fig. 2. Typical size - time history of coal-water slurries

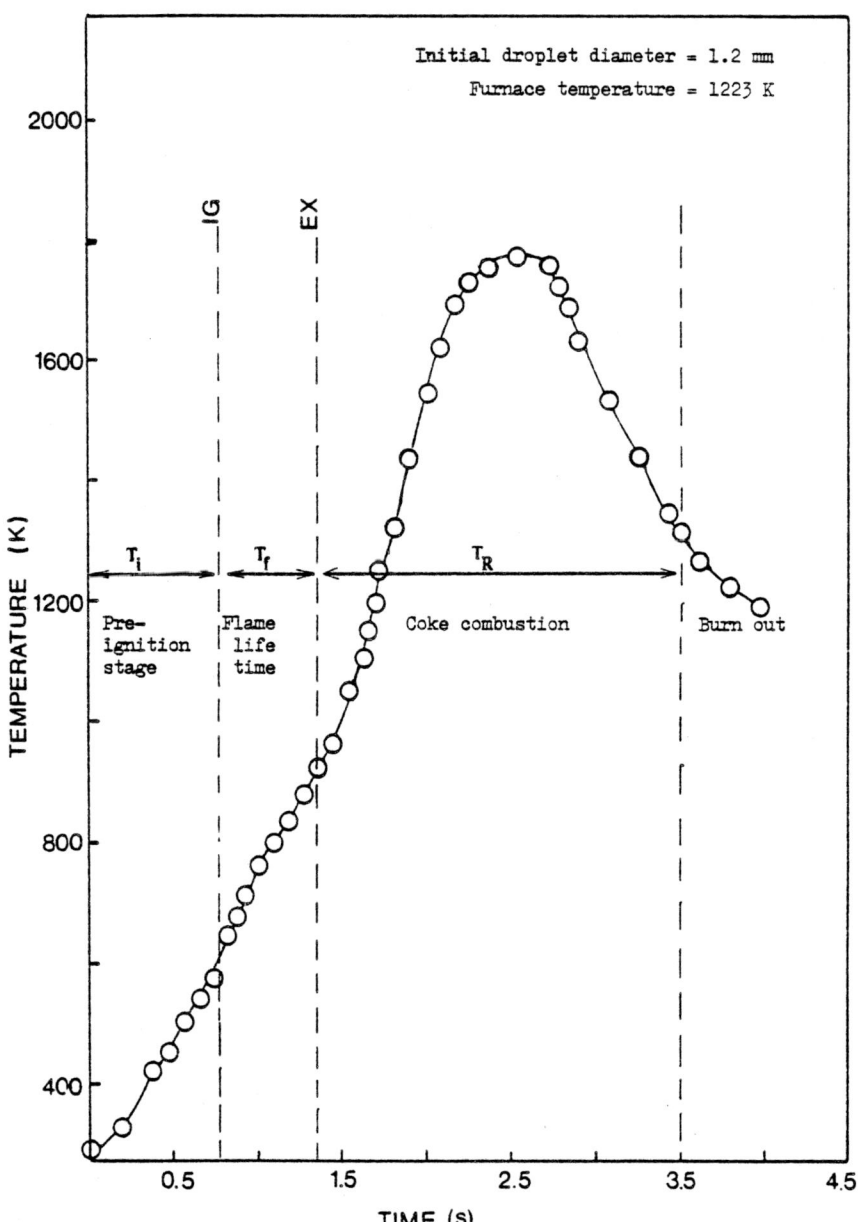

Fig 3. Typical temperature - time history of heavy fuel oil droplet

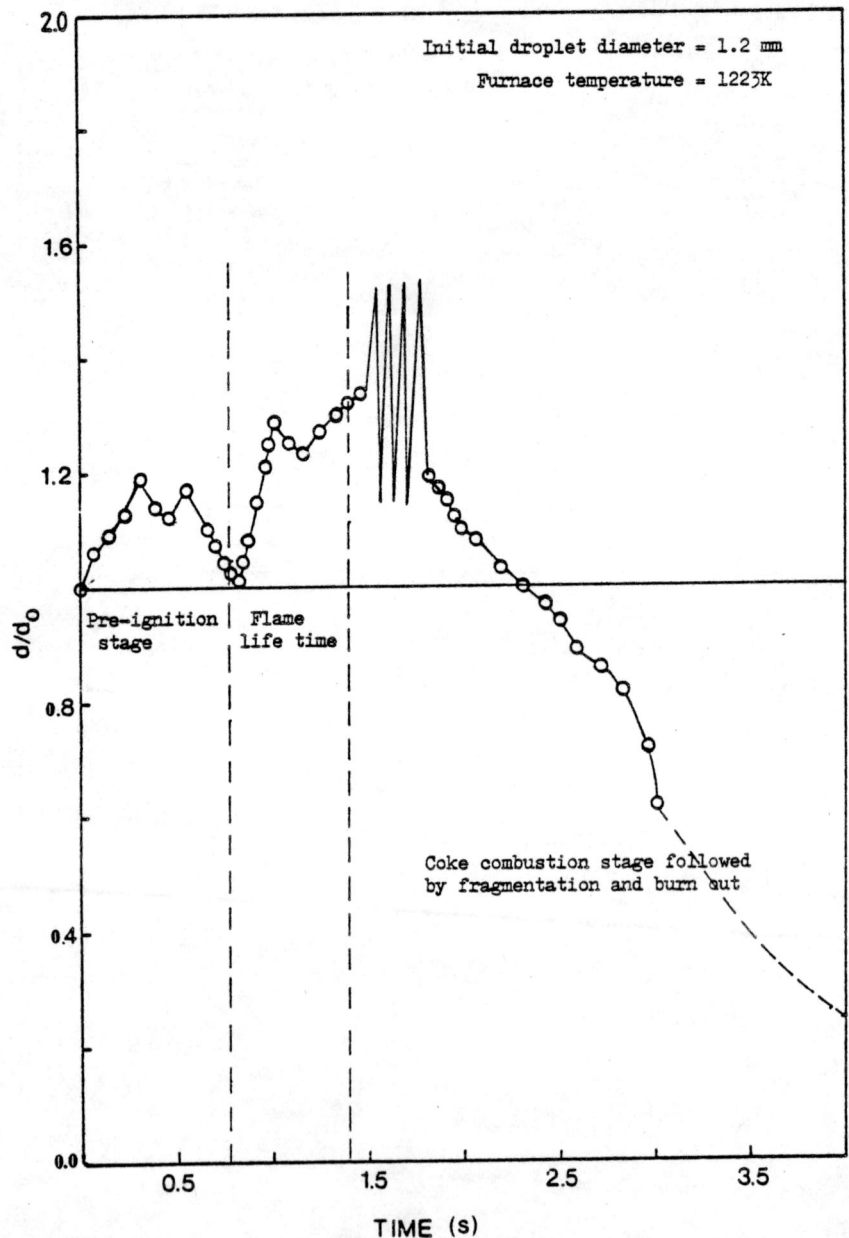

Fig. 4. Typical size - time history of heavy fuel oil droplet

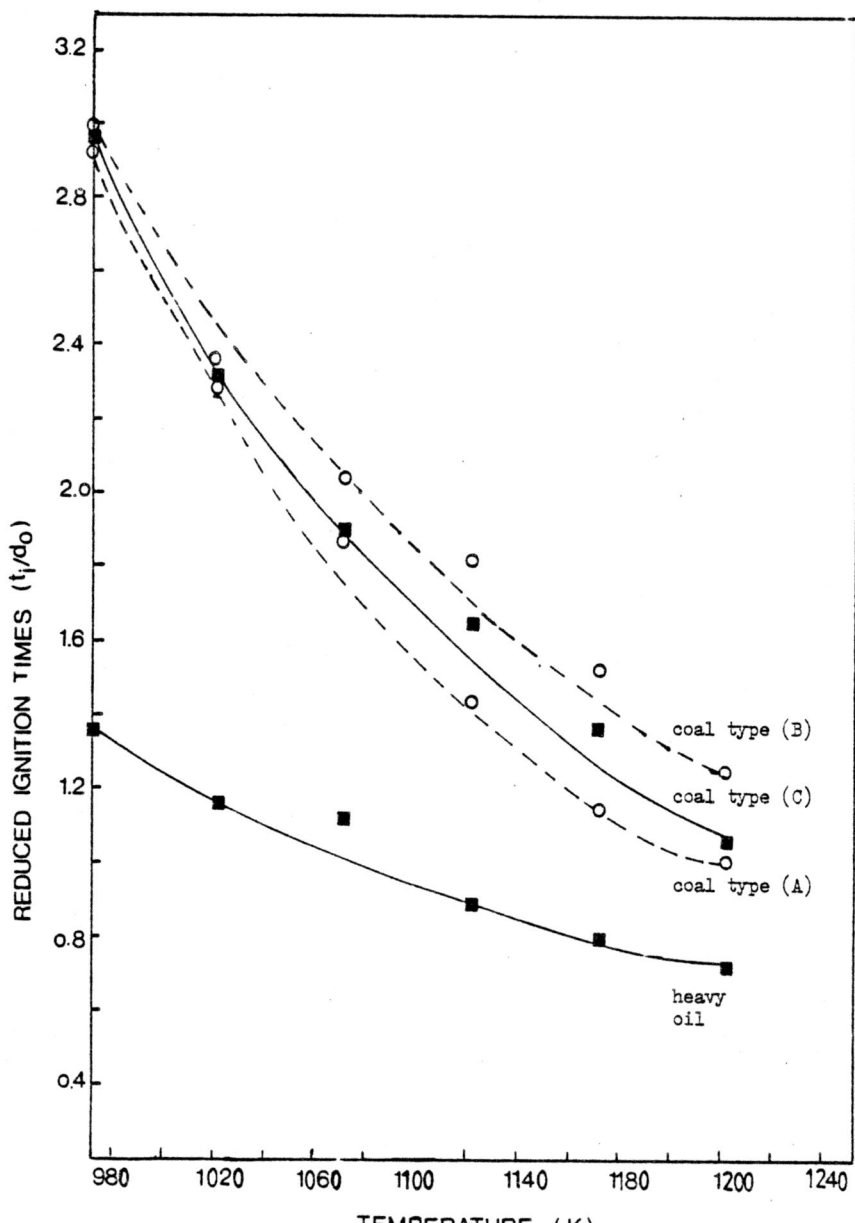

Fig. 5. The variation of droplet ignition delay time, t_i/d_o, with furnace temperature for the three coal-water slurries and the heavy oil

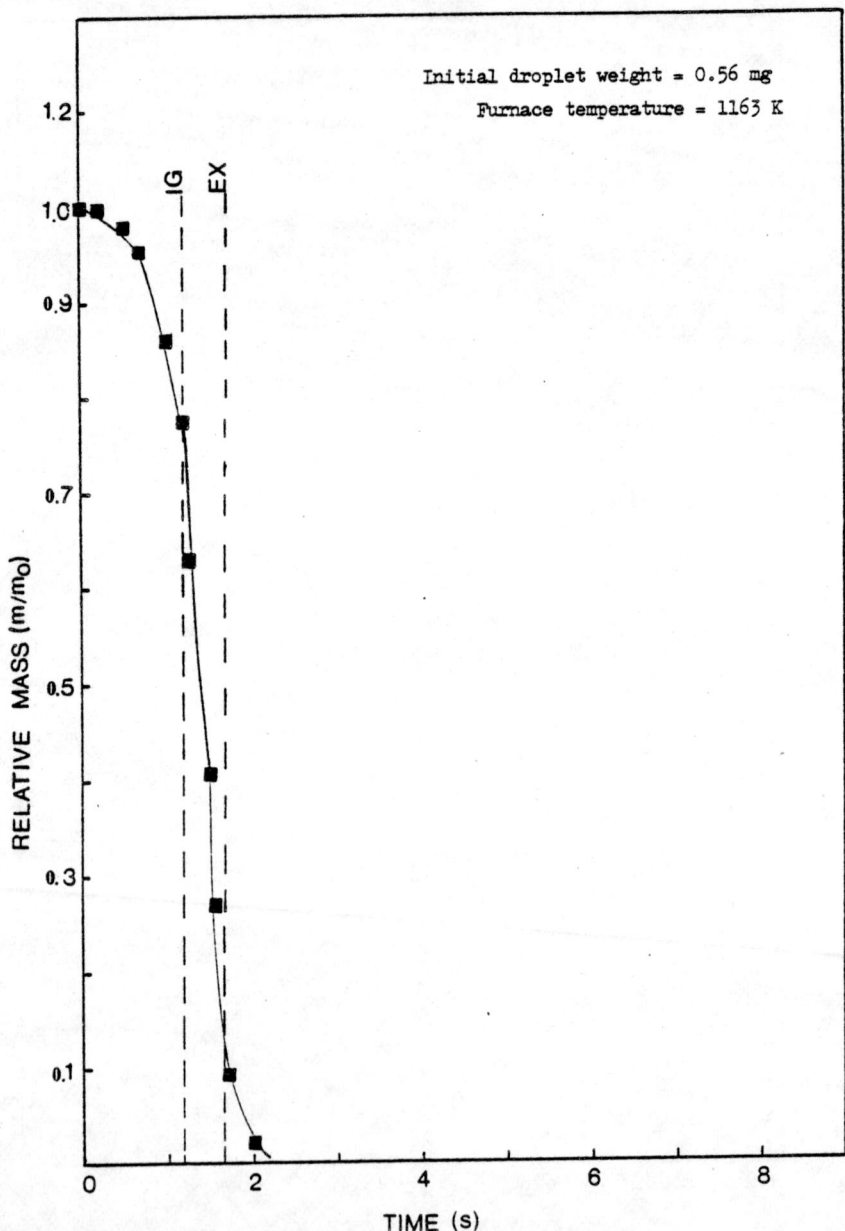

Fig. 6. The variation of relative droplet mass versus time profile for a droplet of heavy fuel oil.

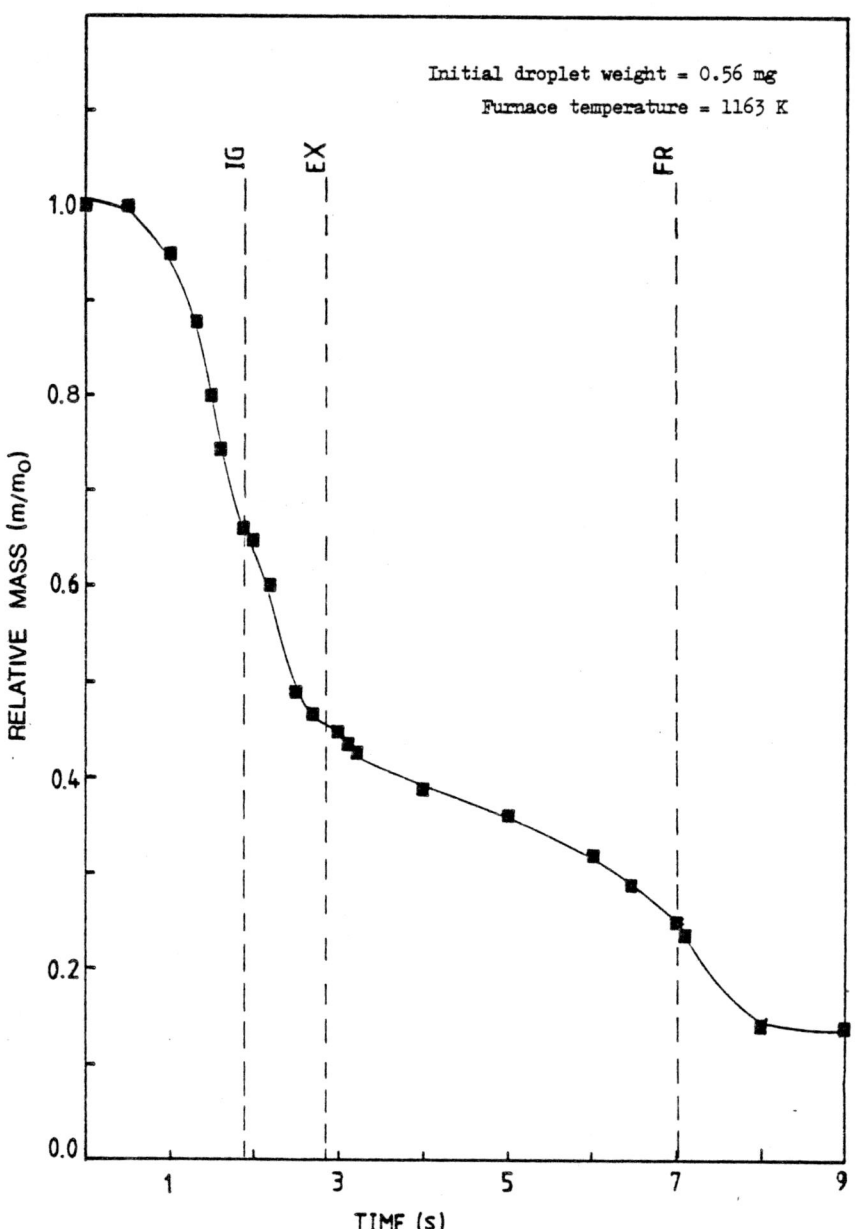

Fig. 7. The variation of relative droplet mass versus time profile for a coal-water slurry droplet containing Type A coal.

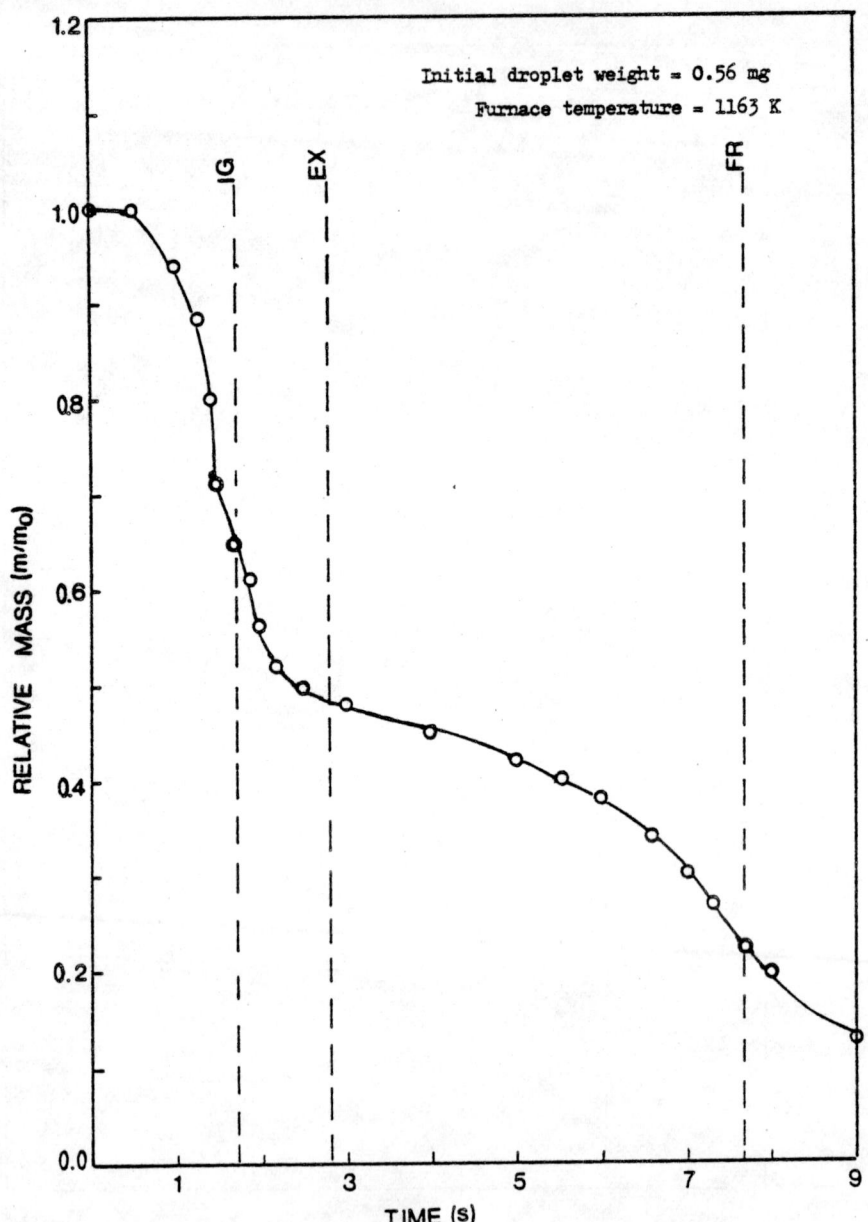

Fig. 8. The variation of relative droplet mass versus time profile for a coal-water slurry droplet containing Type B coal

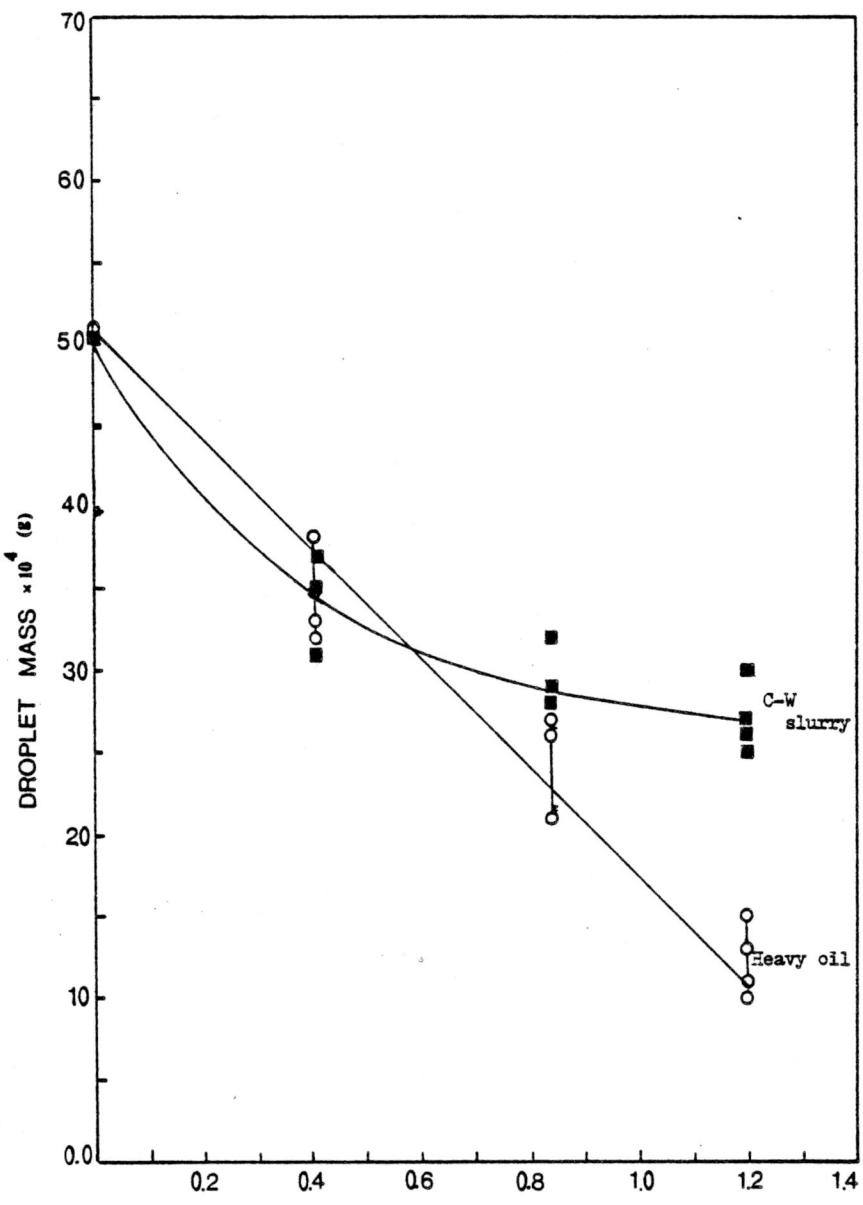

Fig. 9. The variation of the mass of free falling single droplets of
coal-water slurry (type A) versus time for a furnace temperature
of 1123 ± 50 K

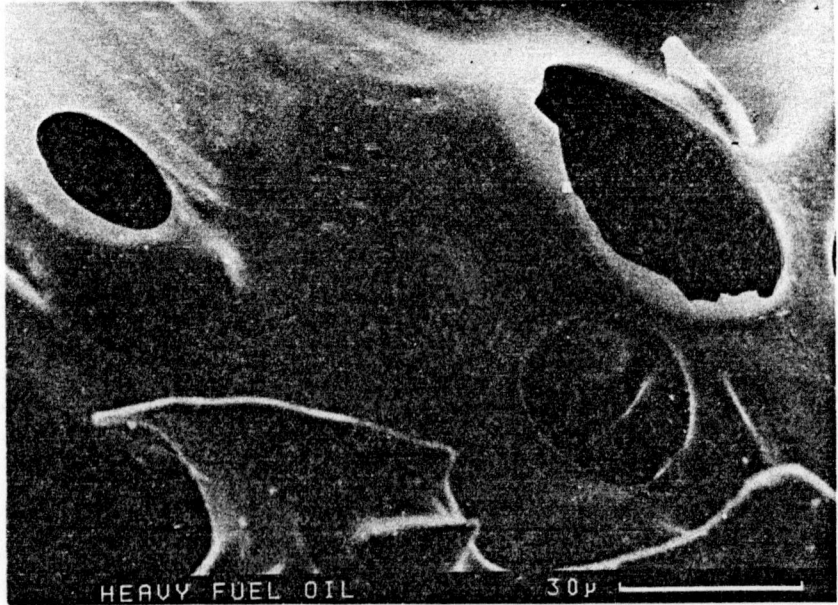

PLATES 1,2. Electron micrographs of residue fragments from heavy fuel oil

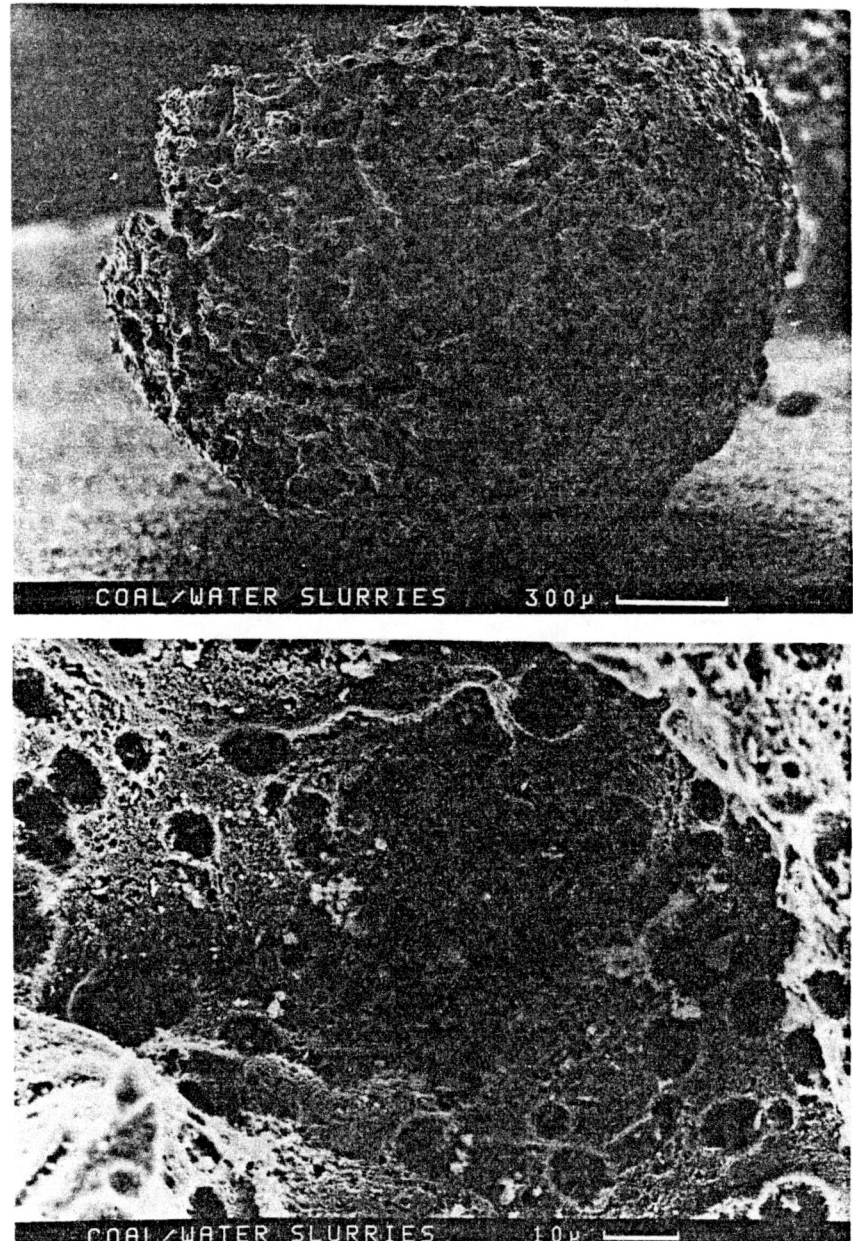

PLATES 3,4. Electron micrographs of residue fragments from coal-water slurries

COMBUSTION OF COAL OIL MIXTURES WITH OXYGEN
FOR INDUSTRIAL APPLICATIONS

COLIN MOORE - BOC,   ·PAUL JENKINS - BSC

The use of oxy-fuel systems using light oils and gas
is an established technique for increasing
productivity, and quality and saving fuel. These fuels
have risen dramatically in price since 1973 and this
paper describes the development of an oxygen heavy
fuel oil burner suitable for coal-liquid mixtures.
Even with propane atomisation, fuel costs about 10% less
than heavy fuel oil alone have been achieved using coal
oil mixtures. The burners were developed in the BSC
test facility at the Welsh Labs.

1. Introduction

BOC has manufactured a range of oxy-fuel burners for several years
and these have been used in a wide range of industrial
applications. The fuels used have principally been light fuel oil
or natural gas and propane, i.e. all very clean and low in sulphur
so their use is compatible with steel making, non-ferrous metal
melting, cement, glass and frit making. The burners have,
therefore, been used in the following furnaces: arc, reverberatory,
kiln and tank, where considerable operating data accumulated over
long periods has optimised the design of the water cooling, fuel and
oxygen supply, and manufacturing materials. The current design of
the 2.5 & 5.0 MW oil burner is shown in Figure 1.

Increasing the proportion of oxygen in the combustion air results in
a dramatic increase in the kinetics of the combustion reactions.
This leads to a significant increase in flame temperature which
gives much higher rates of heat transfer by conduction, convection
and particularly radiation. Furthermore, these higher flame
temperatures provide additional advantages. The increased levels of
dissociation absorb heat of which a high proportion is regained on
contact by the combustion products with the charge and furnace walls.

The increased reaction kinetics also result in increased burning
velocities and reduced ignition temperatures. Both these effects
give a smaller more compact flame with a higher heat release per
unit volume.

CLM-L

The appropriate use of an oxy-fuel burner system allows

1. increase in productivity;

2. improvement in product quality;

3. production flexibility;

4. stabilisation of low grade fuel flames;

5. fuel economy.

These cheap fuels of the 60's and early 70's are today's expensive sources of energy. Fig. 2. shows the change in actual prices of oils, gas, coal and liquid oxygen over the past decade and the probable future trend, and Fig 3 shows the rate of increase of price related to 100 in 1972. It seems to us from these trends that a PF O_2 burner should be economically viable provided that ignition and stable controllable flames could be achieved.

In systems already using coals, the down stream process would already be capable of accepting the ash from the coal. Such process would be steam boilers, kilns and reverberatories. For clean melting, i.e. arc, tank and some reverberatories, the introduction of coal ash could be unacceptable to the process. Modifications to the ash chemistry during combustion could make the resulting ash acceptable and BOC have patents pending on a probable method of achieving compatibility. A further problem in these industries would be their attitude to coal as a fuel compared with oil and gases which are easily handled, monitored and controlled.

From these considerations it was decided to base the oxy-coal burners on the oxy-oil burner shown in Figure 1 and develop the fuel system for carrying pulverized coal in oil.

2. The Test Facility at BSC Port Talbot

BSC have built and operated a burner test unit for several years. It can handle 1.25 MW heat output and has a sophisticated computer control system for input fluids and a comprehensive gas analysis unit with cooled probes for determining the actual behaviour of the flames produced. Experience on other projects using the facility had shown good correlation with results obtained in practice on units up to five times larger. Extrapolation, therefore, into the 5MW range of our existing oxy-fuel burners could be reasonably achieved.

Measurement of gas analysis along and across the flame for CO, O_2, CO_2, H_2 and CH_4 were made during setting up and testing to ensure that combustion was complete. Burner performance was assessed from the wall temperature measurements.

3. The Test Programme

The basic philosophy for burner development was that modifications
to the oxy-oil burner should be as few as possible to enable the
unit to burn commercially available pulverised coal in already used
fluids. To this end, the water cooled jacket and oxygen systems
remain the same as in the oxy-oil burner and only changes in the
easily withdrawable oil tube and nozzle were made.

The coal chosen was a bituminous coal, details of which are given in
Table 1.

TABLE 1

CHEMICAL DETAILS OF COAL USED

Classification	Bituminous
	701
Calorific value	32540 KJ/kg
Volatiles	35.4%
Ash Content	4.6%
Moisture (DAF)	0.8%

Four tonnes of this were commercially pulverized at different
flowrates to give a coarse and fine sample. The size distribution
for these two samples is given in Table 2.

TABLE 2

SIZE DISTRIBUTIONS OF COAL USED

Sieve Size		Percentage Through Sieve	
		Coarse Fraction	Fine Fraction
106 μ		73.6	88.7
75 μ		57.8	76.8
40 μ		40.7	50.7

A heavy fuel oil with the characteristics given in Table 3 was used
for the oxygen oil coal work.

TABLE 3

CHARACTERISTICS OF HEAVY FUEL OIL

Parameter	Value
Viscosity	3500 Redwood Secs
C.V. (Net)	39,800 KJ/Kg
Sulphur	2.8%

The coal-oil mixture contained 40% of coal by weight. The coal concentration is probably somewhat lower than what can be achieved in practice. At the test burner minimum line diameter of 3mm and flows of less than 0.75 Kg/min, blockage at higher solids concentrations are very likely. For the commercial burners sized at, say, 5 and 10 MW outputs, fuel line diameter will be larger for flows of 7.5 to 15 Kg/min. At such flows higher concentrations of coal should be possible.

The burners were fired in the 90% to 110% stoichiometric range and wall temperatures were determined to assess the performance of the system. Data on HFO with and without coal are given in the following sections.

4. Results and Disucssion

4.1 Heavy Fuel Oil - Oxygen

This work was undertaken as a first step to ensure that the first modification of the LFO and oxygen burner produced a stable flame with HFO and oxygen before any coal-oil mixture was tried. Further changes had to be made to the burner before stable combustion was achieved. Atomising media used were steam or propane. The firing conditions are summarised below in tables 4 and 5 and the wall temperture profile for propane atomising is shown in Figure 4..

TABLE 4

Firing Conditions on the HFO-Oxygen System
Using Steam Atomisation

Oil Flow Kg/min		0.73
Steam flow Kgs/min		0.3
Steam by wt. %		40
Oil temp °C		97
Heat release KW		485
O_2 Flow at	m^3/min	1.65
Stoichiometric	1/sec	27.5
Combustion		

TABLE 5

Firing Conditions on the HFO-Oxygen System
Using Propane Atomising

Oil Flow Kg/min		0.68
Propane flow Kg/min		0.047
Propane % of Energy		7.25
Heat release KW		485
O_2 flow at stated	100%	27.5
% of stoichiometric		
combustion 1/sec	90%	24.75
	110%	30.25

On the burner design that evolved, it is possible to begin
firing with oxy-propane to heat up the environment (local or
gross) to a sufficient temperature to maintain a stable flame
of HFO-oxygen. It is also possible to begin on oxy-propane
and change over to oxy-HFO steam, once an adequate
temperature is reached within the furnace.

4.2. HFO-Coal-O_2 Results

This system was fired using coarse and fine fractions of coal
and the firing conditions for the fine fraction tests are
given below for propane atomisation in Table 6.

TABLE 6

Firing Conditions for Coal-Oil Mixture and Oxygen

Mixture composition by wt % HFO		60
Coal		40
Calorific value	JK/Kg	36,900
C.O.M. Flow	Kg/min	0.735
Propane Flow	1/sec	0.4
Propane	% of Energy Input	7.25%
Heat Release	KW	485
O_2 flow at stoich	1/sec	27.5
	m^3/min	1.65
90% stoich	1/sec	24.75
	m^3/min	1.49
110% stoich	1/sec	30.25
	m^3/min	1.8

The temperature profiles from selected tests given in Figures 4, 5 and 6 summarized in Fig. 7. The burners used in these tests have certain novel features which are subject to patent applications by BOC. Copyright also exists for BOC in the burner design.

Where propane was used as the atomising media, the heat input had to be limited to a peak wall temperature of 1580°C. This was 485 KW. At this rate of firing using steam atomising, the wall temperature was reduced some 120°C near the burner and some 30°C about 3m from the burner.

The three fuels used cover a wide range of cost. Table 7 gives the costs of fuel mixtures fired per KWhr at prices quoted in EURA Data for April 1983.

TABLE 7

COSTS OF MIXTURES FIRED

	MIXTURE %		COST
HFO	COAL	PROPANE	P/KWhr
100	-	-	1.19
-	100	-	0.76
-	-	100	1.85
65	35	-	1.04
60	33	7	1.09

5. Conclusions

Analysis of the data also shows:

1) For a given oxygen content, the temperature profiles for HFO, HFO fine and coarse coal are essentially the same.

2) Front and back end wall (and hence flame temperatures) can be controlled within 100°C by controlling oxygen in the range ±10% of stoichiometric combustion.

3) Burner design was considerably simplified by propane atomisation.

4) At 485KW a stable controllable flame can be produced with HFO-coal using propane atomising at about a 10% reduction in cost compared with HFO alone.

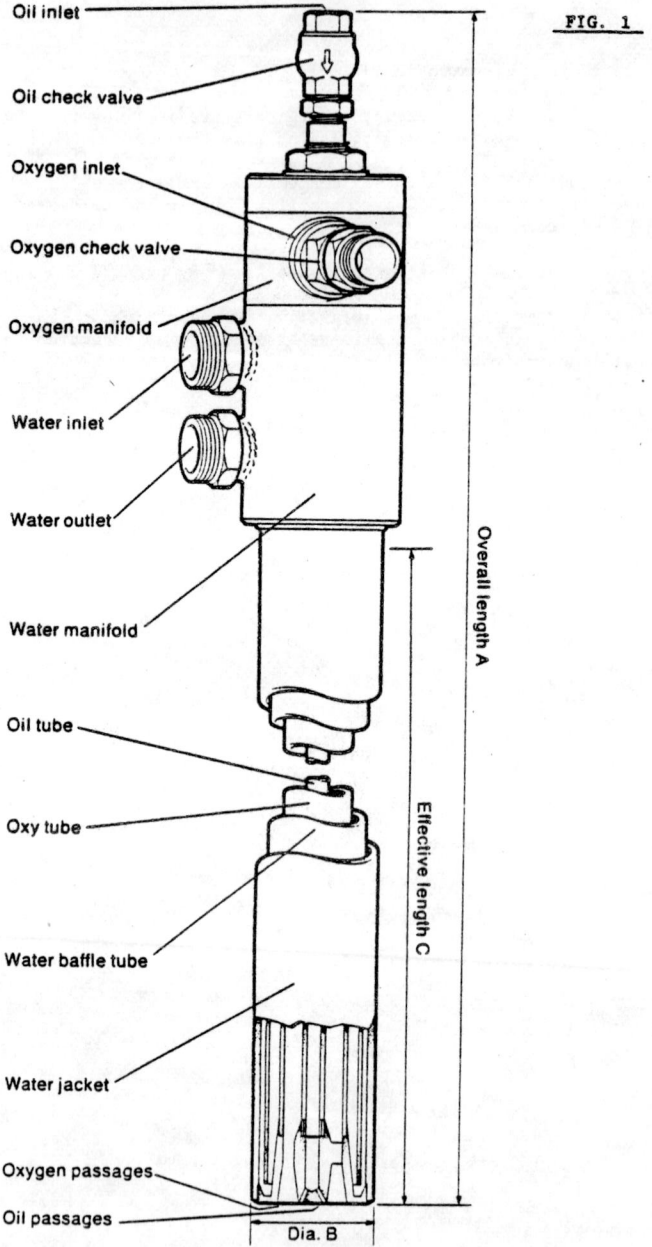

FIG. 1

Oil inlet

Oil check valve

Oxygen inlet

Oxygen check valve

Oxygen manifold

Water inlet

Water outlet

Water manifold

Oil tube

Oxy tube

Water baffle tube

Water jacket

Oxygen passages

Oil passages

Overall length A

Effective length C

Dia. B

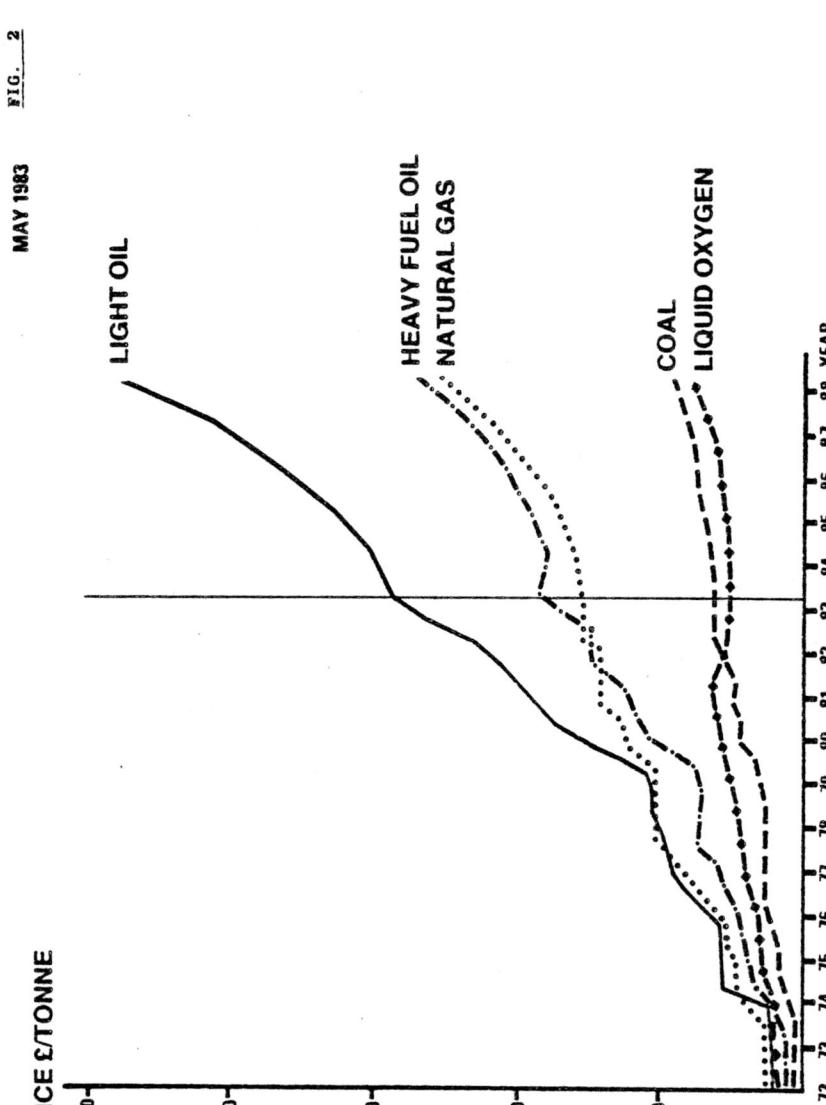

MAY 1983   FIG. 2

FIG. 3   **MAY 1983**

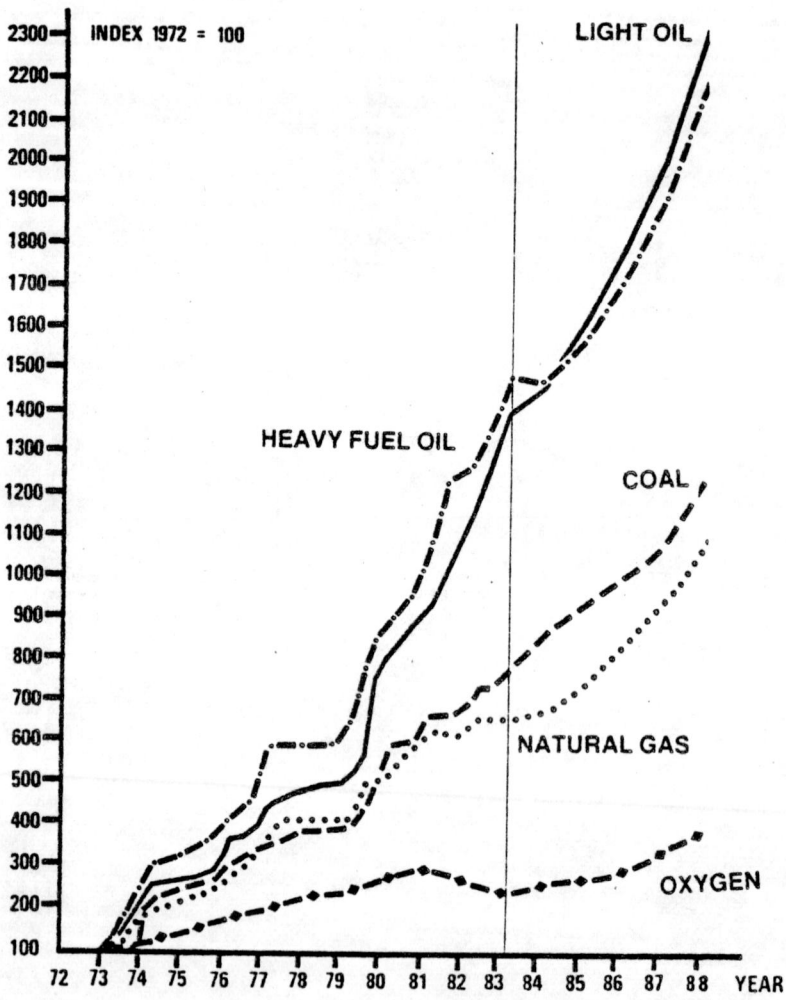

Sources: EURA and Henley Centre

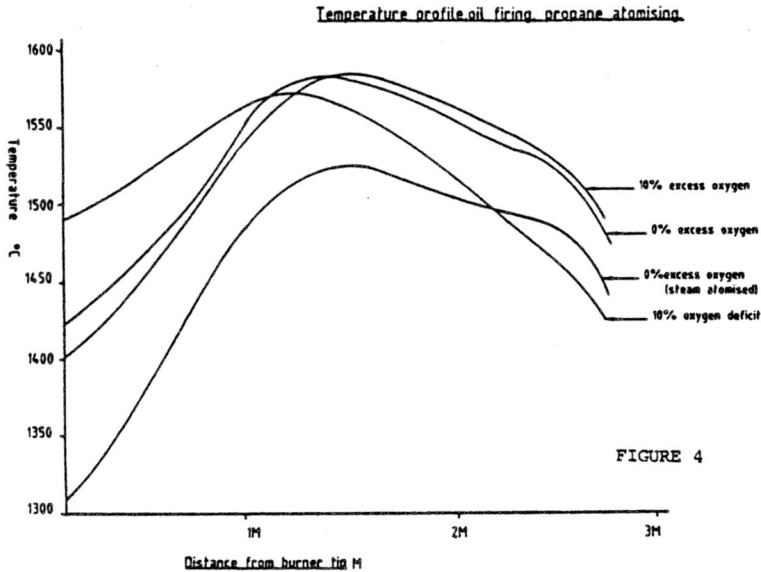

Temperature profile,oil firing, propane atomising.

FIGURE 4

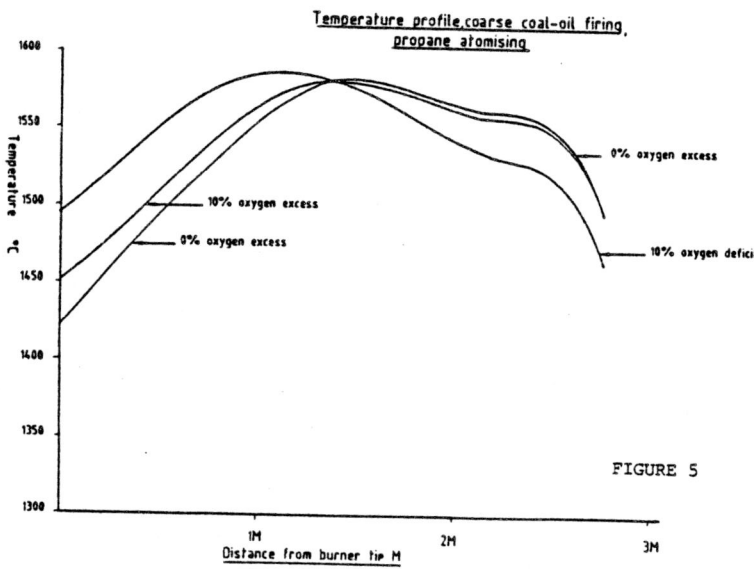

Temperature profile,coarse coal-oil firing, propane atomising

FIGURE 5

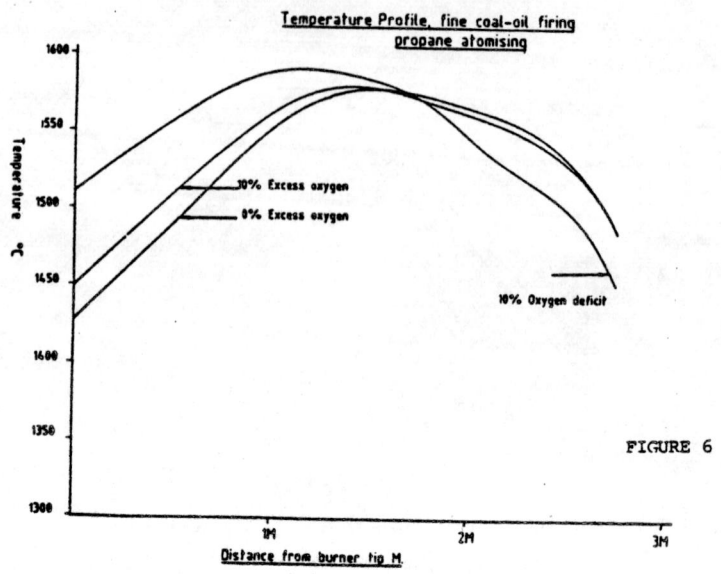

Temperature Profile, fine coal-oil firing propane atomising

FIGURE 6

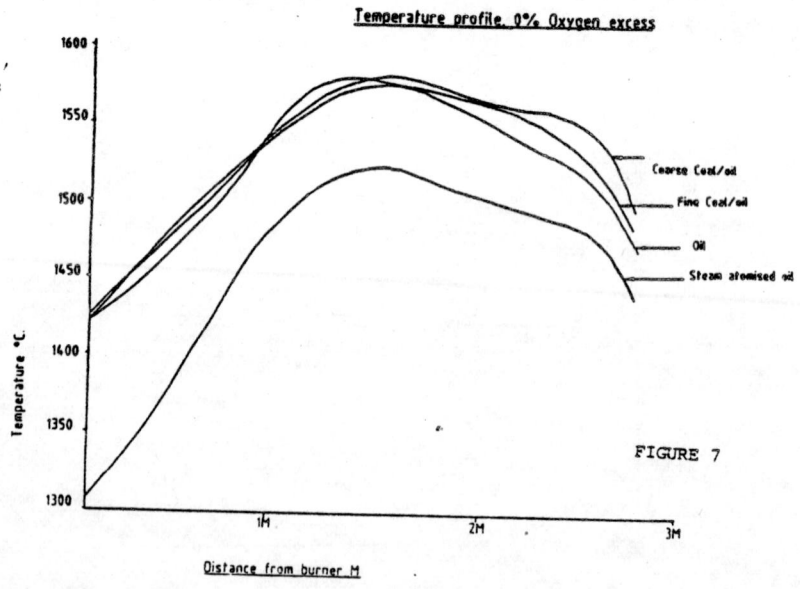

Temperature profile, 0% Oxygen excess

FIGURE 7

TECHNICAL EXPECTATIONS FOR COAL-OIL FUELS
AS A REPLACEMENT IN FUEL OIL FIRED EQUIPMENT

Barrie G. Jenkins[+]   Frank D. Moles[*]   Michael C. Patterson[#]

An experimental programme to assess the technical problems
which are associated with coal-oil fuels has been carried
out by the authors. The results of the experimental study
have led to the development of a mathematical model to
predict flame characteristics. Both of these studies have
been fully reported elsewhere and thus only the salient
points are referred to in this paper. The mathematical
model has been used to make comparisons between residual
fuel oil and coal-oil fuel firing in terms of flame length
or burnout, heat flux and combustion efficiency for various
types and size distributions of coal in coal-oil mixtures.
Conclusions are drawn regarding the changes in process
heating and derating effects associated with various COM
fuels.

INTRODUCTION

The future of coal-oil (COM), coal-oil-water (COW) and coal-water (CWM) fuels
must be considered in terms of their ability to compete economically and
technically with heavy end fuel oils. In the medium term, any conversion of
existing oil fired plant will have to be achieved with only minor modifications
to the fuel handling and combustion system if coal-liquid mixtures are to present a feasible alternative.

In 1977, the British Petroleum Company Limited initiated a collaborative research programme with the Fuels and Energy Research Group at Surrey University
(FERGUS), using the University's 0.15 MW pilot test furnace, to evaluate the
handling and combustion problems associated with micronized coal-oil dispersions. The research programme monitored the combustion quality, emissions
and radiation characteristics of a number of coal types in both 950 and 3500
second fuel oils, and contrasted the results with those from the parent fuel
oil to give a quantitative assessment of the comparability of coal-oil mixtures (1). The results of this work indicated that there were unusual
anomalies which were associated with the rank of the coal used in the mixture,
and further microscopic examination of the combustion residues confirmed the
hypothesis that two competing combustion mechanisms existed. The development
of a mathematical model, based on these conclusions has led to a theoretical
basis by which the prediction of flame characteristics can be made with a
reasonable degree of certainty (2).

+ Airoil-Flaregas Limited, Horton Road, West Drayton, Middx.
* University of Surrey, Guildford, Surrey
# British Gas Corporation, Watson House, Peterborough Road, London

The objective of this paper is to summarize the salient points of both the experimental and theoretical studies, emphasising those areas where further investigation and development are required, and to provide a comparative assessment of the differences between residual fuel oil and coal-oil fuel firing in terms of burnout length, heat flux and combustion efficiency in both hot and cold wall furnace applications.

COM FUEL HANDLING

The successful substitution of COM for residual fuel oil must occur with only minor modifications to the existing fuel handling system, particularly for large fuel consumers, since major conversion costs will probably mean that coal firing, using technically proven and economically predictable pulverized fuel systems, will be more attractive. For the smaller fuel user, large capital conversion costs mean that fixed and fluid bed coal burning units would become economically reasonable options.

The equipment which must be considered when assessing the compatability of any existing handling system for coal-liquid mixtures are

        (a)   valves and filters

        (b)   fuel heating

        (c)   flow measurement

        (d)   pumps

        (e)   burners

A further variable in the system is the method of coal-liquid mixing used. Bought-in stabilized COM or CWM would be delivered into the normal fuel oil tank, and thence taken to the burner by the normal fuel line through all of the above mentioned equipment. In situ mixing of coal and oil would require that the fuel oil handling system remains undisturbed, although operating at reduced capacity, whilst extra equipment would be needed to receive, store, control and mix pulverized coal with the fuel oil immediately before the burner. The latter option is only feasible for large fuel consumption units, or central preparation systems where the economies of scale allow the production of pulverized coal at rates greater than 1 tonne/hour.

Valves and filters

It is advisable to remove all restrictions and convolutions of flow in the system to avoid settling of the coal from the mixture. Most slurries are pseudoplastic Bingham fluids in which the shear stress increases linearly with shear rate, thus increasing the apparent viscosity. During handling trials (1), particular difficulties were experienced with fuel pressure regulating valves, and flow control valves at the lower end of the burner turndown both of which act as severe flow restrictors creating high shear flows. As a general observation, pressure drops through the system were considerably greater for coal-oil mixtures than for the constituent oil at ambient conditions (40-65°C). Similar experience has been reported for coal-water slurries by Barsin (3).

Fuel heating

The stability of COM, COW and CWM fuels is highly dependent on the temperature of the mixtures. Heating of COM to an atomizing temperature consistent with its apparent viscosity led to considerable settling of coal in the fuel supply system, causing blockages and pulsations of fuel to the burner.   The adoption of a preheating temperature a few degrees above that for the constituent fuel oil eliminated bulk settling of the coal.   Initial tests on the research furnace used an immersion type electrical fuel preheater, but it was rapidly discovered that the elevated local film temperatures caused the coal to gasify, and the resultant char to adhere to the element.   The ultimate consequence was total coking of the elements and over-pressure in the preheater drum. The installation of a steam preheater eliminated these problems and gave a greater degree of control to the fuel temperature at the burner.

Flow measurement

The measurement of slurry flows has always been a difficult, and at times in-tractable problem.   The non-Newtonian behaviour and abrasive qualities of coal-liquid slurries prevents the use of orifice flow meters or displacement meters to monitor the fuel flowrate.   Several flow measurement devices were investigated during the course of the tests, showing varying degrees of success. The most practically reliable method is probably the use of a metering pump, although this is an area where further investigations and developments are required.

Pumps

It is probable that any system which is converted to coal-liquid fuel will require an uprated pumping unit to give the increased head requirement mentioned in the previous section on valves and filters.   The types of pump which have been found as most successful with coal-liquid fuel are pneumatic diaphragm pumps, which also act as the fuel pressure regulator, positive displacement screw pumps, and progressive cavity (mono) pumps, all of which operate with relatively low flow velocities and shear through the pump body.

Burners

Selection of the burner for coal-liquid fuels is the single most important factor for successful operation.   A low pressure air atomized burner was used for the combustion tests on the pilot furnace, this design being chosen because of the lack of internal fuel impingement areas, and the relatively large oil-ways for a small burner.   The intention was to eliminate burner erosion and blockage, and this was confirmed by operation where after 250 hours of inter-mittent operation on coal-oil dispersions inspection by the burner manu-facturers indicated no wear in any of the burner parts, or at the atomizing edge.

A number of combustion tests have been undertaken by various workers to evaluate the suitability of twin fluid atomizers for coal-liquid mixtures. Barsin (3) used a modified 'Y' jet to test a 70% coal-water slurry, reporting that the useful life of the atomizer was only a few hours.   Scheffee et. al (4) used a special external air atomized burner for COW combustion which used a high degree of secondary air swirl to improve the combustion efficiency. Tests on a twin fluid steam atomizer (STFA), developed by the CEGB (5), burning COM have been reported by Whaley et. al. (6), the reduced fuel impingement and

low steam requirement of this particular nozzle design giving highly encouraging results to date.

Coking, which can be a problem with simple oil fired burners, occurs rapidly
where atomized coal-liquid fuel is allowed to impinge in the burner quarl or
on the spray nozzle. This can only be eliminated by careful and correct
design of the windbox to ensure an even distribution of secondary air about
the burner, and positioning of the fuel nozzle within the quarl throat to
prevent spray impingement or overheating. It is recommended that aerodynamic
modelling of any proposed installation should be undertaken prior to conversion to coal-liquid firing.

COMBUSTION OF COM

Extensive combustion tests were carried out on 35% coal - 65% oil dispersions
in which the coal type was varied from an anthracitic low volatile coal
(CRC 101), through a medium volatile, high swelling coal (CRC 301) to a high
volatile coking coal (CRC 501). Flame shape and length, as defined by the
0.01% carbon monoxide concentration profile, were measured for each flame
under various excess air conditions and compared with an equivalent fuel oil
flame. Gaseous (CO_2, O_2, CO, NOx, SOx) and solids flue gas emissions were
also measured together with flame emissivity and wall heat flux for all the
studied conditions. The results of this work have been reported previously
(1).

Flame burnout time is generally increased by the presence of coal in oil,
although in the case of the low volatile anthracitic dispersions this effect
was marginal. This observation was in disagreement with experience of pulverized fuel firing of similar coals in rotary cement kilns, where the higher
volatile coals burn more rapidly (7). Close examination of the experimental
results led to the hypothesis that two contrasting combustion mechanisms
existed for coal-oil combustion.

It was proposed that the low volatile coal-oil dispersion burn in the initial
volatile combustion stage as an oil droplet, but during char combustion dis-
integrate into agglomerates composed of coal particle sized cenospheres.
In comparison, higher volatile coals form fused oil droplet size spheres
during volatile combustion, and thus continue to burn as swollen droplet sized
cenospheres in the char phase, see Figure 1. By this mechanism a much
greater surface area is available for char combustion in the case of anthracitic COM, thus reducing the burnout time. This phenomenon was obviously
enhanced by the fact that the COM studied contained micronized (<10μ) coal
particles. Evidence of the existence of these mechanisms was found by
analysis of flue gas solids samples by Dr. P.J. Street of CEGB Marchwood,
using microscopic analysis techniques (8). Similar observations have been
made on agglomeration by Miyasaka and Law (9). Extension of these studies
using single droplet combustion techniques by Lapwood (10) has indicated that
general purpose coals (CRC 802-902) in COM behave in a similar manner to
anthracite, and has attributed the fusion characteristics of COM droplets to
coal softening at the devolatilization temperature (400-550°C), this being
marked for coals of 85-92% carbon content.

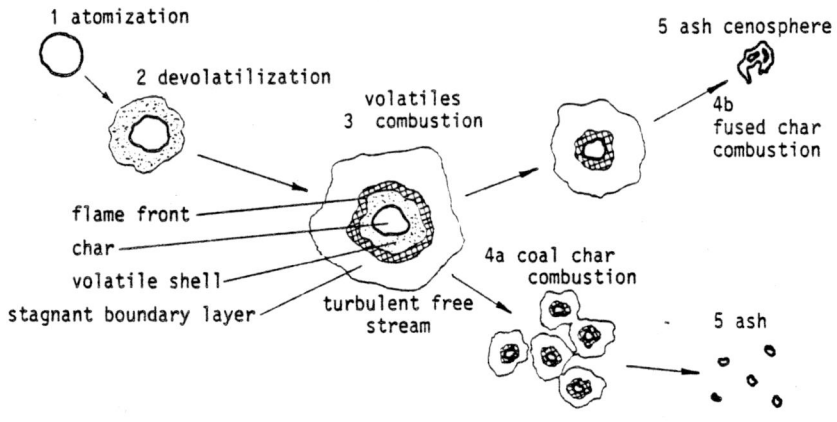

FIGURE 1.   Proposed combustion pathways for coal/oil fuels

In an attempt to qualify these observations theoretically, a mathematic model
has been developed based on the aerodynamic mixing and recirculation of an
enclosed turbulent diffusion jet flame, combined with the burning of oil droplets by mechanisms similar to those controlling pulverized coal combustion.
This has been fully reported elsewhere (2), where a comparison of model predictions and experimental results has shown a highly encouraging degree of
agreement.

RESULTS OF PREDICTIONS FROM THE MODEL

Figure 2 shows a comparison between the measured values of burnout distance
on the test furnace and the predictions of the combustion model.   The measured burnout or flame lengths are based on the axial carbon monoxide extinction point, which is considerably longer than the visual flame length, but
provides a quantifiable condition to which model calculations can converge.
Close agreement is observed between measured and predicted values for excess
air conditions of >5%.   The 950 oil and 501 coal/oil mixture predictions are
based on atomizer droplet size distribution combustion, whereas the 101 coal/
oil mixture predictions use a micronized coal size distribution for char combustion based on a mass median diameter of 5 microns.   It can be seen that
the mid-rank COM flame is approximately 25% longer than the equivalent oil
flame over the excess air range studied, whereas the micronized anthracite
flame is 10-15% shorter.

The size distribution of low and high rank coals in COM is thus seen to have
a significant effect on the burnout characteristics of the flame.   Figure 3
shows model predictions of the importance of the initial coal size distribution in the COM for these coals.   Where the mass median diameter, d, of
the coal is less than 7.5 microns about a Rosin-Rammler distribution index of
1.38, the resultant flame is shorter than the equivalent oil flame, and where

CLM-M

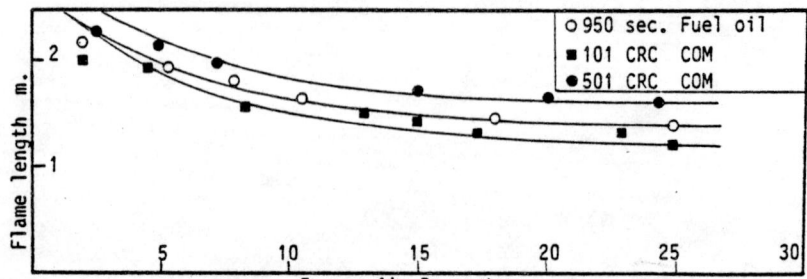

FIGURE 2. Comparative burnout flame lengths for various coal oil mixtures and parent fuel oil

d is less than 14 microns the flame is shorter than a medium rank COM flame. Selection of both the coal type, and the preparation method of the COM are therefore critical in applications where the available combustion space is limited; a condition that would almost invariably apply to combustion chambers initially designed exclusively for gas or fuel oil firing.

Other major considerations affecting both furnace and boiler operation are the heat flux pattern from the flame, and the combustion efficiency. The combustion of COM in any unit will lead to increased stack solids due to the ash in the coal, this being further increased, particularly at low excess air levels, by unburnt carbon char. Typical predictions for 120 micron droplets to burn out at 5% excess air are 1.2 secs. for 950 secs. fuel oil, and 1.8 secs. for a 501 CRC coal/65% 950 secs. fuel oil dispersion.

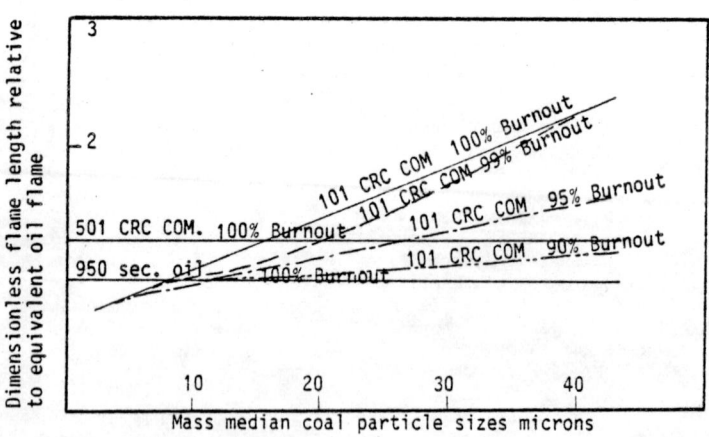

FIGURE 3. The effect of initial coal particle size on COM fuel flame lengths

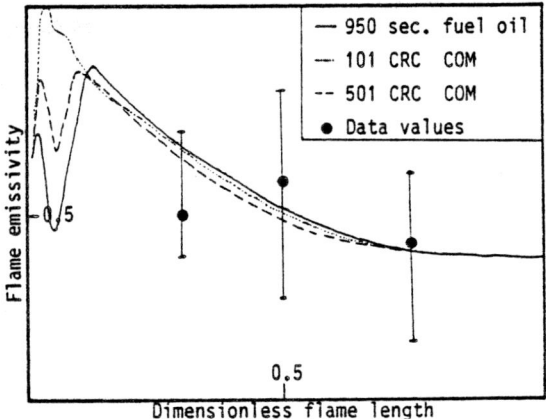

FIGURE 4. Predicted and measured flame emissivities for COM and fuel oil

The presence of particulate coal in the flame increases the emissivity, although the quantitative value is a function of the burner design and air distribution. Figure 4 shows the predicted values of flame emissivity for fuel oil, anthracitic COM, and mid-rank COM from an L.P.A. burner operating at 5% excess air. These compare with measured values from the combustion tests, although it was not possible, due to the mechanical restrictions of the furnace, to monitor the emissivity close to the burner nozzle.

Increased flame emissivity does not imply an increase in overall heat flux from the flame, since the flame temperature is lower. Figure 5 shows the predicted net heat flux from the flame for a 950 second oil and micronized anthracite COM in a hot walled (1430°K) furnace. Also shown are the predicted heat fluxes for the same COM into an equivalent cool (1073°K) and cold (673°K) walled furnace. The distribution of heat from this COM is more 'peaked' than the oil flame and achieves a higher maximum value, however similar calculations (not shown on the graph) for mid-rank COM's gave a flatter, lower heat flux distribution because the delayed combustion reduced the flame temperature to a greater degree than the increased emissivity effect.

Figure 6 correlates the combined effects of furnace temperature, combustion quality and flame length of anthracitic COM against base equivalent oil flame conditions. The flame length of the COM increases initially with reducing furnace temperature down to approximately 900°K, where it then rapidly decreases due to the quenching effect of the surrounding chamber. Combustion quality also decreases with decreasing furnace temperature, this being a not unexpected result. These predictions imply that COM fuels are less likely to be successful in boiler applications, resulting in a probable derating of 7-10%.

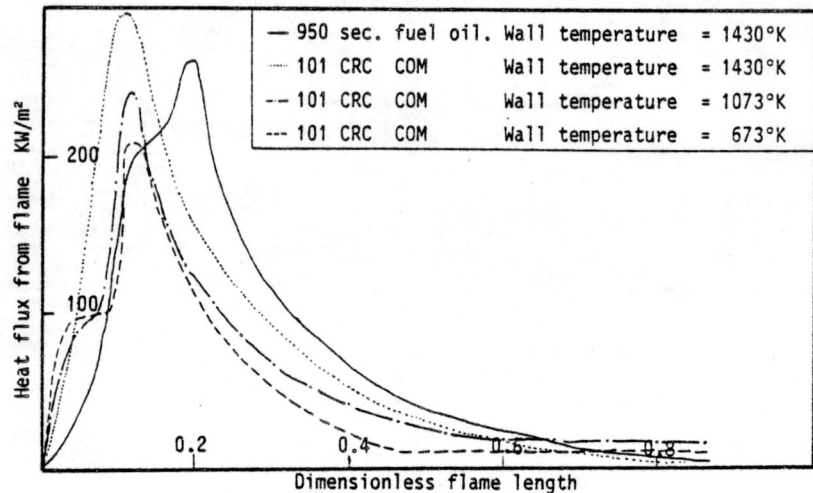

FIGURE 5. Predicted heat flux distribution from COM and oil flames in various furnace environments

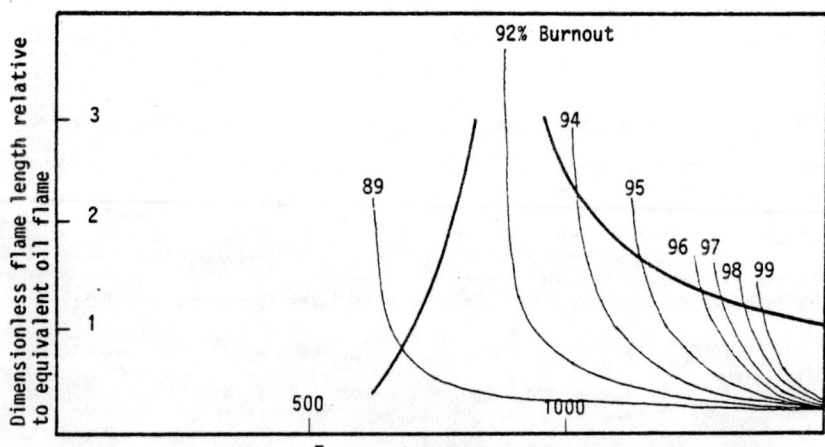

FIGURE 6. Predictions of burnout length and combustion quality for COM fuel as a function of furnace temperature at 5% excess air conditions

CONCLUSIONS

a) Handling and preheating systems for coal/oil fuels must be designed to ensure that flow restrictions, high shear and local high temperature heating are all eliminated. Selection of suitable proprietary equipment will fulfil these requirements.

b) Fuel atomizing temperatures are comparable to those of the constituent fuel oil in COM's.

c) Further development effort is required to arrive at a suitable burner design for coal-liquid mixtures.

d) The type of coal used in COM's has a controlling influence on the combustion characteristics of the fuel, affecting both flame length and heat flux. High and low rank, finely ground (d<10μ) coals in COM's result in shorter 'peakier' flames. Medium rank coals in COM's give longer flames (+25%), and flatter heat flux profiles.

e) Limitations of combustion space in highly rated oil or gas fired furnaces and boilers will require that these units are derated by 7-10% on COM fuels.

1. Alabaf J.S., Patterson M.C., and Moles F.D.
   3rd. Int. Symp. Coal-Oil Mixture Combustion,
   PETC. U.S. Dept. of Energy, 1981

2. Jenkins B.G., Alabaf J.S., Patterson M.C., and Moles F.D.
   Coal : Phoenix of the '80s. Proc. 64th. C.I.C. Coal Symp;
   Halifax, Canada, 1981, Vol. 1, pp 348-356

3. Barsin J.A.
   ibid. pp. 341-347

4. Scheffee R.S., Rossmeissl N.P., Boyd T.J., Henderson C.B. & McHale E.T.
   3rd. Int. Symp. Coal-Oil Mixture Combustion,
   PETC. U.S. Dept. of Energy, 1981

5. Anson D., and Denham R.O.
   British Patent Specification 1470671, 1977

6. Whaley H., Whalen P.J., and Davis F.E.
   6th. Members Conf. of I.F.R.F., Holland, 1980

7. Jenkins B.G., and Moles F.D.
   University of Surrey Rept. CE-RF-57, 1978

8. Lightman P. and Street P.J.
   Fuel, 1968, Vol. 47

9. Miyasaka K., and Law C.K.
   Comb. Sci. Tech., 1980 Vol. 24 pp 71-82

10. Lapwood K.J., Street P.J., and Moles F.D.
    5th. Int. Symp. Coal Slurry Combustion,
    PETC. U.S. Dept. of Energy, 1983

ATOMIZATION CHARACTERISTICS OF COAL-WATER MIXTURES

C. R. Krishna and Thomas Butcher*

The characteristics of the break-up of jets of coal-water mixtures were investigated. It is shown that these behave similarly to liquid (water) jets except for much shorter break-up lengths for the same value of the Rayleigh Parameter. Next, the drop size distributions from an air atomizer were measured. It is shown that the mass mean diameters are more or less the same independent of coal concentration (from 0 to 50 percent) in the mixture at the same ratio of atomizing air flow to fuel flow.

INTRODUCTION

It is well known that proper atomization is a key to satisfactory combustion of liquid fuels. This has led to the development of a number of types of atomizers and also to studies on the fundamentals of the atomization process.[1] With a fuel such as coal-oil mixture (COM), the process of atomization is complicated by the presence of solids in the liquid. However, there has been very little work contributing to its understanding. Only recently has work been started on atomization of coal-oil mixtures.[2,3] Coal-water mixtures (CWM) is a fuel of revived and significant interest.[4] As the liquid (water) is non-combustible, it would seem that the combustion properties of coal-water mixtures would correlate with those of the coal. However, if the water surrounds the coal particles prior to combustion and if the water interacts with the coal, the mixture combustion properties could depend on the atomizer characteristics. There are reported observations that suggest that the coal-water flame appears to be like an oil flame.[5] The reasons for this are not clear, but, if true, this has important implications for coal-water mixtures as a retrofit fuel for furnaces designed to burn oil. The importance of studying atomization has been recognized[4] and the present work represents an attempt towards understanding atomization of coal-water mixtures.

The process of drop formation in liquids can be described roughly as follows.[6] The surface area of the liquid emerging from the atomizer increases until it becomes unstable and disintegrates usually into threads of liquid. Such a thread of liquid is unstable if its length is greater than its circumference and the unstable threads breakdown into rows of drops according to the classical mechanism proposed by Lord Rayleigh.[7] The above description is expected to fit broadly the process of slurry atomization. Atomization, then,

*Brookhaven National Laboratory, Upton, NY 11973.

can be viewed as a combination of processes which can be examined individually. This has been done with liquids to some extent by studying the formation of drops from jets and also the process of breakup of sheets to form
threads of liquid.[8],[9] The goal of the present work is to develop a similar
approach to the understanding of slurry atomization. The breakup characteristics of coal slurry jets were investigated experimentally and compared with
results for liquid jets. Next, the characteristics of an air atomizer were
measured while atomizing water and coal-water mixtures.

JET BREAKUP STUDIES

Brief Review of Past Work

Richardson's chapter in Ref. 8 gives a good survey of the work done till
the date of that publication. An extensive review will not be attempted
here. Only the aspects of jet breakup that provided the framework for the
work reported here will be summarized below.

When a liquid emerges from a nozzle into an ambient fluid, it has been
observed that the jet appears to remain undisturbed for a certain distance and
then breaks up into drops. The distance from the nozzle to the point of disruption is called the continuous length of the jet. This length, L has a
characteristic variation with the jet efflux velocity, V as shown in Figure
1. For a given nozzle and liquid, L increases more or less linearly with
velocity until it reaches a maximum at a critical velocity. Beyond the critical velocity, L decreases as velocity increases over a certain range. The results on the liquid and coal slurry jets to be presented in this paper were
obtained in these two regions.

As Richardson's survey[8] shows, the mechanism of breakup in the two regions are different. In the first region (below the critical velocity), Rayleigh[7] analyzed the breakup as resulting from the interaction of surface
(potential) energy and jet kinetic energy leading to the growth of symmetrical
swelling and contraction (varicosities) in the jet. The surface tension and
density are the primary liquid properties controlling the behavior in this
region, the effect of viscosity being minor. This is expressed in the linear
relationship between (L/d) and the Rayleigh Parameter $V(\rho d/\sigma)^{1/2}$. As we get
closer to the critical velocity, the effects of viscosity become more important and Tyler and Richardson[9] derive a linear relationship between the Rayleigh Parameter at the critical velocity and $(\mu^2/\sigma d\rho)^{1/2}$. In the second region
(beyond the critical velocity), it has been seen[8] that the jet becomes sinuous
and the growth of the sinousitites leads to breakup of the jet. The air resistance to the passage of the sinousities increases roughly as the square of
the velocity and hence the breakup occurs earlier with increase in velocity.
It may be noted that the liquid discharged from an atomizer could be in the
form of a sheet and that mechanisms similar to those for jets have been in-
voked to explain the initial breakup of the sheets.[1]

The brief review above has dealt entirely with liquid jets as no similar
work on solid-liquid mixtures seems to be availale. It will be seen from the
results to be be presented that the slurries show both similarities to and
differences from liquids in jet breakup behavior.

Experimental

A pressure chamber contained the fluids which exited through a tube 0.125
cm ID tube and 6 cm long. The outlet end of the tube was simply cut and
ground flat and smooth. Jet velocity was measured by volume and time

measurements. The jet length was determined by placing a ruler adjacent to the jet and illuminating the jet with a strobe. The jet length was defined as the distance from the tube opening to the first point where a regular break in the jet was seen.

Liquid surface tension was measured by the capillary rise method. The surface tension of the slurries was measured by the classic dropweight method. In this method the drops are formed slowly (20 sec/drop) and some settling could occur in the drop. No difference in the measured surface tension from that of the carrier liquid was noticed and the liquid values were used in the calculations.

Viscosities were measured with a Brookfield LVT viscometer using the concentric cylinder attachement at an approximate shear rate of 63 sec^{-1} for the liquids and coal-water mixtures. The measured properties are listed in Table 1. The coal was a North Dakota lignite ground to about 125 microns mean size and CWM's with different coal loadings were tested.

Table 1
Fluid Properties for Jet Break Up Tests (S.I. UNITS)

	$\rho \times 10^{-3}$	$\sigma \times 10^{3}$	μ	Vc
Water	1	72	.001	1.09
10% Coal-Water Mixture	1.04	72	.0014	1.55
20% Coal-Water Mixture	1.08	72	.0017	1.75
30% Coal-Water Mixture	1.13	72	.0026	2.18
20% Glycerine in Water	1.06	67	.0019	2.00
40% Glycerine in Water	1.11	69	.0037	3.10

Results and Discussion

The jet length during the low speed portion of the break-up curve is shown in Figure 2 for water and coal-water mixtures. This is a non-dimensional representation as in Ref. 8. The data for the water and glycerine in water solutions fall approximately on one straight line in agreement with Tyler and Richardson.[9] The data for the coal-water mixture shows much shorter lengths to breakup than for water and glycerine in water. The jet lengths for the CWM's appear to be independent of coal concentration in the range tested.

The jet critical velocity is defined as the velocity which gives a peak in the length vs velocity plot. A typical plot is illustrated in Figure 1 where results for water and a 20% coal-water mixture are shown. The critical velocity for water in this case is 109 cm/sec and for the 20% coal-water mixture is 175 cm/sec.

The curve of Tyler and Richardson of $Vc(\rho d/\sigma)^{1/2}$ vs $\mu/(\rho d\sigma)^{1/2}$ is reproduced in Figure 3 along with the data for all fluids tested. The data for the water, water-glycerine solutions, and coal-water mixtures all agree fairly well with Tyler and Richardson's curve.

The jets break up into drops which are fairly uniform in size. The size was measured by analyzing photographs for a water jet and a 30% coal-water mixture taken at a sufficient distance from the nozzle. The mean drop size for the water jet was 0.23 cm and the mean size for the coal-water mixture jet was 0.24 cm.

As can be seen from Figures 2 and 3, the behavior of the coal slurry jets is similar in several ways to that of the liquids. The data from the measured drop sizes from the jet show that they are about the same for the liquid and the CWM's and it is roughly twice the jet diameter (roughly in accordance with Rayleigh.[7])

There is an obvious difference between the liquid and CWM jets, however, in terms of the jet length in the first region. The CWM jet lengths are approximately 1/3 of the liquid jet lengths as shown in Figures 1 and 2. According to Tyler and Richardson all data for liquids should lie on the same line in Fig. 2.

Coal-water mixture results show that CWM jets seem to break up by the Rayleigh mechanism as indicated by the linearity of the plot in Figures 1 and 2 and by the resulting drop size being close to that predicted by Rayleigh. However, the reduction in the breakup length compared to the liquids at the same value of Rayleigh parameter could be taken to mean[8] that the ratio of the final amplitude of disturbance causing breakup to the initial amplitude is smaller for the coal-water mixture jets.

The maximum coal loading in these experiments is not very high. While we expect similar behavior up to about 50% coal (similar to the mixtures reported on in the next section), we suspect that heavily-loaded CWM's may behave differently.

ATOMIZATION STUDIES

Brief Review of Past Work

Dombrowski's review[1] is an excellent survey of liquid atomization and the basic picture of drop formation in sprays presented there has been summarized in the Introduction. Work on liquid atomization continues actively. However, there has been very little previous work on the atomization of coal-liquid mixtures. A brief review[10] has been prepared by one of the present authors summarizing the available literature. In view of the early stage of the studies of coal-liquid mixture atomization, the approach taken in the present work was to begin to examine the difference that the presence of particles made to liquid atomization (analogous to the foregoing work on jet break-up). This was done by comparing the characteristics of an air atomizer while spraying water and coal-water mixtures.

Experimental

A schematic of the experimental rig is shown in Figure 4. The coal-water mixture is stored in a vertical slurry tank and the coal is kept suspended by continuous mechanical agitation. The mixture is withdrawn from the tank and pumped to the atomizer by a variable speed positive displacement pump. The exit from the pump can be passed through the atomizer or bypassed to the tank. The laboratory air supply is used for the atomizer and the variable air flow rate is measured by a rotameter. The coal-water mixture flow rate through the atomizer is measured by collecting the flow for a known time. The spray from the atomizer is confined to a collection system which is vented by an exhaust fan to the outside through an entrainment separator. The drop sizes in the spray are measured by a Malvern particle sizer[11] at a fixed distance from the nozzle. The instrument was set to fit a Rosin-Rammler size distribution to the measured drop sizes.

A commercial internal mixing air atomizer (Delavan AIRO 30615-3) designed for liquid spraying was used in the tests. The coal-water mixtures containing 33 and 50 percent by weight coal used in the tests were made without using additives of any kind. The properties of the pulverized coal that was used to make up the mixtures are given in Table 2.

Table 2
Properties of Coal Used in the Atomization Tests

Ultimate Analysis (% by weight)	
C	78
H	5.1
N	1.6
Ash	8
Higher Heating Value	
J kg^{-1}	3.22 x 10^7
Size Distribution:	
% by weight smaller than:	
37 micrometers	7.9
74 micrometers (200 mesh)	69.1
150 micrometers	99.1
250 micrometers	99.9

Results and Discussion

Figures 5, 6, and 7 present some of the key results. The mass mean diameter (MMD) was calculated from the Rosin-Rommler distribution given by $R = Exp [-(d/x)^N]$ where R is the fractional weight of spray above drop diameter d. X and n are the distribution parameters. X denotes the drop diameter for which R=36.8% and N indicates the breadth of the diameter distribution. (Higher values of N represent narrower distributions.) It was found that the ratio of fuel flow to atomizing air mass flow was a significant variable in representing the atomizer characteristics.

Figure 5 shows that at a given air pressure, there is no significant difference in the MMD as the coal loading was changed from 0 to 50% by weight. Predictably, the MMD's are smaller with increased air-fuel ratios. Figure 6 shows that there is a small reduction in MMD when the air pressure is increased. Figure 7 is a plot of N against fuel-air ratios for water and a 33% coal-water mixture. It is seen that at low fuel-air ratios, the N values for the CWM are smaller than for water suggesting that the distribution is wider for the CWM even though the MMD's are quite similar. An explanation for this may be that, at these high air/fuel ratios, some of the coal particles could be outside the water drops after atomization and therefore the instrument measures a composite distribution from those of the water (and coal-water) drops and of the coal particles.

As indicated before, the coal-water mixtures used did not contain any additives or surfactants. While we do not expect surfactants by themselves to alter atomization behavior, they are usually used in coal-water mixtures that have significantly more coal loading than in the mixtures tested here. This combination could affect the atomization characteristics.

REFERENCES

1. Dombrowski, N. and Munday, G. Spray drying. Biochemical and Biological Engineering Science, N. Blakebrogh, Editor, Vol. 2, pp. 209-320, Academic Press, London, 1968.

2. Butcher, T., Pucci, A., and Krishna, C. R. Atomization of coal slurry jets. Proc. 2nd Int. Symp. on Coal-Oil Mixture Combustion, Nov. 1979, Vol. 2, U.S. DOE, Pittsburgh, 1979.

3. Carmi, S. and Ghassemzadeh, M. R. Atomization of coal-oil mixtures. Ibid.

4. Marnell, P. and Krishna, C. R. A program plan for the development of coal-water fuel technology. BNL 29902, May 1981.

5. Marnell, P. Direct firing of coal-water suspensions, state-of-the-art review. Coal Technology'80, Vol. 4, pp. 485-499, 1980.

6. Beer, J. M. and Chigier, N. A. Combustion Aerodynamics, Applied Science, p. 147, London, 1972.

7. Strutt, J. W. (Lord Rayleigh) On the instability of jets. Scientific Papers, Vol. 1, p. 361, Cambridge, England, 1899.

8. Richardson, E. G. Liquid sprays. Flow Properties of Disperse Systems, J. J. Hermans, Ed., p. 266, North-Holland, Amsterdam, 1953.

9. Tyler, E. and Richardson, E. G. The characteristics curves of liquid jets. Proc. Phys. Soc. Lond. 37, p. 279 (1925).

10. Krishna, C. R. Slurry atomization - a brief review. BNL 32952, 1983.

11. Weiner, B. B. Particle and spray sizing using laser diffraction. SPIE, Vol. 170, pp. 53-62.

LIST OF SYMBOLS

d   —   inside diameter of jet tube

L   —   length of jet stream

Ra —   Rayleigh Parameter $= V(\rho\,d/\sigma)^{1/2}$

V   —   jet velocity (V_c — critical jet velocity)

ρ —   Fluid density

σ —   surface tension

μ —   viscosity

JET LENGTH vs. VELOCITY

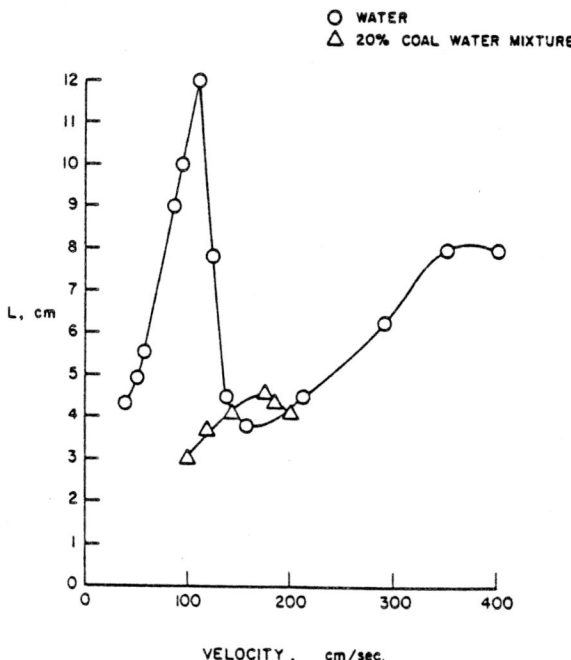

○ WATER
△ 20% COAL WATER MIXTURE

VELOCITY,   cm/sec.

FIGURE 1

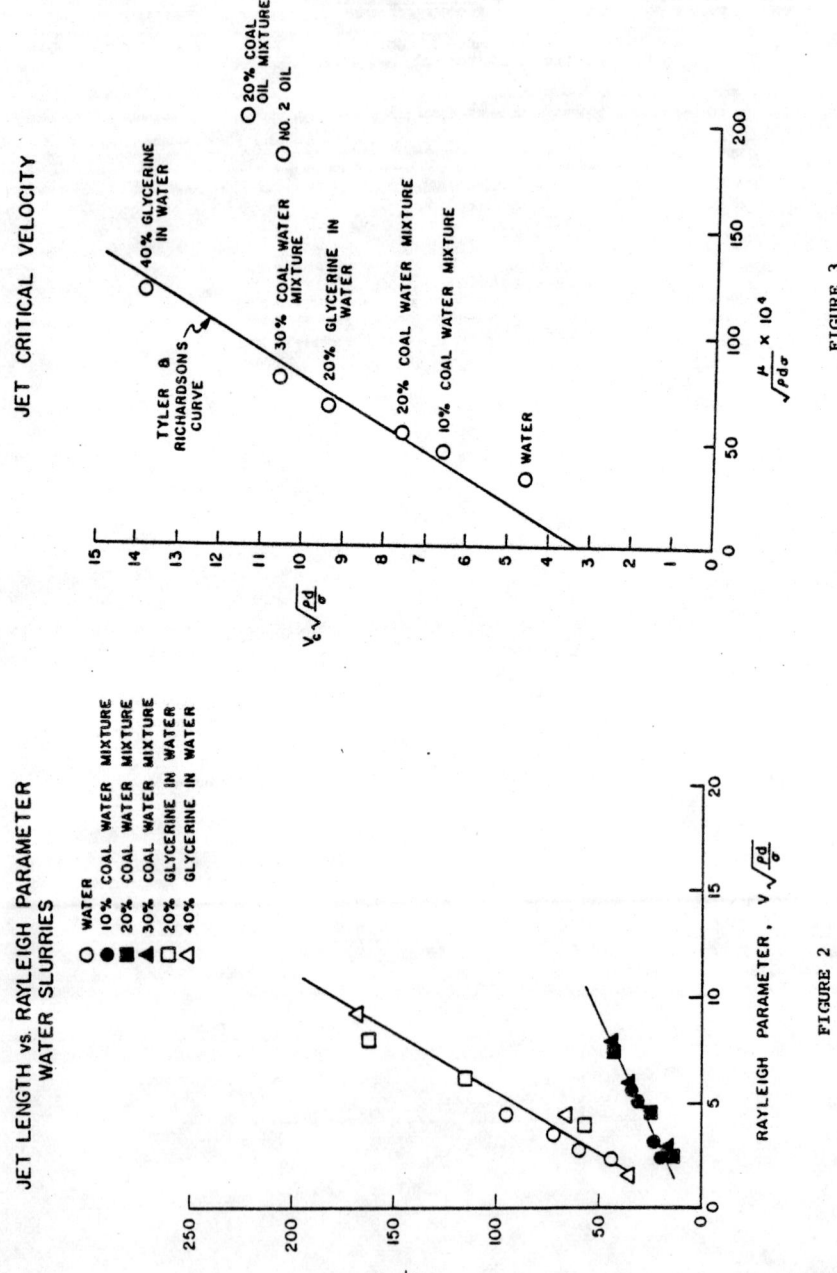

JET CRITICAL VELOCITY

FIGURE 3

JET LENGTH vs. RAYLEIGH PARAMETER
WATER SLURRIES

FIGURE 2

FIGURE 4

ATOMIZER TEST RIG

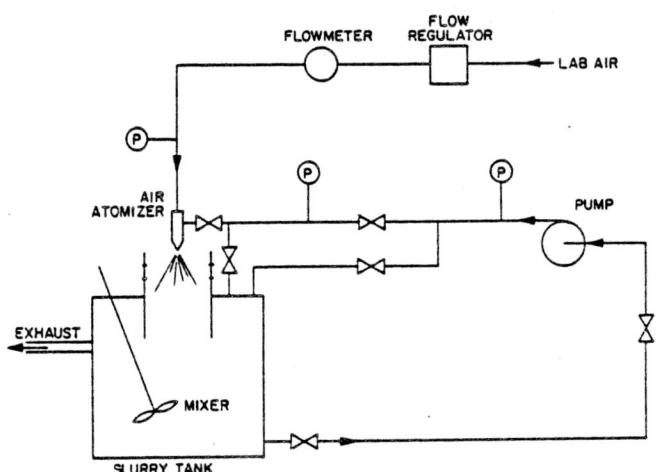

FIGURE 5

EFFECT of COAL CONCENTRATION

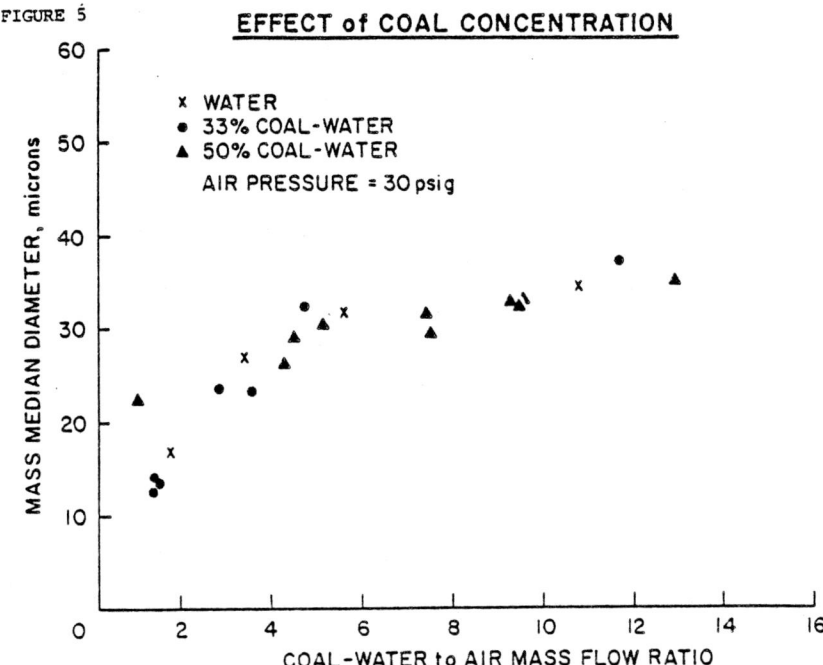

x WATER
● 33% COAL-WATER
▲ 50% COAL-WATER
AIR PRESSURE = 30 psig

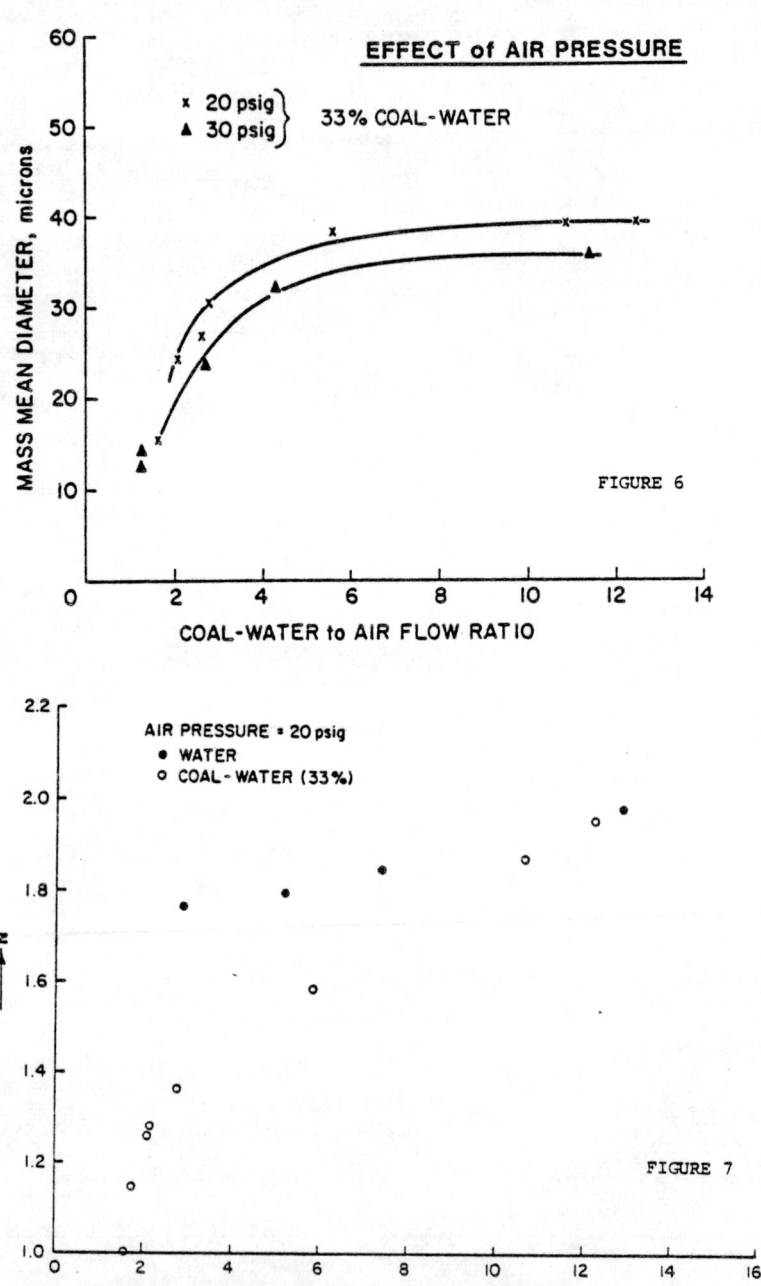

EFFECT of AIR PRESSURE

× 20 psig
▲ 30 psig } 33% COAL-WATER

FIGURE 6

MASS MEAN DIAMETER, microns

COAL-WATER to AIR FLOW RATIO

AIR PRESSURE = 20 psig
● WATER
○ COAL-WATER (33%)

FIGURE 7

N

FUEL TO AIR MASS FLOW RATIO

SMALL AND LARGE THERMAL TEST RIGS FOR COAL SLURRY
BURNER DEVELOPMENT

J.W. ALLEN,   P.R. BEAL,   P.F. HUFTON.

Multi fuel combustion test rigs rated at 88 MW (300 x 10^6
BTU/hr) and 1.5 MW (5 x 10^6 BTU/hr) are described. Some
work on the large rig with coal oil mixtures is discussed
together with the application of this fuel to a 400 MW
steam generating unit.

Preliminary experience in the handling and specification
of coal water mixtures is outlined and early results on
small scale combustion trials are presented. The work
revolves around the design of a burner system to achieve
stable combustion with the coal water mixtures. This
work is continuing and its translation to the large scale
test rig and commercial burner development is discussed.

INTRODUCTION

Thermal test rigs have been used by NEI International Combustion over the past
25 years for the evaluation and development of burner systems for both
industrial and utility boilers. Initially, the rigs were little more than
open brick containers which have evolved to the water-cooled, gas-tight
chamber currently in use to enable the development of low excess air
combustion systems as demanded by market forces and the need for more
effective fuel utilisation. A major step forward in thermal test rig facilities occurred in 1973 with the construction of what is probably the largest
test rig in the Western Hemisphere. The rig design had to meet all of the
known oil burner performance requirements at that time, and also be sufficiently flexible to meet the predicted requirements of the succeeding 10 - 12
years. This rig is still in operation today and is currently undergoing the
next major step change in conversion to enable pulverised fuel, coal slurries
and gas burner systems, sized up to 300 million BTU/hr, to be developed.

A complementary small scale thermal test rig facility, rated at around
5 million BTU/hr was also provided in 1975 to allow fundamental combustion
studies and the development of new ideas in burner design and operation to
take place at a more economical level in comparison with the operation of the
large test rig.

Both these major step changes in the large scale thermal test rig capabilities
have been dictated by market forces. Initially, the need was to develop low
excess oil burners utilising the almost continuously deteriorating quality
fuel oils supplied to the utility boiler industry. The current need is to

NEI INTERNATIONAL COMBUSTION,   DERBY,   ENGLAND.

CLM-N

meet the renewed interest in coal utilisation now that it is realised that fuel oil supplies are not infinite and consequently will be subject to continual increases in price with little or no guarantee as to quality and availability.

Although it is not universal practice to test burner systems for large boilers prior to site installation, the availability of a full-scale thermal test rig enables the development of a tailor-made burner system to suit a particular installation. Customers can see a proposed burner system in operation and any changes required because of alterations in operating procedures or variations in fuel properties can be accommodated. These investigations into altered conditions can be made quickly and economically compared with on-site investigations and without interruption to the customer's operating schedule. Markets for new fuels, such as coal slurries, can be pursued without relying on potential customers to provide full-scale test facilities. In fact, until the firing of coal water and coal oil slurries becomes universally accepted, there should be an increasing demand for off-site demonstrations of the capabilities of these new burner systems. The operation of a full-scale thermal test rig is, therefore, an essential piece of equipment for any burner manufacturer to achieve and maintain a leading position in the supply of combustion systems to the International utility boiler market.

THE LARGE (300 MILLION BTU/HR) THERMAL TEST RIG AT DERBY

General Description of the Rig and its Capabilities

The original design of the large thermal test rig at Derby was ambitious, as it had to meet the requirements of NEI International Combustion to maintain its leadership in oil burner performance in the Western Hemisphere. A major requirement was that of sheer size. The combustion chamber dimensions had to be such that complete combustion could be obtained, upstream of gas sampling positions, with burner systems (firing up to 9 tonnes/hr of fuel oil) designed to produce long narrow flames typical of burners designed for tangential firing applications and those designed to produce more compact large diameter flames typical of front wall firing applications. (Fig.1 indicates the general layout of the test rig and the ancillary supply and analysis equipment.)

These considerations resulted in a combustion chamber of internal dimensions 21.34 m (70 ft) long by 5.49 m (18 ft) square cross section. Cooling of the combustion chamber is achieved by means of static water sandwiched between inner and outer steel skins on the side walls and end wall of the chamber remote from the burner. The centre section of the burner wall is built entirely of refractory to enable easy changing of burner configurations. The roof is also entirely water-cooled and the hearth covered with a layer of refractory pebbles on a bed of sand.

Oil storage is provided by two lagged and steam heated storage tanks, each of 45460 litres (10,000 gallons) capacity. Oil is transferred from the tanks, via a low pressure transfer pump, to a primary pumping and heating circuit capable of delivering oil at a pressure of 44.8 bar (650 lb/in^2) at up to 13640 kg/hr (30,000 lbs) and a viscosity of 70 Redwood No.1 seconds (16 cs, 80 SSU). A second pumping and heating circuit was installed, at a later date, enabling an oil delivery pressure up to 83 bar (1200 lb/in^2) and oil temperatures up to 200°C (392°F) to be achieved. This secondary pumping heating system was used particularly in a study of the combustion of fuel oils containing a high percentage (up to 14%) of hard asphaltenes.

Combustion air is supplied to the burner windbox by a four-stage axial flow fan capable of supplying 37.77 m^3/sec (80,000 ft^3/m) of combustion air at a maximum pressure of 44.8 mbar (18 ins water gauge). Noise levels from the fan are controlled by axial flow silencers immediately upstream and downstream of the fans and air flow rate is indicated by a venturi meter. Coarse air flow control is via the number of fan stages brought into operation and fine control is achieved by a remotely controlled butterfly damper.

All the flows to the rig are controlled from an operating console situated at the end of the combustion chamber remote from the burner. From this position the rig operator can observe an end view of the flame through a glass porthole. A comprehensive gas analysis system is also incorporated with recording local to the control console.

During the operation of the burner, gases are sampled from the 2.74 m (9 ft) diameter 18.29 m (60 ft) high refractory lined stack, at a point some 10.7 m (35 ft) above ground level located in the stack.

At this same point a platform has been erected to enable the isokinetic sampling of the flue gases for determination of the solids burden. The gas analysis instrumentation provides a continuous record of O_2, CO, CO_2 and NOx in the flue gases throughout a test. A smoke density meter and facility for determining the Bacharach smoke number is also available. Observation and sampling ports are provided along the side wall of the combustion chamber to enable photographs of the flames and internal flame samples to be taken as required.

Steam for oil heating and atomisation is available from a package boiler of 4994 kg (10,980 lb) per hour steam capacity at a delivery pressure of 17.24 bar (250 lb/in^2).

Most oil burner test work is carried out using cold combustion air. However, combustion air preheat can be achieved by means of duct burners located in the combustion air supply ducting after the axial flow fans. With this system in operation, oxygen has to be supplied to the combustion air stream to maintain a 21% O_2 content in the combustion air at the burner windbox.

In order to provide a complete burner test facility, comprehensive laboratory facilities and expertise are available to provide chemical and physical analysis of fuels and particulates and also isothermal test facilities to assess the quality of atomisation of the various atomiser designs under test.

Conversion of the Rig to provide a Coal and Coal Slurry Firing Capability

The object of the conversion exercise is to maintain the existing oil firing capability and provide the facility for firing pulverised fuel, coal slurries (e.g. coal water or coal oil mixtures) and gas at similar maximum heat input rates of around 300 million BTU/hr. Both light and dense phase systems for the firing of pulverised fuel are incorporated and the coal slurry firing facilities are designed to handle coal water mixtures containing up to 75% coal and coal oil mixtures with up to 50% coal. A typical British East Midlands steam raising coal of 15% ash content (dry basis) and 7220 Kcals/hr (13,000 BTU/lb) calorific value, was used as the basis for the design calculations. (Fig.2 shows the general layout of the test rig after the conversion to provide this multi-fuel burner development facility.)

The problems to be overcome in the conversion exercise were the deposition and collection of ash, both within and outside the test rig. Preheated primary and secondary air was required to assist coal and coal slurry combustion.

Handling systems for dense phase and lean phase pulverised fuel and for coal slurries, were required.

Local Authority environmental requirements restrict solids discharge from the rig stack to 72 kg/hr (158 lb/hr). Therefore, in order to cope with the maximum coal firing rate of 10 tonnes/hr with a 15% ash coal, the waste gases leaving the rig, at about 1000°C (1830°F), must be conditioned and cleaned before being dispersed to the atmosphere. The system decided upon for the gas conditioning and cleaning comprises a high pressure hot water waste heat boiler followed by a multi-cyclone dust collector. An induced draught fan after the dust collector conveys the cooled clean gases into the stack and thus to the atmosphere. The incorporation of the waste heat recovery system allows the provision of pre-heat to the combustion air and primary air (conveying lean phase pulverised fuel) via high pressure hot water heat exchangers located in the appropriate air ducts. Surplus heat from the waste heat boiler is dissipated via a series of forced draught dump coolers located remotely from the test rig and operating in the closed circuit mode in line with the waste heat boiler. Water is supplied from the waste heat boiler to the heat exchangers at a temperature of 218°C (425°F) and pressure of 30 bar (440 lb/in2) providing 121°C (250°F) pre-heat for the primary air lean phase pulverised fuel conveying and 177°C (350°F) pre-heat for the secondary combustion air. The system of duct burners and oxygen injection is retained in order to provide increased secondary (or combustion) air pre-heat as required.

In addition to collection of ash outside the combustion chamber, it is anticipated that up to 50% of the total ash content of the coal could be collected inside the chamber. To facilitate the removal of this ash and any spillages of unburnt pulverised fuel, the existing refractory pebble floor has been replaced by a water-cooled floor of similar design to the side walls as mentioned in the general description of the rig. As the existing mode of-operation, for the development of low excess air oil burners, is to be maintained a water-cooled door has been provided in the rig back wall enabling combustion gases to be diverted either directly, or via the waste gas conditioning system, to the existing stack.

Coal conveying is to be based on a dense phase system either feeding directly to a purpose-designed burner system or into a pre-heated primary air system giving the required air dilution to provide a lean phase pulverised fuel firing facility.

The dense phase system comprises a pressurised double-blow tank unit fed from a 20 tonne pulverised coal storage silo. Nitrogen purge facilities and continuous temperature monitoring of the storage silos are provided to ensure safety during the storage and handling of fine coal products. The dense phase system will allow pulverised fuel to be conveyed in relatively small diameter pipelines, compared to the more conventional lean phase firing systems, which could be an important consideration when boiler conversions from oil to coal firing are contemplated and existing access to the boiler fronts is limited. Fuel flow rates from the dense phase system will be monitored by load cells located in the blow tank system.

The slurry feed system comprises a storage tank which can be stirred and heated continuously and a mono-pump with facilities for flushing out the complete system after any particular firing exercise. Slurry flow will be monitored by a "Doppler Effect" flow meter, with the slurry continuously circulated around the pump and storage tank and taken off as required to the burner system.

A facility for continuous data logging of fuel flows, gas flows, gas analysis

and temperature is provided and this microprocessor controlled unit will also
provide a central control over all the operating parameters, giving a continuous visual display of the levels of the parameters throughout a test run.

Attention has also had to be given to the sound levels from equipment to be
provided in the conversion work. For the protection of operating personnel,
the sound pressure level from any individual item of plant will not exceed
90 dBA at 1 metre. The maximum sound power level from the complete plant is
limited to 117 dBA, ref. 10^{-12} watts, with a boundary condition of 65 dBA at
200 metres.

THE SMALL (5 MILLION BTU/HR) THERMAL TEST RIG AT DERBY

Fundamental combustion studies and particular aspects of burner systems
development work can be carried out quickly and economically on this small rig.
The rig is linked to the same gas analysis system as the large rig and is
equipped for oil, gas and coal slurry firing. Additionally, a combustion air
pre-heater is available capable of delivering combustion air at up to $425^{\circ}C$
$(800^{\circ}F)$. The general layout of the small rig and its ancillary equipment is
shown in Fig.3.

The combustion chamber comprises a water-cooled mineral wool lined steel
cylinder 4 m (13 ft) long by 1.1 m (3.75 ft) diameter, with a 5.2 m (17 ft)
refractory lined stack equipped with gas sampling and temperature measurement
points. Observation ports are provided on the combustion chamber axis, close
to the burner and in the chamber rear wall. Compressed air and steam are
available for fuel atomisation purposes as are pumping and circulating trains
for oil fuels and coal slurries. At a 5 million BTU/hr rating this rig is
capable of burning 125 kg/hr (275 lbs) of gas oil and 205-230 kg/hr (450-500
lbs) of coal water slurry or their equivalent.

WORK ON COAL SLURRY UTILISATION

General [1]

Over the past ten years there has been a considerable increase in interest in
the possibility of replacing fuel oil, used for power generation, with coal.
Much of this interest has concentrated on the use of coal slurries, rather
than pulverised coal, in an attempt to minimise the capital conversion costs,
as it may be possible to use existing fuel oil storage tanks, pipelines and
pumps for handling coal slurries.

Coal oil mixtures containing up to 50% finely ground coal have been produced
and fired successfully. However, even a 50/50 mix is only equivalent to
38-39% oil replacement on a calorific value basis. Recent work has concentrated on coal water mixtures which contain about 70% finely ground coal and
30% water to produce a viscous but pumpable slurry and offer the possibility
of 100% fuel oil replacement.

Fig.4 shows the viscosities of heavy fuel oils, coal oil mixtures and coal
water mixtures. At present, the viscosity of the latter cannot be reduced by
heating, because of the possible instability of the additives used in slurry
production. However, as work progresses on coal water mixture utilisation,
this probably may be overcome. This means that care must be taken with burner
design if steam is to be considered as an atomising medium for coal water
mixture burners. [2,3]

These points are developed further in the discussion on firing the coal slurries.

Experience with coal water slurries on the small scale (i.e. firing 200-250 kg/hr) has enabled the compilation of a general specification for coal-water mixture properties and associated handling equipment.

The coal water mixture should have a maximum particle size of 200 μm and the coal volatile matter should be a minimum of 25% (dry basis). There should be little or no settling during transportation and any settling which does occur should be easily overcome by a simple pumping and recirculation system. Burner designs should be based on external mix atomisation avoiding sudden changes in direction and diameter in the coal-water mixture flow system which tend to deposit coal from the slurry and are points of excessive wear in atomiser components. Air atomisation is preferred to steam because of the possibility of breakdown of the coal water slurries above 60°C. Coal-water slurries should be fed to the burners via a continuous recirculation system and in the case of intermittent rig work facilities should be provided for water flushing of all lines and valves after each run. As coal-water mixtures can freeze at 0°C, some provision for mild heating, subject to temperature limitations mentioned above, should be provided where low temperature conditions can occur.

It is possible that some of these restrictions can be widened when operating on the large scale, particularly the atomiser design in relation to viscosity. The use of steam as an atomising agent, in a twin barrel burner gun system, may be possible on the large scale.

Using coal slurries to replace fuel oil will require attention to be paid to ash deposition and corrosion/erosion effects throughout the combustion and off gas collection system. A major factor which may influence the use of coal-water mixtures is the possibility of them being derived from a coal beneficiation process which usually treats coal in a finely divided form. The removal of ash and particularly sulphur from the parent coal will minimise the problems of furnace fouling and waste gas cleaning mentioned above. In addition to coal oil and coal water mixtures, coal methanol or coal/methanol/water mixtures have also been studied. These mixtures may offer advantages over coal water in improved initial flame stabilisation, combustion characteristics and thermal efficiency. [4,5]

Coal Oil Mixture (COM) Firing

A development exercise was carried out on the large test rig in 1980/81 aimed at firing a 50/50 COM prior to a demonstration project on a utility boiler in the USA. The COM for this exercise was prepared on site, immediately prior to use. Table 1 gives some results on the range of coal oil mixtures investigated. Mixtures of a 4% ash coal (ground to 80% through 200 mesh) and 3500 secs. heavy fuel oil were utilised, fired at a maximum rate of 6136 kg/hr (13,500 lb/hr).

The development exercise established that COM could be fired successfully, but excess air levels required to maintain efficient combustion would be similar to those required for pulverised fuel firing rather than oil firing, i.e. in the order of 20-25% excess air.

Burners of similar design were used on a COM demonstration project on a 400 MW steam generator at a Florida Power and Light Company station in the USA. [6] Because of the use of increased excess air levels, with a 50/50 COM, the maximum load was limited to 355 MW because of limitations on the ID fan margins (the flue gas weight from 50/50 COM firing is approximately 30%

greater than that from 100% oil firing). Fig.5 shows the boiler efficiencies obtained at various load factors and COM concentrations. The decrease in unit efficiency for all COM tests was due to the higher excess air levels employed and the increased combustible losses. The boiler efficiency varied from 91% at 100 MW to 89% at 400 MW with 100% oil firing whilst COM boiler efficiency was 0.7 to 1.5 percent lower over the same load range.

Other factors which had to be taken into consideration were the provision of furnace wall and retractable soot blowers, the erosion of convection heat transfer surfaces and the removal of particulates at the boiler to avoid ID fan erosion and meet environmental requirements.

Generally speaking, this project demonstrated the technical feasibility of burning COM in steam generators designed for oil firing.

Coal Water Mixture (CWM) Firing

At the time of writing, work has been carried out solely on the small thermal test rig, concentrating on the problems associated with the stability of CWM flames.

Possible difficulties in handling the highly viscous CWM have been discussed earlier in the paper. The problem is greatly increased because of the small scale of operations and the associated physical size of slots and/or orifices in the atomisers employed. In order to minimise blockage problems, on the small scale, air is used as an atomising agent and the burner unit employs a straight through CWM flow, avoiding narrow orifices and sudden changes in cross section (Fig.6).

With this atomiser, viscosity reduction prior to final atomising via "swirl air" is obtained by the use of "shear air" to aerate the CWM and produce an apparent reduction in viscosity. Fig.7, calculated using a technique developed for the viscosity blending of the fuel oils, shows this effect. [7]

Viscosities of CWM are usually reported, depending upon the type of viscometer used, at shear rates ranging from 30 to 100 reciprocal seconds, viscosity reducing as the shear rate is increased. It has been suggested that shear rates through a large scale atomiser may be in the order of thousands of reciprocal seconds,[8] thus producing the effective viscosity reduction required for satisfactory atomisation. In fact, reported work on CWM combustion has not mentioned problems arising as a consequence of the high viscosity CWM. Some isothermal work, using glycerol as the viscous liquid substitute for CWM, has indicated that effective atomisation is possible with the type of atomisers used in the small scale work. Mean droplet sizes in the range 40-60 μm were reported from atomisers with various "shear hole" to "swirl hole" diameter ratios.

CWM usually comprises 70% coal, 30% water plus small fractions of "stabilising" and "viscosity control" additives. The major problem in combustion is the stabilisation of the flame close to the burner nozzle. In order to achieve this, rapid evaporation of the water and rapid release and ignition of coal volatiles is required. High volatile coals (minimum 25% VM on dry basis) help to achieve this initial stability. However, the burner system configuration and aerodynamics controlling the transfer of heat to the CWM spray root are of major importance. Two methods of increasing the initial spray heating rate are available, the use of refractory quarls and both the internal and external recirculation of flame gases. Pre-heated combustion air also assists flame development, but is of secondary importance to the radiative and convective heat transfer processes mentioned previously.

Burner configurations used in small scale work are shown in Fig.8. The quarl shapes were installed on the front wall of the water cooled small scale rig described earlier.

With an axial swirler, of full air register diameter, using combustion air pre-heated to 250-300°C, a stable flame was obtained from system C, both systems A and B requiring gas support to maintain flame stability with a flame front close to the burner nozzle.

Early work on system A assessed the effect of atomiser design on combustion and results are reported in Table 2. Cold combustion air was used at this stage and the effects of "apparent viscosity" reduction on improved combustion were evidenced by an increase in stack temperature with reduction in viscosity for a given fuel input rate. With this system, using pre-heated combustion air, stable flame fronts were established some 4 - 6 ft from the burner nozzle. The use of hot refractory, pre-heated to at least 800°C, in close proximity to the CWM spray, enabled stable unsupported flames to be achieved.

At the time of writing, experiments are under way to minimise the use of the hot refractory effect which, although possible in industrial boiler/furnace operations, should be kept to a minimum in larger water tube boiler plant used for steam raising.

On the larger scale, radiation from the bulk of the flame to the spray root will be greatly enhanced. Successful firing at a rate of 80 x 10[6] BTU has been reported using a small refractory cone to give initial flame support. [9]

A 4 tonnes/hr trial is planned for the large test rig using a burner design based on that proved successful for COM firing and a swirler/quarl system arising from the current small scale work.

For easy application to utility boilers, particularly in the retrofit situation, the use of refractory as an aid to flame stabilisation will have to be minimised. Successful ignition and flame stability in a cold combustion chamber will also have to be demonstrated.

CONCLUSIONS

A demonstration of coal oil mixture firing on the large thermal test rig was transferred successfully to full scale commercial operation.

Current small scale test work, on coal water mixture handling and combustion, will be developed along similar lines.

The availability of a full scale combustion test facility enables custom built commercial burner systems to be developed quickly. Implications of changes in operating requirements can be studied and incorporated relatively inexpensively before burner systems are installed on site.

REFERENCES

1. H.M. Lee — An Overview of Proposed Slurry Technologies and their Cost Savings. Seventh International Technical Conference on Slurry Transportation - March 1982.

2. P.F. Hufton, R.S. North — Coal Oil and Coal Water Mixture Firing. Paper presented to CEA - October 1982.

3. J.W. Allen, P.R. Beal, P.F. Hufton — Large (300 million BTU/hr) and Small (5 million BTU/hr) Thermal Test Rigs for Coal and Coal Slurry Burner Development. 185th American Chemical Society Meeting, Seattle - March 1983.

4. R.P. Rusakan — Coal Oil/Coal Water Slurries .... Now Its Coal Methanol Power - October 1982.

5. Y.W. Pan, G.T. Bellas, R.B. Snedden, B.E. Wieczenski and J.T. Joubert — Exploratory Coal Water and Coal Methanol Tests in Oil Designed Boilers. Pittsburgh Energy Technology Centre 1982.

6. A.D. Schmidt, J.L. Friedmet — Full Scale Tests Firing Coal Oil Mixtures in a 400 MW Steam Generator. Third International Symposium on Coal Oil Mixture Combustion 1982.

7. H.M. Spiers (Ed.) — Technical Data on Fuel 6th Edition 1961.

8. G.A. Farthing, R.P. Daley, S.J. Vecci, T.M. Sommer — Chemical and Physical Properties of Highly Treated Coal Water Mixtures. 185th American Chemical Society Meeting, Seattle - March 1983.

9. R.W. Bonio, D.A. Smith, M.J. Rini, R.C. La Flesh — Coal Water Slurry Burner Technology Development. 185th American Chemical Society Meeting, Seattle - March 1983.

Test No.	Coal %	O_2 %	Stack Solids lb/10^6 BTU	Carbon in Ash %	Atomising Steam lb/hr @ PSI
1	30	1.5	-	-	1074 @ 90
2	30	1.3	-	-	1074 @ 90
3	36	1.3	1.56	76	1074 @ 90
4	40	3.5	0.69	58	1074 @ 80
5	44	3.5	0.72	51	1312 @ 115
6	0	1.0	0.22	-	1074 @ 90
7	0	0.5	0.31	-	1074 @ 90

(- = not measured)

Total fuel firing rate approximately 13,500 lb/hr = 222 x 10^6 BTU/hr

Tests 4 and 5   (Flame length 24 - 26 ft
                (Flame width 14 - 15 ft
                (NOx   245 - 250 ppm

Tests 6 and 7   Oil firing only

TABLE 1

Results from Coal Oil Mixture Combustion Trials - Large Test Rig

TABLE 2 - BURNER SYSTEM A (FIG. 12)

2.1 Nozzle C - Shear Hole/Swirl Hole Area Ratio 0.50

Coal Water Mixture (CWM) feed rate			
kgs/hr (lbs/hr)	110 (242)	148 (325)	192 (400)
CWM feed pressure bar (psi)	0.6 (8)	0.9 (12.5)	1.2 (16)
Atomising air pressure bar (psi)	4.1 (60)	4.1 (60)	4.1 (60)
Atomising air ratio			
wt air : wt CWM	0.48	0.36	0.29
Heat input to test rig % MCR	58	76	92
Heat input ratio CWM : gas	6.8	9.2	11.3
Excess O_2 % in flue gas	1.6	1.5	1.4
Flue gas temperature °C	779	805	820
Apparent aerated CWM viscosity cS	300	350	400

2.2 Nozzle D - Shear Hole/Swirl Hole Area Ratio 0.98

Coal Water Mixture (CWM) feed rate			
kgs/hr (lbs/hr)	110 (242)	148 (325)	182 (400)
CWM feed pressure bar (psi)	0.6 (8)	0.9 (12.5)	1.2 (16)
Atomising air pressure bar (psi)	4.1 (60)	4.1 (60)	4.1 (60)
Atomising air ratio			
wt air : wt CWM	0.48	0.36	0.29
Heat input to rig % MCR	58	76	92
Heat input ratio CWM : gas	6.8	9.2	11.3
Excess O_2 % in flue gas	1.5/4.4	1.5/4.7	1.5/4.6
Flue gas temperature °C	803/795	806/806	820/817
Apparent viscosity aerated CWM cS	100	150	250

2.3 Nozzle E - Shear Hole/Swirl Hole Aoea Ratio 2.0

Coal Water Mixture (CWM) feed rate			
kgs/hr (lbs/hr)	110 (242)	148 (325)	182 (400)
CWM feed pressure bar (psi)	0.6 (8)	0.9 (12.5)	1.2 (16)
Atomising air pressure bar (psi)	4.1 (60)	4.1 (60)	4.4 (60)
Atomising air ratio			
wt air : wt CWM	0.48	0.36	0.29
Heat input to rig % MCR	58	76	92
Heat input ratio CWM : gas	6.8	9.2	11.3
Excess O_2 % in flue gas	1.5/4.0	1.5/4.0	1.5/4.0
Flue gas temperature °C	792/820	846/862	925/937
Apparent viscosity aerated CWM cS	60	100	150

2.4   Unsupported Flame (System A)

       Flame front half way along furnace length.

325 lb/hr	Coal Water Mixture (CWM)
20 - 25%	Excess air
970°C	Stack temperature
0.35	Atomising air ratio

2.5   Unsupported Flame (System C)

653 lb/hr	Coal Water Mixture
270°C	Combustion air preheat
20 - 25%	Excess air
0.2	Atomising air ratio
96.6%	Combustion efficiency

TABLE 2

Results of Preliminary Small Scale Thermal Tests on
Coal Water Mixture Atomisers

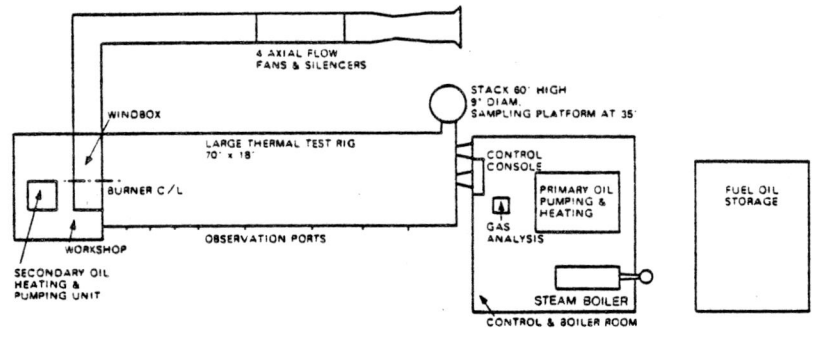

FIGURE 1 GENERAL LAYOUT OF LARGE THERMAL TEST RIG FOR LOW EXCESS AIR OIL BURNER DEVELOPMENT

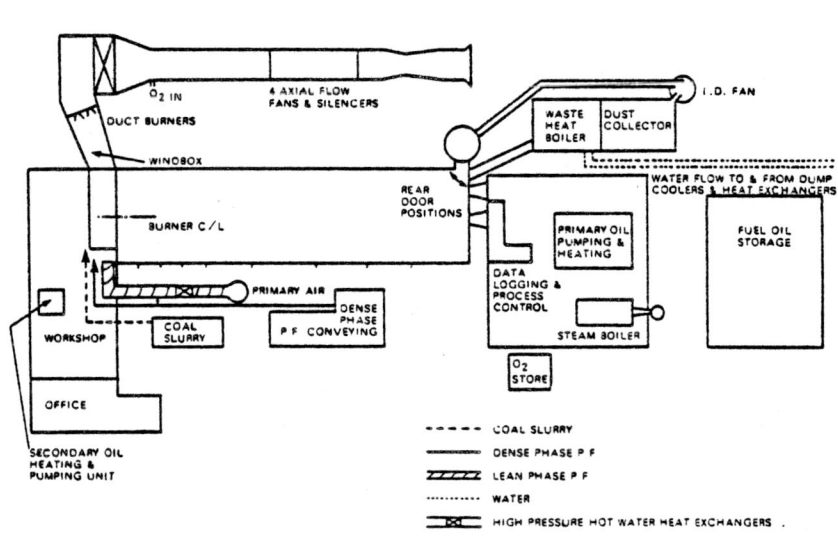

FIGURE 2 GENERAL LAYOUT OF LARGE THERMAL TEST RIG FOR MULTI-FUEL BURNER DEVELOPMENT

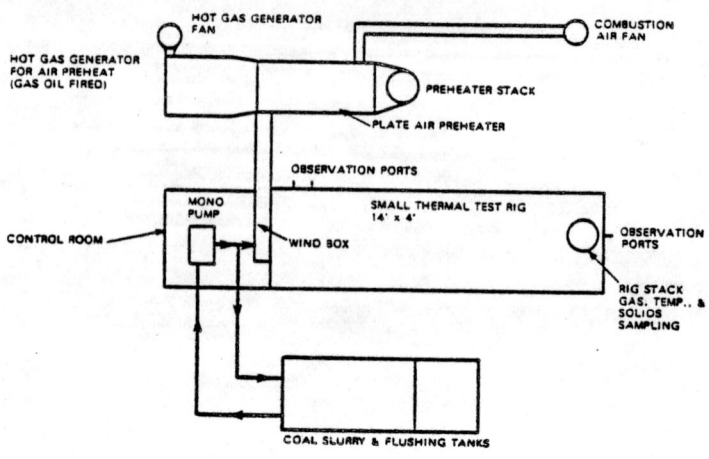

FIGURE 3 GENERAL LAYOUT OF SMALL SCALE THERMAL TEST RIG

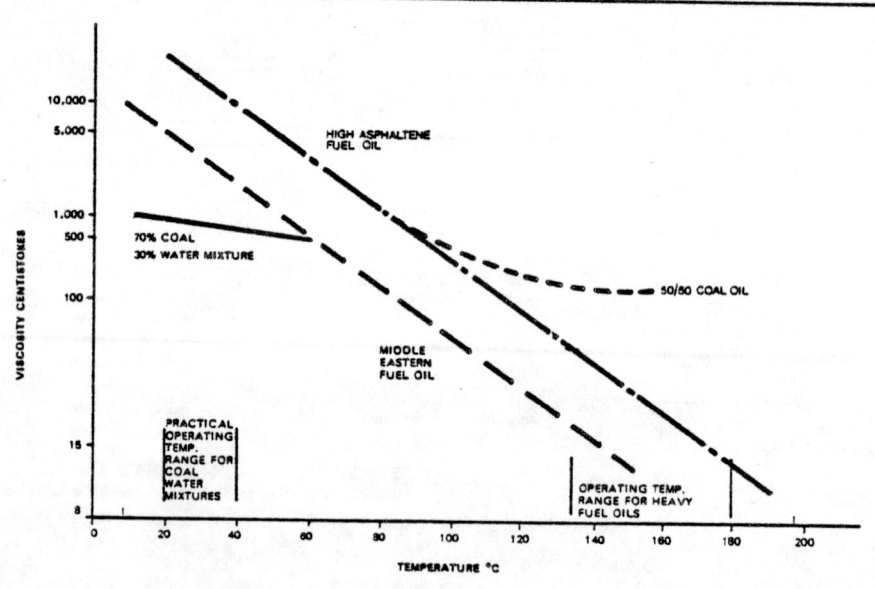

FIGURE 4 COMPARISON OF COAL WATER MIXTURE & HEAVY FUEL OIL VISCOSITIES

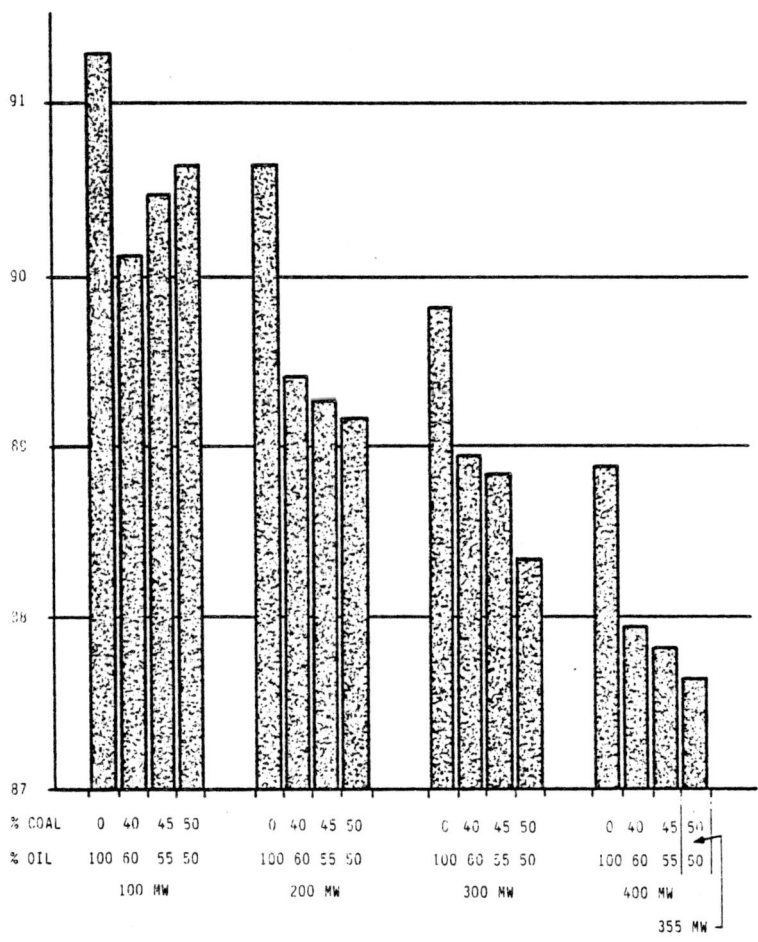

FIGURE 5 BOILER EFFICIENCY

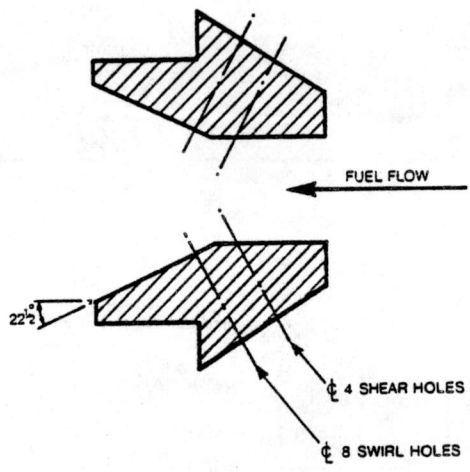

FUEL FLOW

22½°

¢ 4 SHEAR HOLES

¢ 8 SWIRL HOLES

FIGURE 6   DIAGRAM OF COAL WATER MIXTURE BURNER NOZZLE

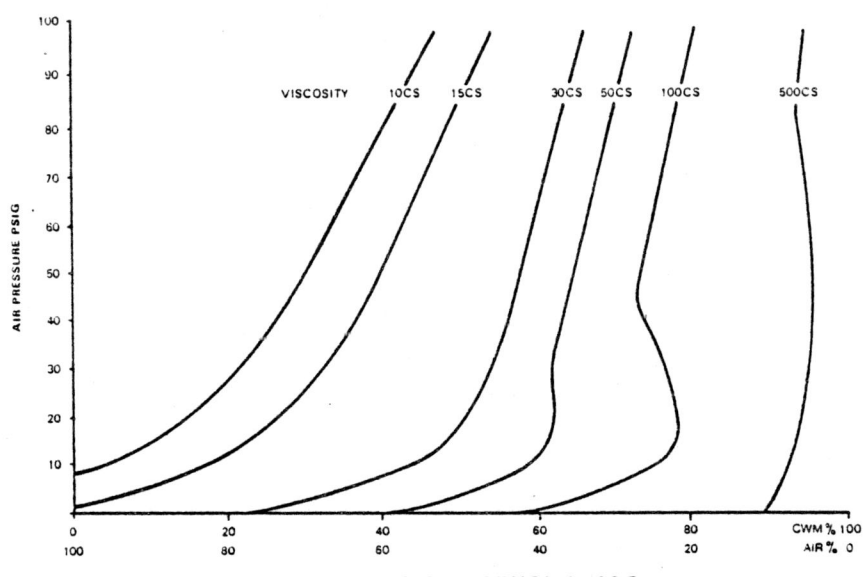

FIGURE 7   VISCOSITY VARIATION, BY AERATION, OF COAL WATER MIXTURES AT 20°C

CLM-0

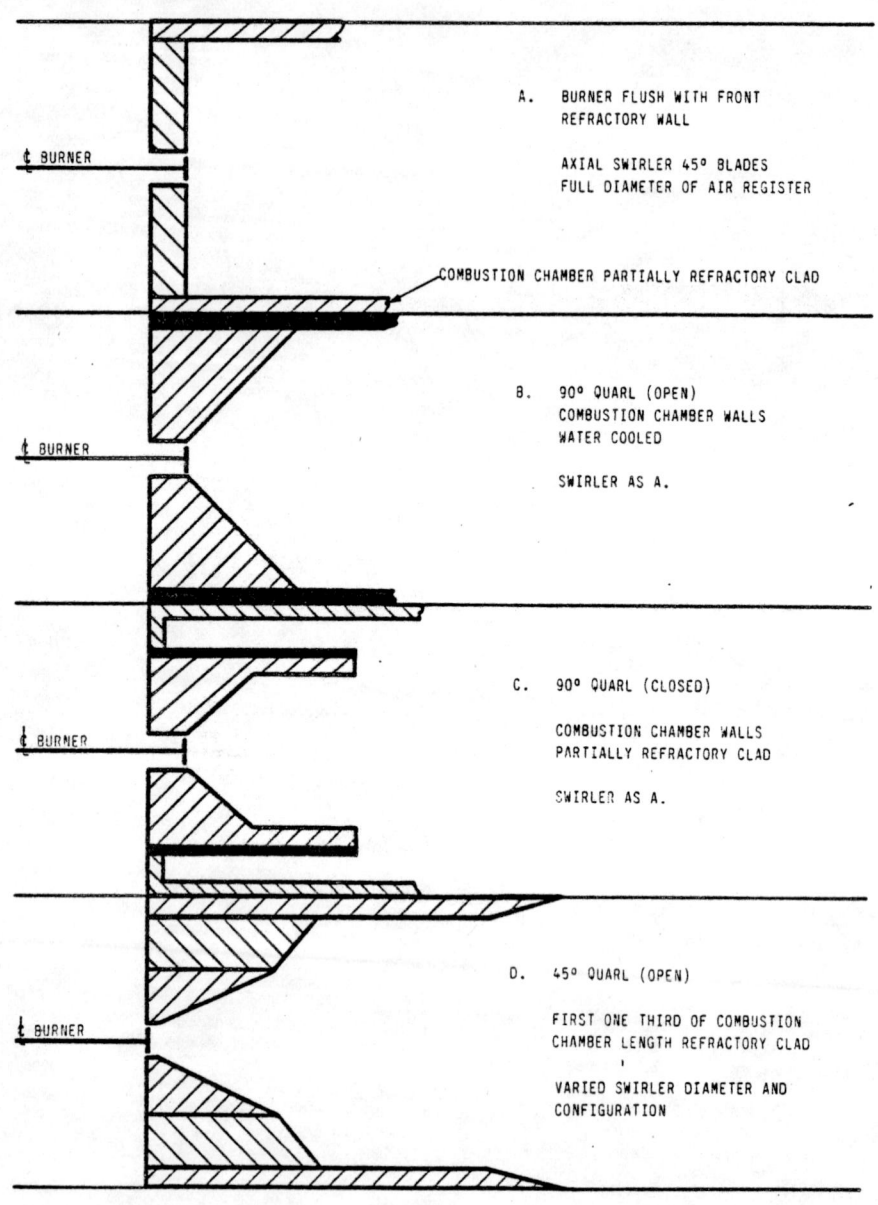

FIGURE 8 BURNER QUARL SYSTEMS SMALL SCALE RIG

BURNING COAL-WATER MIXTURES IN PRESSURISED FLUIDISED BEDS.

A.G. Roberts*, K.K. Pillai*, S.N. Barker*, and J.W. Byam**.

The use of coal-water mixtures (CWM) in combined-cycle
plant incorporating pressurised fluidised-bed combustors
(PFBC's) is discussed. The potential reduction in costs
compared with the current method of preparing and feeding
coal are considered qualitatively, and it is shown that
the reduction in cycle efficiency is less in a combined
cycle than in a simple steam cycle (about 0.8% compared
with 1.5% for a CWM containing 30% water.

Earlier tests were carried out with a CWM produced from
beneficiated finely-ground coal. Recent tests with a
CWM containing 30% water produced from a high-ash, high
sulphur coal with a size 3mm x 0 are described. When
burned in a small PFBC at 16 atm pressure, high combustion
efficiencies and sulphur retentions were obtained. The
work suggests that the use of CWM could still further
increase the attractiveness of combined cycle PFBC plant.

INTRODUCTION

A typical combined cycle incorporating a pressurised fluidised bed combustor
(PFBC) is shown in Fig. 1. The hot gases from the fluidised bed drive the
gas turbine and steam is generated in tubes immersed in the PFBC. The latter
is thus acting as a supercharged boiler. The ratio of steam turbine output
to gas turbine output is mainly influenced by the amount of excess air, which
is usually arranged to be between 20 and 100%. PFBC combined cycles have
been identified as having the potential of producing electric power more
cheaply than by any of the alternative advanced concepts.

Coal handling, processing and feeding costs amount to about 10% of the
estimated total plant costs. In this respect PFBC is similar to most types
of plant which burn or gasify coal.

The method of handling coal usually considered for PFBC's is to dry and crush
the coal to a size typically 3mm x 0 and to feed it pneumatically into the
fluidised bed through a number of nozzles. The steps in the process are
shown schematically in Fig. 2. It is the system used in all existing pilot-
scale PFBC's and considered in all projected commercial designs. It suffers
from some drawbacks, however:

     1)   Drying the coal to zero surface moisture (which is essential
         for reliable operation in pneumatic conveying systems) involves
         high capital cost. If waste heat cannot be fully utilised, then

*   NCB Coal Utilisation Research Laboratory, Leatherhead, England.
**. United States Department of Energy, Morgantown, U.S.A.

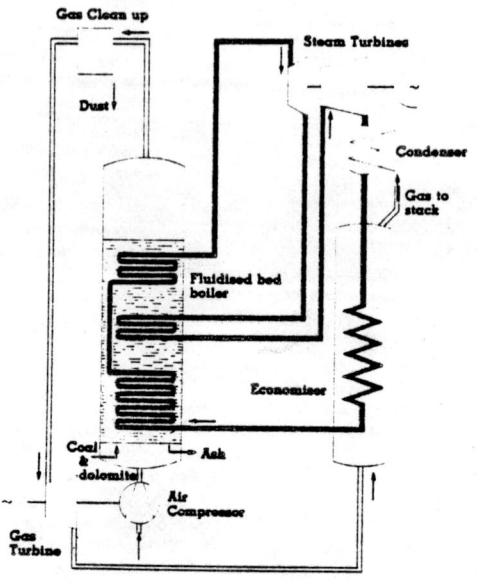

Fig. 1.   Typical PFBC combined cycle

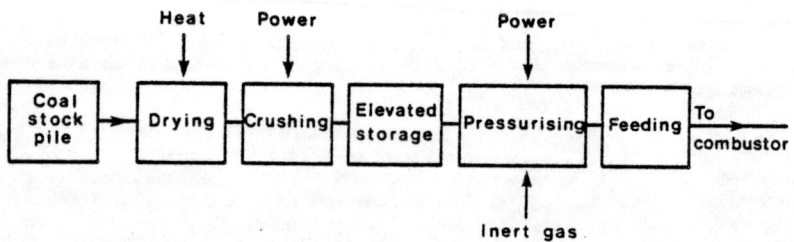

Fig. 2.   Preparation and feeding system for crushed coal

the use of auxiliary fuel is necessary, with a consequent loss of cycle efficiency.

2)   Crushing the coal involves capital cost and some energy consumption (although the latter is relatively small).

3)   Pressurising the coal requires a lock hopper system which involves capital cost for pressure vessels and also energy consumption in supplying an inert gas for pressurising the system.

Clearly, the competitive position of PFBC would be still further enhanced if a more attractive coal feeding system were available.

USE OF COAL-WATER MIXTURES IN PFBC

The use of coal in the form of a pumpable coal-water mixture (CWM) is one possibility for reducing costs.   The main elements of a system for feeding a PFBC with CWM are shown in Fig. 3.   Compared with the conventional system of Fig. 2, there are the following advantages:

1)   Elimination of the drying system (although coal crushing or grinding is still necessary)

2)   Elimination of pressure vessels

3)   Elimination of inert gas production and the losses involved in pressurising lock hoppers.

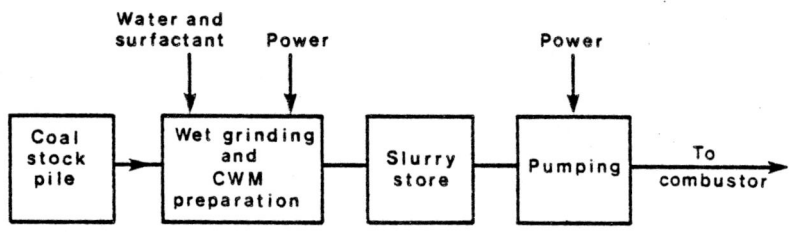

Fig. 3.   Preparation and feeding system for CWM

There are additional advantages when it comes to distributing coal over the area of a fluidised bed.   A power plant of 250 MWe would require a combustor cross-section of about 70 m^2 and a coal consumption of about 100 tonnes/h. Current practice suggests that this would require a minimum of about 30 feed nozzles.   Obtaining uniform flow to each point with pneumatic coal feeding would involve relatively expensive control equipment, whereas the flow of CWM can be controlled more easily.   Furthermore, a CWM system could be expected to have a more rapid response to changes in firing rate and less sensitivity to pressure fluctuations in the combustor.

Effect on cycle efficiency.   Clearly, feeding extra water into the combustor results in some loss of cycle efficiency.

Some results obtained by the NCB's ARACHNE flowsheeting computer package* are given in Fig. 4 for a gas turbine with a 10 : 1 pressure ratio, and for a 10% ash coal.   It will be seen that a CWM of 30% water content results in a loss of about 0.8 percentage points on cycle efficiency.   This is in contrast to a conventional steam cycle, in which a CWM of 30% water content would result in a loss of about 1½ percentage points.

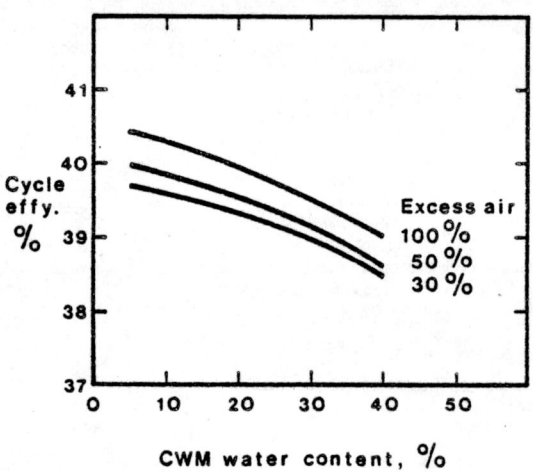

10 bar combined cycle PFBC plant
10% ash in coal
Stack temperature : 280°F

Fig. 4. Effect of water content of CWM and excess air on cycle efficiency

The difference between a combined cycle and a conventional steam cycle is that some of the energy used to convert the water into steam is recovered in a

* Further details of ARACHNE are given in another paper to this conference:-
"The assessment of coal water slurries in coal conversion processes by computer program" by Ford, Holmes, Pillai and Reed.

combined cycle. Since a gas cycle operates at a higher temperature than a
steam cycle, conversion of the recovered energy to shaft power can be
accomplished at a higher efficiency in the gas turbine and combined cycle
than in the conventional steam cycle. The energy associated with the steam
escaping via the stack is irretrievable.

When utilising CWM, the total plant output decreases slightly as the stack loss
increases. The gas turbine power output increases as the mass flow increases,
but not sufficiently to compensate for the reduction in steam turbine power.
CWMs of high water content might cause problems in increasing the swallowing
capacity of the gas turbine, but this is not likely with moisture contents of
30% or less.

Comparisons of power output and generation efficiency for PFBC systems fed
with dry coal or CWM are summarised in Table. 1. The calculations are based
on a gas turbine producing 70 MWe when the system is burning dry coal.

Table 1.   Effect of CWM on Power Output and Cycle Efficiency

		"Dry" coal	CWM
Pressure ratio		10	10
Excess air	%	30	30
Gas turbine output	MWe	70	81
Steam turbine output	MWe	347	326
Total output	MWe	417	407
Generation efficiency	%	39.8	39.0
Total water with coal	%	5	30

The ash content of the coal has a slight effect on the cycle efficiency. This
is because the ratio of water to solid matter must be maintained in order to
keep a stable mixture. The effect of increasing the ash content is to require
a larger ratio of water to combustible coal, i.e. for the same thermal input
a larger quantity of water has to be fed with a higher ash coal. Thus,
increasing the ash content from 10% (as in Fig. 4) to 20% would increase the
effective water content of the CWM by a factor of 1.13.

Availability of CWM . CWMs will probably be produced commercially from fine
pulverised coal having a controlled size distribution, in slurries containing
25 to 35% water. Because the anticipated market is for repowering existing
oil – or gas-fired boilers, the coal is usually "beneficiated" by reducing
the ash and sulphur contents. Such a CWM would probably be delivered by
pipe-line or barge, thus eliminating on-site coal stockyards and raw coal
handling equipment. The cost of producing such a CWM is significant and, in
general, probably outweighs its advantages so far as a PFBC combined-cycle
plant is concerned, although there may be instances where off-site preparation
is attractive.   However, beneficiation is unnecessary for a PFBC and the
fuel can be much coarser than in the CWMs currently being produced.   Coarse
slurries containing 30% water with a top size of 3mm have been produced in the
past[1] on a semi-commercial scale for firing a cyclone-fired boiler.   It may
be possible to produce a slurry containing dolomite or coal and dolomite, thus
simplifying the dolomite feeding system as well.

TEST PROGRAMME

In view of the potential savings in captial and operating costs for coal
preparation and feeding, it was decided to investigate the performance of
CWMs with regard to such factors as combustion efficiency and sulphur
retention. A programme is being carried out as part of the US DOE PFBC
programme at CURL.   Some early work in a 12 inch diameter combustor at a
pressure of 6 atm has already been reported [2]. The CWMs were produced
from fine coal (50% less than 20 micron) of low sulphur content by commercial
manufacturers, and the results indicated excellent combustion efficiency
and sulphur retentions. The programme has recently been continued by
burning a slurry of coarse coal at higher pressures than hitherto, and the
results of these tests are described in this paper.

TEST FACILITY

The facility is shown in Fig. 5 and details of the combustor in Fig. 6 . The
combustor consists of a refractory-lined vessel providing a fluidised bed
with a cross-section of 305 x 305 mm.   The fluidised bed and freeboard
together occupy a height of 4.8m.   Immersed within the bed is a tube bank
comprising a large number of cooled and uncooled tubes.   Air is admitted
through a sparge-type distributor.

Hot combustion gases proceed from the freeboard through a cyclone dust
collector and then along a duct in which are installed the dust sampling
systems and the main gas sampling points.   The gases are then quenched by
water sprays and passed through a pressure let-down valve.   The sparge-type
air distributor consists of four headers containing 16 stand-pipes arranged
on 70mm square pitch.   Each stand-pipe, which is closed at the top, has ten
holes arranged in two rows in the top circumference to give a total flow
area equal to 0.77% of the combustor cross-section.   Coaxial within each
stand-pipe is a small-bore tube conveying propane to the tip of the stand-
pipe.   In this way a pre-mixed air/propane supply is provided at the air
nozzles for combustor start-up.

The tube bank is composed of 25mm dia. elements which can be either water-
cooled tubes or uncooled bars.   The number of elements used in any
particular test depends on the nominated operating conditions.   For example,
the bed height required to immerse the maximum 119 water-cooled elements is
2.7m and the consequent cooling load enables the combustor to be operated
at about 20 atm, 1 m/s and 30% excess air or at lower pressure and higher
velocities.   Conditions requiring a lower rate of heat extraction can be
obtained by using fewer cooled elements.   The tube geometry can be
maintained merely by interchanging cooled for uncooled elements, and the
tube geometry itself can be altered by removing some of the 119 elements.

Bed level is controlled by removing excess bed material continuously in
gravity flow past the sparge tubes of the distributor.   Bed material is
cooled by passing over a steam-cooled heat exchanger and the rate of bed
removal is controlled by the speed of a rotary metering valve which is
located below the combustor.

COAL-WATER MIXTURE

The coal was a Pittsburgh No. 8 coal having the analysis given in Table 2.
It was prepared by sieving from existing stocks of dried coal, mixing with
water in a cement mixer and storing in 550 gallon tanks fitted with stirrers.

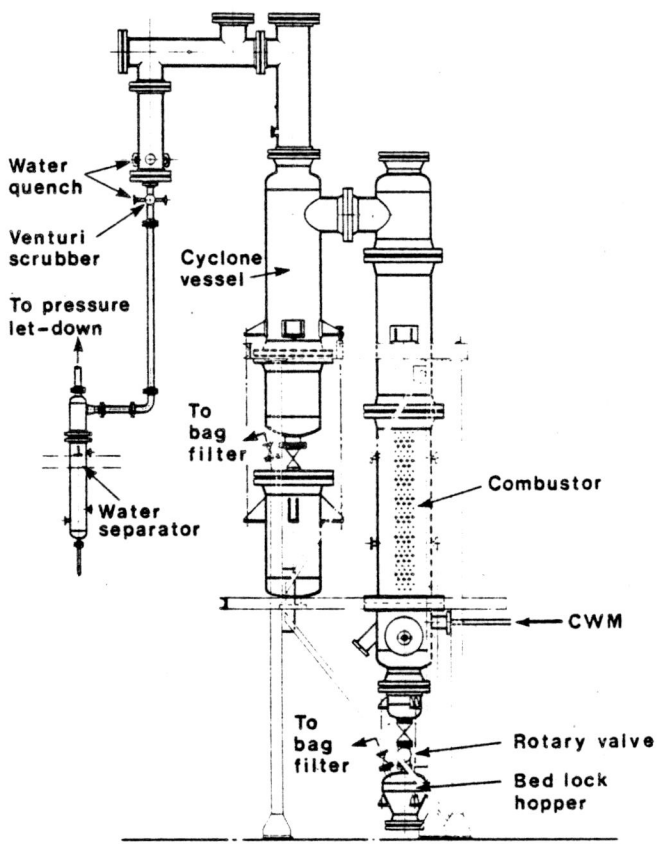

Fig. 5.   Diagram of test facility

No surfactant was used.   From there the slurry was pumped as required into a feeder tank using a diaphragm pump.   The main pressurising pump was a progressive cavity pump manufactured by Moineau, which pumped slurry directly from the feed tank into the injection nozzle.   The latter (Fig. 7) was a water-cooled tube of 48mm dia, extending 50mm into the combustor at a level of 160mm above the distributor.   A small amount of air was injected near the outlet of the nozzle to break up the slurry stream.

A schematic diagram of the pumping system is shown in Fig. 8.   Two nozzles were installed, but only one was used in the tests described here.   Although

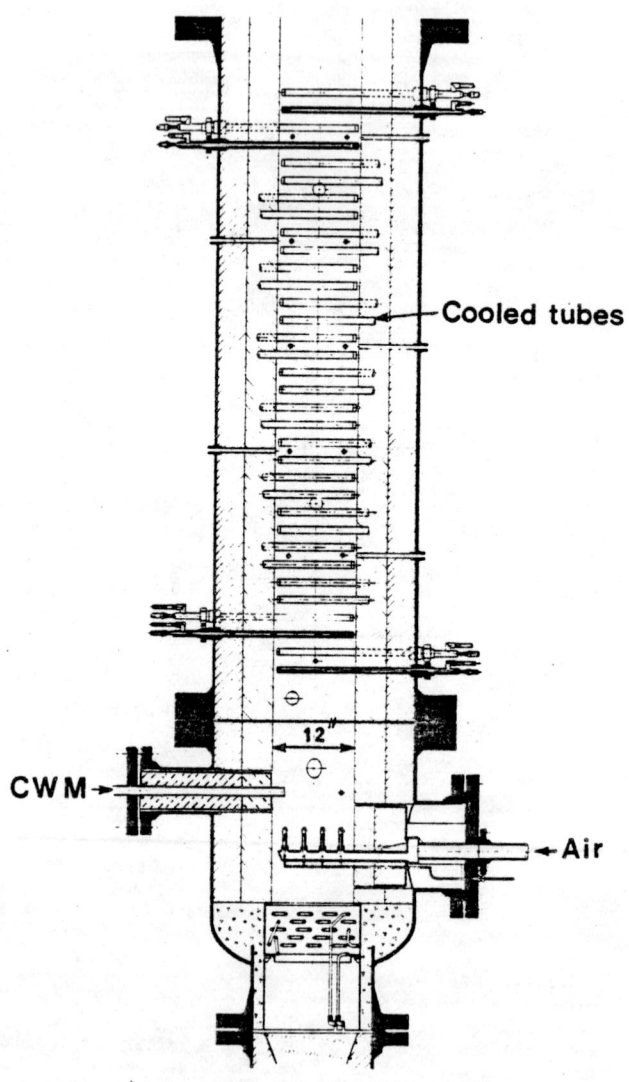

Fig. 6.   Diagram of combustor

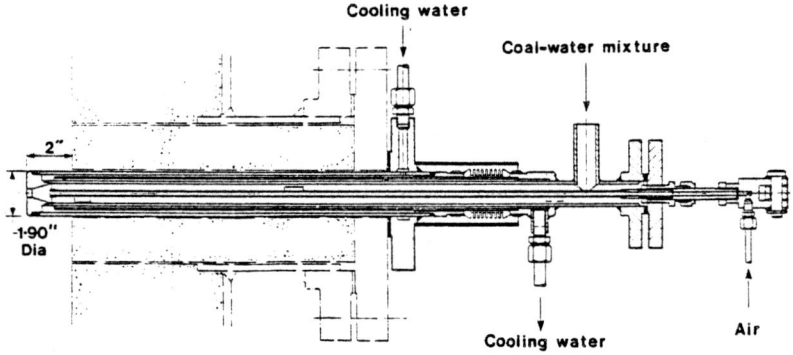

Fig. 7.   CWM injector nozzle

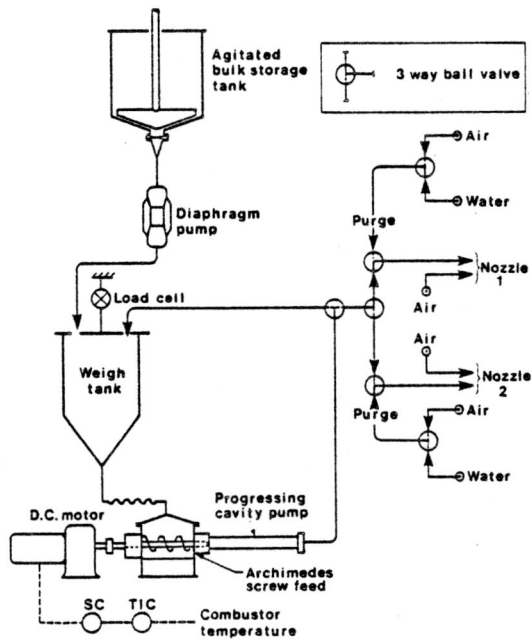

Fig. 8.   Diagram of CWM pumping system

some care was taken in producing the correct size distribution for the coal, it was felt that more liberties could have been taken with the coarse coal than is possible when producing slurries from fine coal.

Table 2. Details of coal-water mixture

Water content 30%

On-dry basis:
Ash content	13%
Volatile content	36%
Sulphur content	4%
Calorific value	29,200 kJ/kg

Size Distribution:
+3mm	-	0.1%
+0.85mm	-	30%
+0.15mm	-	65%

RESULTS

Operating conditions, together with the major results, are summarised in Table 3. The main variables examined were bed temperature, fluidising velocity and the amount of dolomite used. The values of bed temperature, velocity, bed height and excess air were typical of those proposed for commercial plant. The pressure was 16 atm throughout.

Table 3. Operating conditions and results

Coal	:	Pittsburgh No. 8
Sorbent	:	Plum Run (Ohio) dolomite
Pressure	:	16 atm
Bed height:		2.7m

Temperature	°C	750	750	850	850	900	950
Fluidising velocity	m/s	1.2	1.2	1.2	0.9	0.9	1.2
Ca/S mol ratio		2.3	2.9	1.7	2.1	2.2	2.0
CWM feed rate	kg/h	136		200	160	168	197
Excess air	%	90		50	40	30	35
CO in exhaust	vppm	50		20	30	<10	<10
NO$_x$ in exhaust	vppm	230		200	200	200	160
SO$_2$ in exhaust	vppm	225	175	225	110	125	100
Combustion efficiency	%	98.7		99.5	99.6	99.5	99.8
Sulphur capture	%	83	86½	89	95	94½	95½
NO$_x$ emission	g/MJ	0.27		0.15	0.14	0.13	0.11

Combustion efficiencies, sulphur capture and NO_x emissions were identical with those obtained in the same combustor and at the same conditions when the coal was dried, crushed and fed pneumatically. The following comments, therefore, apply to operation with either dry crushed coal or CWM:-

1) Combustion efficiency was about $99\frac{1}{2}\%$ at a bed temperature of $850^{\circ}C$, increasing to nearly 100% at a bed temperature of $950^{\circ}C$ and decreasing to about $98\frac{1}{2}\%$ at $750^{\circ}C$.

2) At $850^{\circ}C$, 90% of the sulphur could be captured in the solids by adding dolomite at a rate to give a Ca/S mol ratio of about 1.7. The amount of dolomite would have to be increased at lower bed temperatures but could be decreased at higher bed temperatures in order to achieve the same sulphur capture.

3) The emission of oxides of nitrogen depended principally on the level of excess air, varying from about 0.11 g/MJ at 30% excess air to about 0.26 g/MJ at 90% excess air. The changes in excess air in the tests shown in Table 3 were largely due to changes in bed temperature (and therefore heat input). Past experience has shown that NO_x emissions are influenced by excess air rather than temperature.

One of the main features when burning CWM was the extremely steady operation and ease of control. These are expected characteristics of liquid fuels. Although feeding crushed coal by pneumatic means into a pressurised reactor has become a reliable technique over recent years, the improvement in steadiness obtained when pumping CWM was clear in the tests described here. CWM is a homogeneous material and from an experimental point of view (and possibly also from a commercial one) variations (with time) of particle size, ash and sulphur contents - all of which are normal problems when working with dry crushed coal - are minimised.

The effect of adding air into the injection nozzle was to break up the CWM into globules of up to 6mm in size. This is believed to be an important feature in the subsequent combustion mechanisms. Examination of samples of solids taken from the bed showed that at least some of these globules had been transformed into carbon agglomerates of about the same size. The overall effect, together with the influence of the extra water, was that burning rates were somewhat slower than with dry crushed coal.

DISCUSSION

The experimental work has shown that high combustion efficiencies (greater than 99%) can be expected in PFBCs and that sulphur retention appears to be as good as that obtained when burning pneumatically-conveyed crushed coal. As might be expected with a "liquid" fuel, control, feeding and distribution are considerably simpler and cheaper.

From a process point of view, therefore, the use of CWM appears to have many advantages. Ultimately, its adoption will depend on the cost of production and transport and on how much a client is prepared to pay for a "convenience" fuel. If the CWM is prepared centrally and delivered to site (e.g. by barge or pipe-line) it is likely to be a beneficiated, finely-ground product since this is the product which will be generally available for repowering existing oil and gas-fired boilers. It will tend to be expensive for PFBC applications, although the earlier testwork (2) showed that this type of CWM can be burned satisfactorily. The work reported here has confirmed that neither beneficiation nor fine grinding are necessary in fluidised combustion. Slurries of 3mm x 0 coal have been produced in the past (1) for other purposes and there seems to be every reason to believe that the use of CWM could still further increase the attractive potential of PFBC combined cycle plant.

REFERENCES

(1)   Kelcec, G., Olivadoti, P.P., & Duzy, A.F.
        ASME paper 62-PWR-3 (Reprinted in Coal Utilisation, 1963 (March), 15)

(2)   Roberts, A.G., Pillai, K.K., Barker, S.N. & Carpenter, L.K. "Combustion
        of 'Run-of-Mine' coal and coal-water mixtures in a small PFBC'.
        7th International Conference on Fluidised Bed Combustion, Philadelphia,
        1982.

ACKNOWLEDGEMENTS

This experimental work was carried out as part of the United States Department of Energy Contract DE AC21-80MC14129.   Any views expressed in this paper are those of the authors and not necessarily those of the US DOE or the NCB. The work was carried out under the direction of H.R. Hoy, O.B.E.

COAL OIL FUELS AS AN OPTION IN CEGB POWER STATIONS.

K.J. Matthews* and R. Conolly**

*   Central Electricity Generating Board, Marchwood Engineering Laboratories, Marchwood, Southampton, SO4 4ZB, U.K.
** Central Electricity Generating Board, Scientific Services Department, South Western Region, Bristol, BS13 8AN.

Three separate full scale power station trials on a 120 MW_e boiler at Padiham Power Station using coal-oil dispersions (CODs) manufactured by B.P. are described and the main results are presented. Small scale (0.5 MW_t) laboratory investigations, using a wider range of coals in the dispersions, have also been carried out together with some wear loop testing. Results from all of this work have been used to assess the technical feasibility of using coal-oil dispersions in CEGB power stations. It is concluded that COD could be used as an auxiliary fuel on p.f. fired power stations, but that substantial derating of modern highly rated CEGB oil designed boilers would be necessary on conversion to COD firing. However, where heat release rates are appreciably lower, operation without derating would be possible for selected COD formulations.

INTRODUCTION

The incentive for the CEGB to invest capital in converting boilers firing oil to use other cheaper fuels is dependent on many factors. The situation is changing as new oil-fired and nuclear plant is being commissioned at a time when demand is not rising. This is illustrated in Figure 1(a) which shows that nuclear and coal-fired power stations, operating on base load at load factors between 60 and 90%, provide most of the capacity required to meet the demand which peaks at just over 40GW. The large modern oil designed stations operate on a two shift basis, with load factors of 10 to 30%, to provide peak demand power. The older oil-fired units, which were converted from coal-firing in the 1960s, are not used very much now and, after the present closure programme, their combined capacity will be only just over 1GW. When considering the potential for replacing oil by coal-oil fuels in the CEGB systems it is obvious that these older converted stations do not appear very attractive in view of their limited further life and low load factors. These load factors are certain to be further reduced by the introduction of 5GW of nuclear, 2GW of coal-fired and 3GW of oil-fired plant over the next five years.

The additional modern oil-fired plant, bringing the total to almost 10GW, is clearly uneconomic to run at a time when oil is relatively expensive. There is consequently a great incentive to investigate alternative ways of operating these units, particularly using cheaper fuels. A number of different possibilities have been assessed including the use of low calorific value gas (available as refinery waste or produced from coal in gasifiers), cyclone firing and firing with liquid coal based fuels.

The present paper describes work carried out on coal-oil fuels within the CEGB to determine the technical feasibility of using these fuels in the

Board's power stations.   In this respect it is worth noting that oil consumption in the large modern oil and mixed-fired stations, see Figure 1(c), is currently above 4Mt p.a. and may increase as new plant is commissioned. The potential fuel cost saving would therefore be substantial if a significantly cheaper fuel were available.   On the other hand, the total fuel bill for older converted stations is small and reducing rapidly.

The quantity of oil used in coal-fired boilers is also shown in Figure 1(c).   This oil is used for light-up, flame stabilising (particularly when burning low grade fuels or wet coal) and modulation between mill groups.   The quantities involved have been reduced substantially in recent years because of the increase in fuel costs, but the current level approaching 1Mt p.a. is not likely to be reduced significantly by further improvements to the operating procedures.   In fact, the quantities used in this application could increase as the commissioning of AGR plant will tend to increase the load following requirement in the p.f.-fired tranche.   Applications of cheaper auxiliary fuels in p.f. plant could therefore result in significant savings.

COAL-OIL FUELS

It is not within the scope of this paper to reiterate the history of the development of coal-oil fuels.   However, it is worth noting that the stimulus for the present work came initially from a request from BP that the Board should consider the potential of their recently developed and patented coal-oil fuel within the CEGB system.   The result was a collaborative programme of trials on a 120MW$_e$ boiler and at 0.5MW$_t$ on a laboratory combustion facility.   The main results of all this work are reported in this paper.

The BP patent[1] gives details of their process which involves grinding the coal in the oil until the coal particle size is less than 10 microns. This produces a stable dispersion without the use of additives which are necessary to stabilise mixtures of p.f. sized coal in oil[2].   The resulting product is known as a coal-oil dispersion (COD) to distinguish it from conventional coal-oil mixtures (COMs).

The main difficulty in handling coal oil mixtures and dispersions results from their high viscosities.   Because of their pseudoplastic nature the apparent viscosity depends on shear rate and the values quoted in Figure 2 are at a shear stress of 100 N m^{-2} which is typical of fairly short transfer lines[3].   These data may not, of course, be very relevant to the situation in power station burner components where shear rates are much higher.   Coal-oil fuels are also thixotropic at pumping temperatures so that the viscosity will decrease along a pipeline.   However, at firing temperatures the fluids become Newtonian.   It should be noted that, in general, dispersions are more viscous than mixtures for a given coal loading.   Also, to reduce the viscosity to conventional RFO firing viscosity of about 15 cP, the fuel must be preheated to a much higher temperature.   In practice, the maximum bulk liquid temperature which can be used is limited by dispersion stability considerations.

Two quite different coals were used in the COD's fired on the full scale boiler.   One of these COD's was also fired on the 0.5 MW$_t$ combustion facility in a test programme which also included three COD's especially made by BP using coals specified by MEL.   The latter were chosen to cover a range of ash contents and abrasive mineral content.   Details of the analyses and properties of these fuels are given in Table 1.   The analyses of the ash contents are noted in Table 2.

TRIALS

Full Scale Boiler.

The full scale trials were carried out on the No.1. boiler at Padiham P.S. in Lancashire. This boiler is a single drum, natural circulation, reheat unit rated for 108 kg s^{-1} of steam generation. Built by Babcock & Wilcox in 1961, it operates at 110 bar with a final steam temperature of 543ºC. Originally designed for coal firing with 3 cyclone burners, the boiler was converted in 1974 to fire residual fuel oil. A schematic is shown in Figure 3.

In the conversion the cyclones were removed together with the target and screen tubes. The lower section of the front was rebuilt to accommodate 15 Hamworthy steam atomising burners. These are arranged in 3 rows of 5 and are each of 2 t h^{-1} capacity with a 6:1 turndown. Primary air, heated if necessary by a separate steam air heater, is supplied from a bus main using up to four primary air fans. Secondary air is fed from a conventional windbox arrangement with individual sleeve dampers to isolate the flow at minimum firing rate. A refractory quarl separates the two air streams. Flame stabilisation is by means of an axial flow swirler mounted on the end of the oil gun carrier tube. Fixed blade secondary air swirlers are used to define the overall flame shape.

Three separate load carrying trials have been carried out at Padiham. The first two were each of about two hours duration during which time approximately 50 tonnes of fuel were burnt. The most recent trial, in February 1983, was much more extensive, a total of about 5000 tonnes of COD being burnt during the thirteen days of testing under various operating conditions. Prior to each trial, a considerable amount of single burner test work was also undertaken to assess the performance of different atomiser types.

The results from the first trials, now referred to as Padiham 1, have been published[4] in detail. Extensive measurements were made during the trials to obtain a direct comparison between COD and residual fuel oil (RFO) firing. During this trial the boiler load was limited to 94MW$_e$ by the throughput of the standard Hamworthy steam atomised Y-jet burners at the maximum permissible rail pressure. The COD used was made from Pleasley coal, see Table 1, which contained ash with a relatively high iron content and a low ash fusion temperature, see Table 2.

Measurements were made of flue gas composition and dust burdens, ash deposition rates at the superheater and furnace heat flux levels. Boiler efficiency was calculated by the losses method at just above the smoke point which was determined by monitoring the carbon monoxide produced at various levels of excess oxygen.

The second trial, referred to as Padiham 2, was carried out at full load of 120MW$_e$. A direct comparison was again made with RFO firing and similar measurements were made to those listed above for Padiham 1. Apart from the higher load, the main differences from Padiham 1 were the incorporation of a more refractory Herrington coal in the COD (see Table 2) and the use of the symmetric twin fluid atomiser (STFA) which was developed by the CEGB[5] and had previously been used in Canada[6] for firing COM.

235

Since the Padiham 2 trials the design of the STFA atomiser has been optimised for COD on the MEL spray rig. This enabled the third trials series, Padiham 3, to be undertaken with a better atomiser performance and consequent improvement in combustion burnout. The Padiham 3 trials were much more elaborate and detailed results will be reported elsewhere. The main objective was to ascertain the effect of long term COD operation on boiler performance over a period of one month. To do this the boiler was extensively instrumented to enable all pressure, flow and temperature readings needed to complete heat balances on all of the boiler elements to be logged on two data loggers. In addition, the effectiveness of sootblowers was determined both in the furnace chamber and in the convective passes. As a check on the heat balance, the furnace mean temperature in the plane slightly below the furnace exit, see Figure 3, was measured by a sonic pyrometry system which has been developed by MEL[7]. The Padiham 3 trials were carried out using the same Herrington coal in the COD as had been used in Padiham 2, but the coal content was slightly lower, see Table 1.

Laboratory Experimental Facility.

Because of the cost involved in mounting full boiler trials, it was decided at the outset that some laboratory scale trials would be beneficial. This made it possible to test dispersions made from a variety of coals. In particular, coals were chosen with a wide range of ash contents and also with a wide range of quartz and pyrites contents, see Table 1.

The $0.5MW_t$ combustion facility at MEL, see Figure 4, has been used for many investigations into the effect of oil composition and additives on combustion performance[8]. It comprises a 0.5m diameter water cooled refractory lined duct fired from one end by a parallel flow oil burner. The air supply can be preheated and sampling can be done via probing ports at 0.3m intervals along the duct. Dust burdens can also be measured further downstream and the flue gases can be washed to provide samples for analysis of final burnout conditions.

In the laboratory work, four CODs were used, including the Pleasley COD used in Padiham 1. The main emphasis of work was on deposition rates at the end of the duct where the temperature was similar to full scale furnace exit gas temperatures. Dust burdens and flue gas washings also enabled combustion burnout to be estimated and three burner types were fired. The bulk of the work was done with either a steam atomised Y-jet tip or a self-stabilised air atomised oil burner (SSAAOB) design developed by MEL[9]. A Monarch pressure jet burner proved to be unsatisfactory owing to frequent burner blockage. In order to estimate the effect of the abrasive nature of the coal ash, the CODs were also circulated through a loop in which 90° and 45° pipe bends and various valves were monitored for wear. The pumps used during this work were a Viking internal gear pump and an Imo triple screw pump. It should be noted that the latter was not considered to be suitable for COD operation.

Results.

Handling.

The CODs used throughout this work all had broadly similar pumping and handling characteristics. As expected, pumping difficulties were encountered unless the COD temperature was above about 70°C and the pressure drop on the suction side of the pump was minimised. In practice, this latter requirement meant that both transfer and main pump strainers had to be removed during the Padiham 3 trials. Similarly, start up of the wear loop was found to be impossible when the high pressure drop Micromotion flow-meter[10] was fitted on the suction side of the high pressure pump.

The CODs used in all of the work were completely stable so long as temperatures of about 140°C were not exceeded for extended periods. It is difficult to avoid doing this in some heaters and a finned tube heating coil in the storage tank used during the Padiham 3 trials was found to be fouled with thick COD containing about 45% of coal at the end of the trials. On the other hand, some samples of COD are still stable after three years storage at ambient temperature at MEL.

The wear loop work has confirmed the results reported by Kenison[11] that pipework will not be subject to wear when transporting COD at conventional power station flow velocities. This is not surprising as the high viscosity of COD results in very low Reynolds numbers, see Table 3, and the consequent laminar flow clearly restricts particulate contact with the walls.

No atomiser wear was detected during the short (about 20 hours) $0.5MW_t$ combustion trials, but both of the high pressure pumps showed quite severe wear. However, the wear on the Viking pump after some 1000 hours running on COD did not seriously affect its performance. With some modifications to incorporate external bearings, this type of internal gear pump would probably be acceptable. Kenison[11] has concluded that the best option would be a progressive cavity eccentric screw type pump with a flexible stator. A pump of this design was used as the low pressure transfer pump on the Padiham 2 trials.

During the longer term Padiham 3 trials one of the triple screw transfer pumps was taken out of service after 600 hours operation on COD, but inspection showed that little wear had occurred. The main station high pressure triple screw pumps were still functioning adequately, apart from minor seal leaks, at the end of the trial. During this longer term trial of about 180 hours actual firing, there were signs of wear on some STFA atomiser inserts, but no deterioration in performance was detected. This wear is not considered to be a problem as it would be possible to manufacture these inserts in tungsten carbide or some other wear resistant material.

It may be concluded that, apart from the changes discussed above, a conventional RFO system is suitable for handling COD.

Combustion.

In all of the trials at full ($25MW_t$ per burner) load, the ignition and flame stability characteristics of COD were similar to those of the RFO and were therefore satisfactory. This is a result of the relatively high proportion of RFO in the COD which gives rise to rapid initial devolatilisa-tion to give ignitable gaseous fuel-oxidant mixtures in the root of the flame despite the presence of coal particles. Experience on the laboratory scale showed the same similarity between COD and RFO, although on this scale it was difficult to obtain flames properly stabilised by internal recirculation with either fuel. However, although the flames stood off from the burner by about a stabiliser diameter, they were stable and gave consistently repeatable results.

One useful measure of the combustion performance of the fuel is the amount of excess oxygen that is required to avoid smoke production in a given burner system. Production of smoke at low oxygen levels is closely related to the production of carbon monoxide which is more readily measured. Carbon monoxide versus oxygen curves, as measured at the reheater outlet at Padiham, are plotted in Figure 5.

The full line labelled Padiham 2 was obtained at nearly full load burning Herrington COD on unoptimised STFA burners. The dotted line, from Padiham 3, shows how burner optimisation improved the smoke point. A similarly improved performance was obtained with a set of specially designed Hamworthy tips.

These results indicate that COD can be burnt at low excess air levels without smoke production. However, this does not mean that the carbon burnout at these low oxygen levels will be satisfactory. Indeed, extensive tests during the Padiham 3 trials showed that it was necessary to use more than 2% excess oxygen to reduce the unburnt carbon in dust below the level considered to be safe to avoid fire hazards in precipitators. A similar reduction in carbon in dust was obtained at 1.5% excess oxygen by using between 5 and 10% water as emulsion in the COD. This technique could be a useful alternative to increasing combustion air flow.

Carbon loss was the most significant factor in the boiler efficiency calculation, see Table 4, in both the Padiham 1 and 2 trials, which were carried out at just above the smoke point. The table indicates how optimising the burner prior to the Padiham 3 trial and increasing the combustion air reduced the unburnt carbon losses from 1.45% to 0.11%. The burner optimisation was aimed at controlling carbon loss by reducing the maximum droplet size produced and the success of this was evident from the dust sample particle sizes shown in Table 5.

Microscopic examination of the dust samples obtained in all of the trials at full scale and on the 0.5 MW_t rig indicated that the coal particles in each droplet aggregated during the devolatilisation phase. Coke spheres formed were similar in size to the original droplet sizes. Coal particle aggregation during devolatilisation has also been observed in single suspended droplet work at MEL[12] and the extent of this behaviour has been related to the rank of the coal. The potential advantage of very fine milling of the coal was clearly, from the combustion point of view, negated by this behaviour. However, studies[13] at the International Flame Research Foundation have shown that greater burnout over a wide range of operating conditions could be achieved with a COD compared with a COM prepared from the same coal.

Apart from the loss of efficiency associated with high carbon-in-dust levels, the effectiveness of electrostatic precipitators would be impaired. Dust resistivities above about 10^9 ohm cm are normally required for efficient dust removal. Values obtained during the CEGB trials range from 10^4 ohm cm at about 130-160°C during Padiham 1 (carbon-in-dust $\sim$ 25%) to 10^6 ohm cm (carbon-in-dust $\sim$ 60%) at higher load during Padiham 2 and, typically, 10^{10} ohm cm (carbon-in-dust < 15%) during the Padiham 3 trials.

It should be noted that high carbon-in-dust levels would also be undesirable from the fire hazard point of view. This is a potential problem throughout the boiler system but particularly so in the air heater and economiser and precipitator hoppers.

Ash Deposition.

Ash deposition rates have been measured at the superheater level, see Figure 3, at Padiham during all of the trials.   In addition, some measurements have been made in the furnace chamber and in the primary superheater. Deposition studies in the $0.5MW_t$ facility were at the end of the horizontal duct where temperature levels were similar to those found at the furnace exit level at Padiham.

The results from the superheater measurements at Padiham are summarised in Table 6.   There were several relevant differences between Padiham 1 and Padiham 2.   The higher load would result in a higher gas temperature and velocity at the probe and these factors alone could explain the increase in deposition rate despite the lower ash content of the fuel.   Data from the $0.5MW_t$ work using four different coals in the CODs, see Figure 6, indicated a statistically "significant" dependence of deposition rate on gas temperature in the relevant temperature ranges for the Pleasley COD and "probably significant" dependence for Butterwell and Ashington CODs.   Furthermore, despite a variation of ash contents from 3 to 13%, the deposition rates and the extent of dependence on temperature, between 0.27 and 0.44 g m^{-2} h^{-1} deg.C^{-1}, were all very similar.   This has been associated with the ash compositions shown in Table 2.   For instance, those ashes with high iron content, Butterwell and Pleasley, have lower total ash content.   The presence of iron tends to enhance slag formation, especially in reducing conditions, by lowering the viscosity of glasses and the melting point of silicates.

Butterwell and Pleasley ashes are also high in calcium which can affect the physical nature of the deposit and Pleasley ash is especially high in alkali metals which aid the formation of sintered low density deposits.   The nett result was that the deposition rates measured in this work were more dependent on gas temperature than on ash content and composition.

The same dependence may be expected at Padiham and the deposition rates measured during the first two full scale trials may be associated with differing gas side conditions.   Inspection of Figure 6(b) for Pleasley COD indicates that the full scale ash deposition rate of 0.014 kg m^{-2} h^{-1}(Table 6) occurred on the rig at a gas temperature of just above 1200°C.   This is perhaps a little higher than would be expected at Padiham at the superheater level when generating 94MW, as gas temperatures of this order were measured with a suction pyrometer during the Padiham 2 trials at 120MW.

The probe surface temperature can also have an influence on deposition rates.   Using the regression equations plotted in Figure 6, the data have been referred to 1300°C and plotted against probe temperature in Figure 7. The two low ash CODs show similar statistically "significant" or "probably significant" (when the unsteady flow data are eliminated in the case of the Pleasley COD) increases in deposition rate with probe temperature.   The data from the higher ash CODs, on the other hand, show no statistically significant trends.

However, experience on full scale boiler plant using long probes confirms the surface temperature dependence and this effect is evident in the Padiham trials results where data from the Padiham 3 trials were obtained at significantly higher probe temperatures than had been used during the Padiham 2 trials (see Table 6).   The deposition rates, see Table 6, were two orders of magnitude higher and this could reflect the change in the physical nature of the deposits, particularly their stickiness, at the higher surface temperature. These rates are, perhaps, more appropriate to the partly ashed-up boiler where the exposed surface temperature has increased as the deposit has built up.

Eventually, if the deposit builds up to a thickness where the surface is above the ash fusion temperature, melting occurs. During the Padiham 3 trials, superheater deposits built up to about 150 mm in length and at this length, their extremities were seen to be molten. However, this caused no operational problems as the basic deposit was friable and tended to break away whenever the boiler cooled down. Also, during all of the Padiham trials, slag did form on the refractory burner quarls and other refractory surfaces in the furnace, but this again did not cause any operational problems.

The effect of ash deposition in the boiler on the heat pick-up in the various elements is illustrated by the change in furnace gas temperature with time recorded by sonic pyrometry close to the furnace exit during the Padiham 3 trials. As the furnace wall deposits built up, the furnace gas temperature increased gradually to a value approximately 100 deg.C higher than the clean furnace value. Superheater temperatures also tended to rise and this necessitated a significant increase in spray desuperheating. However, the resultant modification to heat pick-up patterns suggested that it would not be necessary to sootblow more frequently than weekly when operating on a two-shift basis. Full details of the analyses of the fouling data and the associated water side heat pick-up variations will be reported elsewhere.

DISCUSSION

As far as the CEGB was concerned, the incentive to carry out this work in collaboration with BP was to ascertain whether COD could be used as a fuel in CEGB power stations. It should be noted that BP had more international market applications in mind and conditions in CEGB power stations are somewhat different in some respects. The following discussion is generally only relevant to the CEGB situation.

P.F. Fired Power Stations.

Replacing the oil used in p.f. fired power stations by COD should not be a problem from the ash deposition and handling point of view, because these units are obviously designed to handle the ash present in coal. As the flame ignition and stability characteristics are similar to oil, the only potential combustion problem is associated with the possibility of high carbon-in-dust levels during start-up leading to air heater and precipitator hopper fires. Improvements to the atomisation performance during the most recent trials at Padiham suggest that a more satisfactory atomiser design is now available, especially since the long term wear characteristics of the STFA atomiser can be improved by using tungsten carbide inserts.

Wear characteristics of other items of plant are such that these items must be chosen carefully. BP[11] are confident that they can specify suitable pumps, meters and valves, and their work and the wear loop work at MEL suggests that there will be insignificant wear on pipework.

When costing a possible conversion to COD, the main capital expenditure will be for alternative pumps, heaters and meters. Any pipework modifications are likely to be limited to minimising the suction heads to facilitate start-up. The commercial viability of the scheme will depend on whether the sum of the amortised capital costs and the additional pumping and heating costs associated with the higher viscosity of the fuel are offset by the fuel cost saving. The latter factor may depend on whether the fuel is made locally on site or is bought in from a central production facility as envisaged by BP.

The present situation is that the work reported in this paper has provided sufficient technical and operational information to suggest that COD could be used in p.f. fired power stations if this were found to be economically attractive.

Oil-Fired Power Stations.

The Padiham boiler used for the COD trials is typical of the CEGB's few remaining boilers which were converted from coal to oil firing between 1955 and the early 70s. At first sight these units would appear to be ideal for conversion to COD firing having been designed originally for coal firing. On the other hand, nearly all of the Board's oil-fired plant operational by 1985 will be the 10GW of modern highly rated plant. Because of the introduction of new nuclear units, these oil-fired boilers will still be operating on very low load factors unless alternative cheaper fuels are available. However, there are many features incorporated in the highly rated designs which make coal firing, either directly or as a liquid based fuel, difficult.

Figure 8 demonstrates the most significant problem graphically. It is based on results presented by Borio and Hargrove[14] in a study by Combustion Engineering commissioned by EPRI. Borio used two measurements of firing intensity to relate various coal-oil mixture demonstration trials to the conditions in modern US oil-fired boiler designs and concluded that the demonstration trials to that date were on plant with relatively low firing intensities. Despite this, in both the Sanford and Salem Harbour cases, load was limited by furnace slagging. Borio concluded that derating by 20 - 60% would be necessary for the other US oil designs even when using a coal with an initial deformation temperature under reducing conditions above 1480°C.

The US oil designs are seen to be very close in firing intensity to our modern coal fired design at Drax! The values for our modern oil-fired boilers are much higher, reflecting the tightness of these designs. On the other hand, it is apparent that Padiham is, in this respect, a very reasonable test boiler to provide data for US (and the similar Japanese) oil-fired designs.

The nominal heat release rate is a good indicator of the furnace exit gas temperature (FEGT) for a clean furnace as it reflects the heat input in relation to the surface area available for heat extraction. The FEGT for the US designs is typically up to about 1300°C which is considerably lower than the 1450°C in the CEGB's modern oil-fired designs.

When firing a fuel containing a significant amount of ash, the effect of ash deposition on the walls of the furnace chamber is to increase the wall surface temperature and reduce heat transfer. The resulting increase in FEGT was estimated in a study commissioned by BP[15] to be 46 deg.C for a 120MW boiler designed with an oil-fired FEGT of 1195°C. This is somewhat lower than the 100 deg.C increase measured at Padiham by sonic pyrometry during the Padiham 3 trials. It should be noted that the reduction in furnace steam generation may be offset by the increase in attemperator flow necessary to control superheater steam temperatures.

The relevance of FEGT is that it affects the rate of ash deposition on the convection tubes at the furnace exit plane and beyond, especially if the FEGT is above the temperature at which the surface of the ash particles is sticky. BP concluded that derating could be avoided in their study boiler by using coal with a higher IDT and they quoted readily available sources with IDT's up to 1440°C. The CEGB's oil-fired designs, with clean furnace FEGTs of about 1450°C present a much more difficult problem. It must be remembered that these quoted FEGTs are mean values and in reality there is a range of gas temperatures at the furnace exit extending perhaps 150deg.C.either side of the mean. In order to avoid derating a clean boiler it would therefore be necessary to use coals with satisfactory deposition characteristics at temperatures above 1600°C to prevent the formation of slagging deposits. Theoretical studies indicate that the FEGT varies with load to the power 0.25, so that reduction of the FEGT to the level experienced during the Padiham trials would imply derating by about 50%.

Derating implies longer furnace and boiler residence times and this would certainly be beneficial from the point of view of combustion burnout. Results from Padiham 3 suggest that adequate burnout can be achieved with current burner technology if sufficient excess oxygen is used. Other coals, on the other hand, could have superior burnout characteristics and may not require increased oxygen.

Borio et al[14], using a limiting velocity of $21.3m s^{-1}$, predicted for COM that load would be restricted by convection pass erosion in three of the US oil designs they studied. However, microscopic examination of ash particles from the COD combustion work reported here has indicated that erosion should not be a problem because of their small size.

Assuming that it would be acceptable to run these large modern units in a derated mode,they would still require considerable modification to cope with the ash. Sootblowers and ash removal hoppers would be essential and it would be necessary to provide precipitators to clean up the flue gas. Ash removal and disposal facilities would also be required. The capital cost of these major modifications and the additional operating costs, primarily the precipitator energy and extra pumping and heating, would have to be justified against the fuel cost saving.

The quantity of fuel involved for a derated 660MW unit on two shift operation would be of the order of 1000t/day. For a complete power station a 2Mt p.a. plant would be required. If this were built on site the extra capital costs associated with the COD plant and its coal supplies would have to be considered. If the COD were supplied from external sources, the uncertainty in the long term availability of the fuel and its cost would have to be allowed for. Taking into account the difficulties discussed above, the incentive for the CEGB to pursue conversion to a fuel which would only provide 30% of the heat input from coal and result in substantially derated plant must ultimately depend on the fuel cost advantage.

It should be noted that it may be possible to convert oil-fired plant in the rest of the world, where furnace designs are more liberal, to COD firing without derating and more cost effectively. Experience during the Padiham 3 trials has shown that similarly rated boilers can be operated successfully at full load provided sufficient ash removal facilities are available. This experience, to some extent, contradicts the predictions of Borio et al[14], presumably because of differences in fuel properties and detailed plant design. On the other hand it is in line with the experience of the Florida Power Corporation[16] at their 120MW Bartow No.1. unit which has a similar plan area firing intensity to Padiham (see Figure 8). This plant has been burning an "ultrafine" coal-oil mixture very successfully on a commercial basis since May 1982 during which time more than 50,000 tonnes of fuel have been used.

ACKNOWLEDGEMENTS

Much of the work reported in this paper was carried out at Padiham P.S. and our thanks are due to the Station Manager, Mr. J.T. Simpson, and his staff for their part in preparing for and carrying out the trials. Many of our colleagues in the North Western Region and at Marchwood Engineering Laboratories have contributed to this programme, particularly in obtaining and analysing test data. Our thanks must also go to BP, who provided all of the COD fuel used in this work, and to their staff who took part in many stimulating discussions.

This paper is published with the permission of the Central Electricity Generating Board and the British Petroleum Company plc.

REFERENCES

1. CAIRNS, R.J.R., British Patent Specification No.1. 532 193. "Coal Oil Mixtures", August, 1978.

2. BURROWS, S., Process Engineering, p.72, June, 1981.

3. VEAL, C.J., WALL, D.R., and GOSZEK, A.J., "Stable Coal/Fuel Oil Dispersions", 2nd.Int.Symposium on Coal Oil Mixture Combustion,Nov.1979.

4. CONOLLY, R., GADBURY, P., JACQUES, M.T. and MATTHEWS, K.J., "The Combustion of Coal in Oil Dispersion - Direct Comparison with Firing of Residual Fuel Oil in a 120MW$_e$ Utility Boiler", 4th Int.Symposium on Coal Slurry Combustion, May, 1982.

5. SARJEANT, M., "Blast Atomiser Developments in the CEGB", Paper 4-4, 2nd. Int.Conf. on Liquid Atomisation and Spray Systems, June, 1982.

6. WHALEY, H., WHALEN, P.J. and DAVIES, F.W., "Utilisation of a Beneficiated Coal-Oil Mixture in a Small Utility Boiler", 6th Members Conference of the Int.Flame Research Foundation, May, 1980.

7. GREEN, S.F. and WOODHAM, A.U., "Rapid Furnace Temperature Distribution Measurements by Sonic Pyrometry", 7th Members Conference of the Int. Flame Research Foundation, May, 1983.

8. CUNNINGHAM, A.T.S. and JACKSON, P.J., "Operation Methods of Reducing Unburnt Carbon in the Combustion of Fuel Oil", CEGB Note R/M/N942, 1977.

9. ANSON, D. and SARJEANT, M., "A Self-Stabilised Air Atomised Oil Burner", CEGB Note No.R/M/N909, 1976.

10. PLACHE, K.O., "Coriolis/Gyroscopic Flow Meter", Mechanical Engineering, p.36, March, 1978.

11. KENISON, R.C.,"Erosive and Abrasive Wear in Heavy Fuel Oil Equipment Handling Coal-Oil Dispersion", Inst.Chem.Eng.Conference, "Design 1982", Sept. 1982.

12. LAPWOOD, K.J., STREET, P.J. and MOLES, F.D., "An Examination of the
     Behaviour and Structure of Single Droplets of Coal-Oil Fuels During
     Combustion", 5th Int.Symposium on Coal Slurry Combustion and Technology,
     April, 1983.

13. GOULD, P. and WHITEHEAD, D.M., "Combustion of BP's Coal-Oil Dispersions",
     Santa Barbara Conference, "Combustion of Tomorrow's Fuels", November,
     1982.

14. BORIO, R.W. and HARGROVE, M.J., "Coal-Oil Mixture and Coal-Water Mixture
     Fuels for Steam Generators", 4th Int.Symp. on Coal Slurry Combustion,
     May, 1982.

15. GADBURY, P., "Theoretical Boiler Performance When Firing BP Coal-Oil
     Dispersions", 3rd Int.Symp. on Coal-Oil Mixture Combustion, April, 1981.

16. HIGGINS, M.E., "Commercial Operation of the Paul L. Bartow Unit 1 on
     Coal-Oil Mixture", 5th Int.Symp. on Coal Slurry Combustion and
     Technology, April, 1983.

Fuel	Grime-thorpe COD	Ashington COD	Butterwell COD	Pleasley COD	Herrington COD	
Trials	MEL	MEL	MEL	MEL Padiham 1	Padiham 2	Padiham 3
Gross Cal. Value. (MJ/kg)	35.24	34.32	37.70	38.20	38.86	39.16
Analysis % wt						
Carbon	74.8	72.7	81.0	81.1	84.6	82.8
Hydrogen	8.7	8.5	8.9	8.6	8.4	8.8
Oxygen	1.5	2.1	2.5	3.1	3.0	2.0
Sulphur	3.1	2.8	3.0	2.2	2.5	1.7
Nitrogen	0.5	0.5	0.6	0.8	0.9	0.8
Sodium	0.005	0.006	0.004	0.048	0.16	0.2
Vanadium	0.004	0.004	0.004	0.007	0.007	0.007
Chlorine	0.4	0.06	0.01	0.18	N.A	N.A
Ash in COD	11.6	13.0*	3.1	4.5	2.33	3.3
Quartz in COD	1.70	1.00	0.14	0.30	0.08†	0.08†
Pyrites in COD	0.85	0.40	0.44	0.25	0.24†	0.24†
Coal in COD	33.4	35.3*	34.6	40.5	40.0	37.5
Density (kg/m³)	N.A	N.A	N.A	1066	1101	1085
Viscosity (Pa s) @ 70°C and 100 Pa shear stress	N.A	N.A	N.A	2.00	2.83	1.21

* Including 6.3% Kaolin.

† Data derived from other samples of Herrington coal.

TABLE 1:   ANALYSES OF COAL-OIL DISPERSIONS.

Fuel	Grime-thorpe COD	Ashington COD	Butterwell COD	Pleasley COD		Herrington COD
Trials	MEL	MEL	MEL	MEL	Padiham 1	Padiham 3
SiO_2	58.4	51.7	44.9	44.6	37.9	45.6
Al_2O_3	27.4	31.3	27.6	22.3	19.5	34.5
CaO	1.0	1.8	3.5	5.85	5.0	3.5
MgO	1.6	0.7	1.4	0.39	0.43	0.6
Fe_2O_3	8.9	4.4	14.9	17.8	18.9	7.5
TiO_2	1.0	0.55	1.17	0.79	0.67	1.5
Na_2O	0.69	2.5	0.24	2.6	1.43	ND
K_2O	3.9		1.3	2.55	2.4	1.0
Mn_3O_4	0.07	0.05	0.07	N.A	0.15	ND
P_2O_5	0.16	0.05	0.18	N.A	0.44	0.5
SO_3	ND	0.52	1.3	N.A	9.0	4.1
V_2O_5	N.A	N.A	N.A	N.A	0.29	0.4
Cr_2O_3	N.A	N.A	N.A	N.A	N.A	0.2
NiO	N.A	N.A	N.A	N.A	N.A	0.2
PbO	N.A	N.A	N.A	N.A	N.A	0.1
Ash IDT (°C)	1245	1385	1270	1150	1080	1500
Hemis-phere T (°C)	1390	1455	1385	1210	1180	1575
Flow T (°C)	1500	1495	1430	1330	1500	>1600

TABLE 2:   COD ASH ANALYSES.

	Flow Rate kg s^{-1}	Diameter m	Velocity m s^{-1}	Re$_{COD}$	Re$_{RFO}$
MEL WEAR LOOP					
$\frac{3}{8}$" NB Pipe	0.14	0.009	1.86	9	-
$\frac{1}{2}$" NB Pipe	0.14	0.013	1.04	7	-
GRAIN P.S.					
Burner Rail	10.2	0.150	0.56	430	5800
Burner Supply	1.7	0.040	1.35	270	3600
Burner Bar Annulus	1.7	0.042–0.060	1.18	50	700
Burner Orifice	0.1	0.003	4.17	210	2900
0.5 MW$_t$ RIG.					
Supply Pipe	0.014	0.013	0.11	8	100
Burner Bar Annulus	0.014	0.021–0.013	0.06	3	35
Burner Orifice	0.014	0.001	17.83	90	1200

TABLE 3: VELOCITIES AND REYNOLDS NUMBERS IN THE MEL WEAR LOOP AND OTHER PLANT.

TRIAL	PADIHAM 1		PADIHAM 2		PADIHAM 5000		
Fuel	COD	RFO	COD	RFO	COD	COD	COD
Load MW$_e$	94	86	117	120	94	120	120
Excess Oxygen %	0.9	0.5	1.0	0.6	1.2	1.7	3.0
LOSSES (%):-							
Dry Flue Gas	5.24	4.30	5.60	4.47	5.65	5.56	6.73
Flue Gas Moisture	5.70	6.37	5.82	6.21	5.95	5.92	5.96
Unburnt Carbon	0.67	0.07	1.45	0.05	0.60	0.76	0.11
Unburnt CO	0.06	0.11	0.03	0.06	0.14	0.01	0.01
Air Moisture	0.07	0.07	0.09	0.09	0.07	0.09	0.10
Radiation etc. Assumed.	0.64	0.70	0.5	0.5	0.64	0.5	0.50
Boiler Efficiency %	87.6	88.37	87.97*	88.6.*	86.94	87.12	86.43

* During the Padiham 2 trials a feedheater was out of service giving rise to efficiencies 0.2% higher than normal.

TABLE 4: BOILER EFFICIENCIES.

Trial Conditions			Mass Median Particle Diameter μm		
Fuel	Load MW	Excess Oxygen %	Coulter Counter	Sieving	Microscopy
PADIHAM 1					
COD	94	0.9	59	67	-
RFO	86	0.5	27	-	27
PADIHAM 2					
COD	117	1.0	52	102	104
RFO	120	0.6	34	-	39
PADIHAM 3					
COD	94	1.2	-	43	-
COD	120	1.7	-	47	-
COD	120	3.0	-	46	-

TABLE 5: DUST PARTICLE SIZES.

Trial Conditions			Deposition Rate, Mean or Range kg m^{-2} hr^{-1}	Probe Temperature Range ^oC	Mean Dust Burden g m^{-3}	Mean Carbon in dust %	Ash to Precipitators %
Fuel	Load	Excess Oxygen					
PADIHAM 1							
COD	94	0.9	0.014	400 - 670	1.7	24.7	73
RFO	86	0.5	0.001		0.3	68.4	-
PADIHAM 2							
COD	117	1.0	0.021	425 - 780	2.9	59.9	40
RFO	120	0.6	0.010		0.2	88.0	-
PADIHAM 3							
COD	94	1.2	0.6 - 1.4	600 - 820	2.0	34.2	51
COD	120	1.7	2.0 - 3.2	650 - 865	2.3	36.2	58
COD	120	3.0	1.1 - 2.3	545 - 780	1.5	6.7	43

TABLE 6:   ASH DEPOSITION DATA.

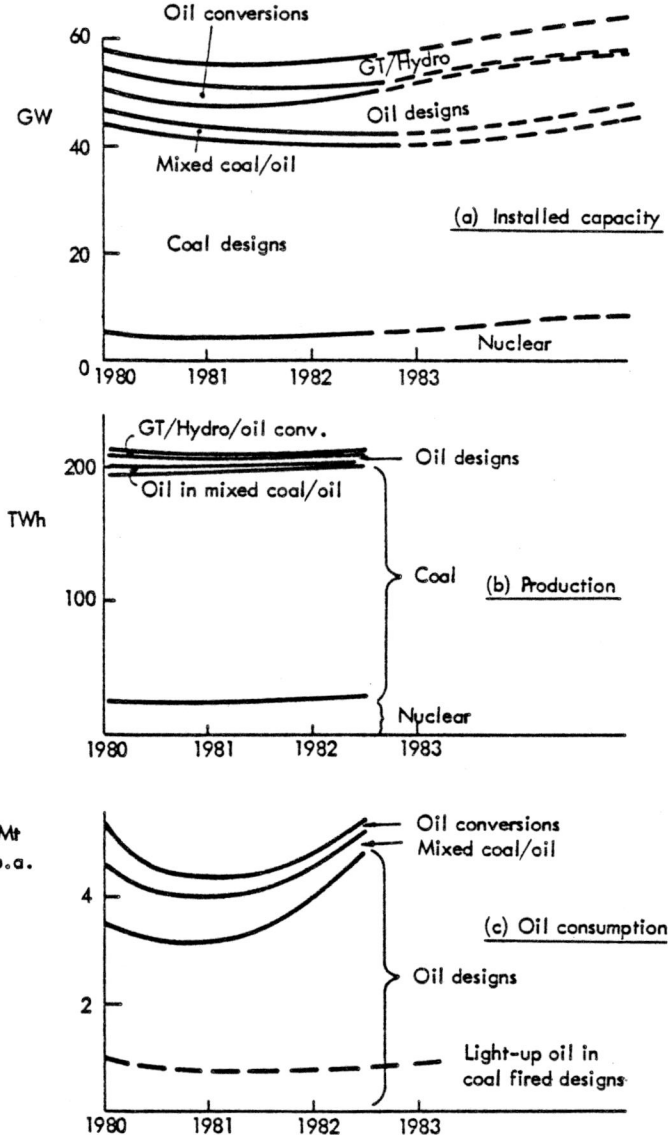

FIGURE 1. CEGB PLANT DATA

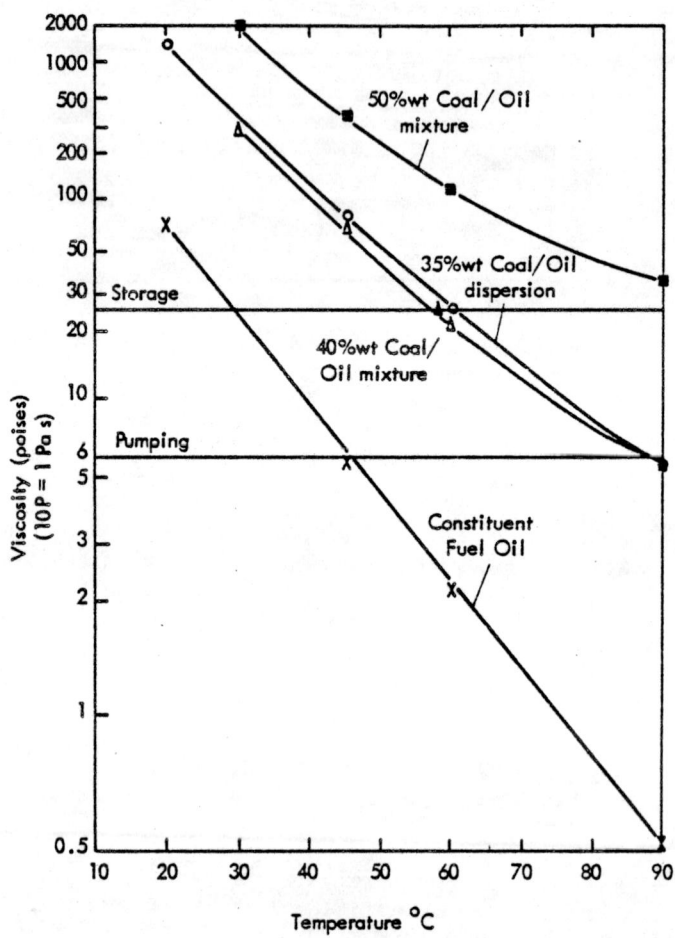

FIGURE 2. VISCOSITY vs TEMPERATURE RELATIONSHIPS
(reproduced from Veal et al[3]).

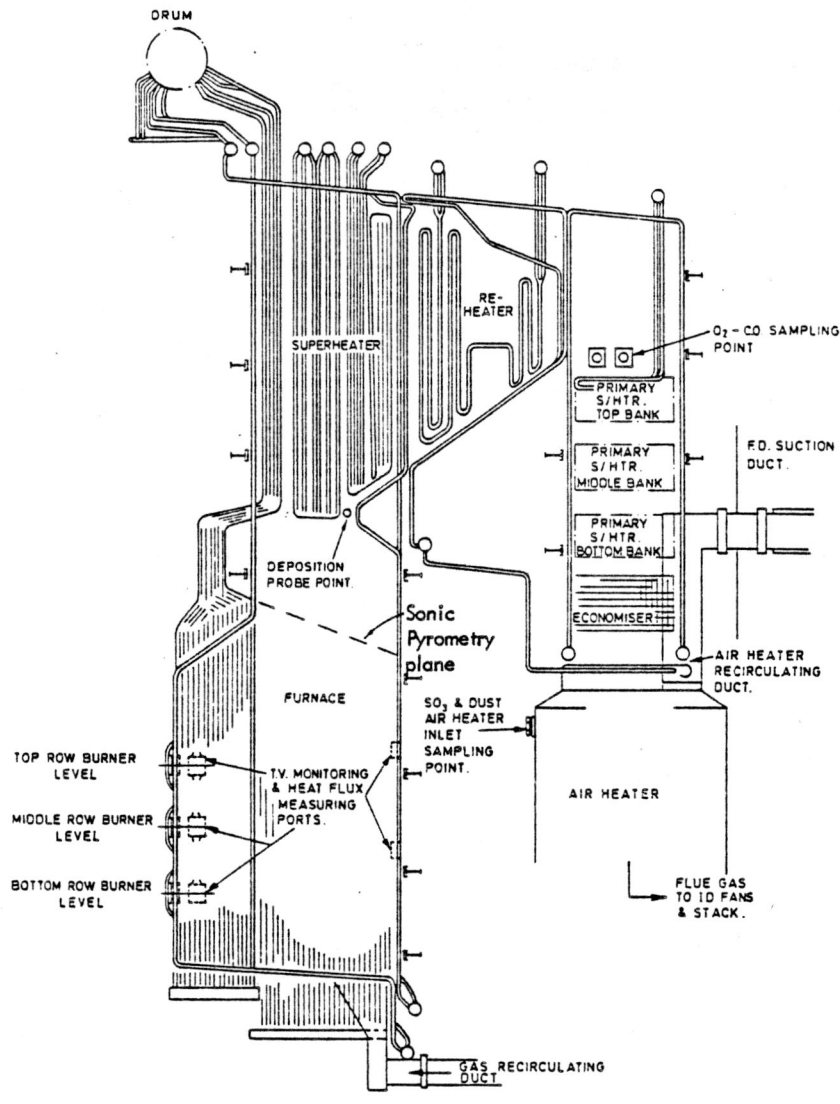

FIG. 3. SCHEMATIC OF BOILER SHOWING SAMPLING AND
MEASURING POINTS.

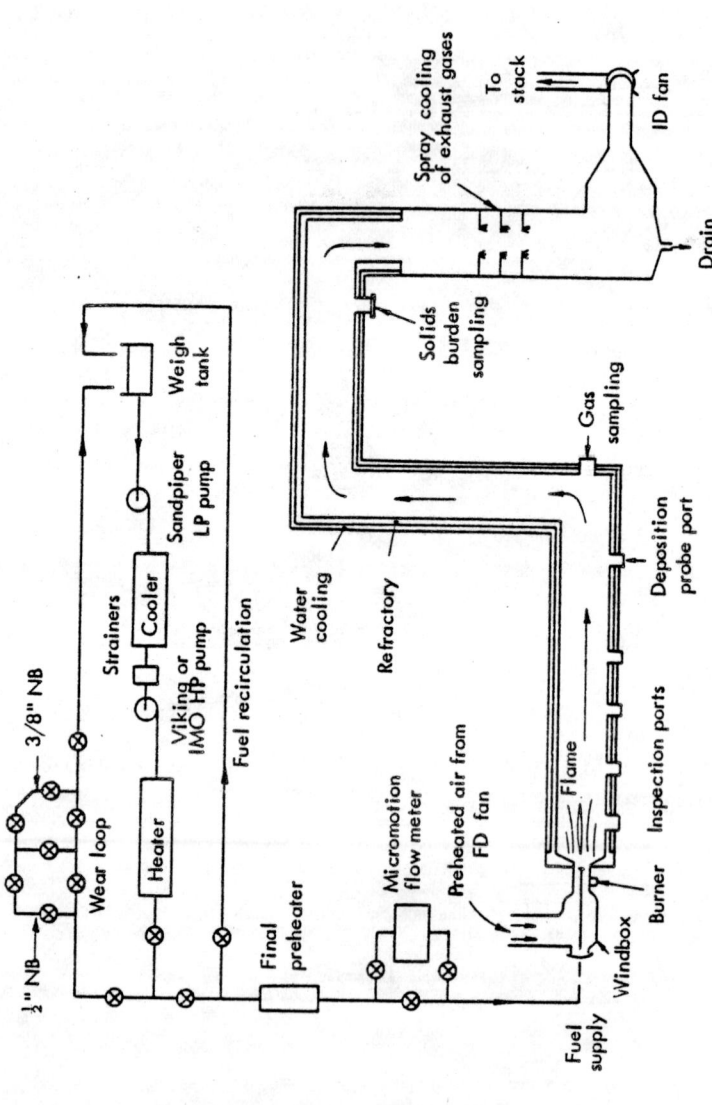

FIGURE 4. 0.5MW$_t$ COMBUSTION FACILITY AND WEAR LOOP SCHEMATIC LAYOUT

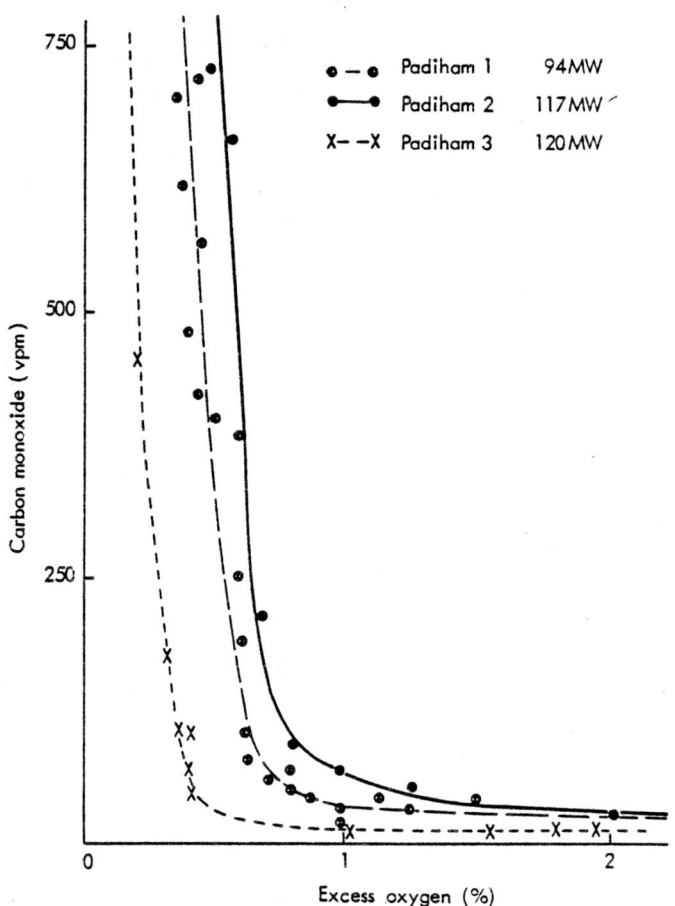

FIGURE 5. VARIATION OF CO WITH EXCESS O_2 AT
REHEATER OUTLET.

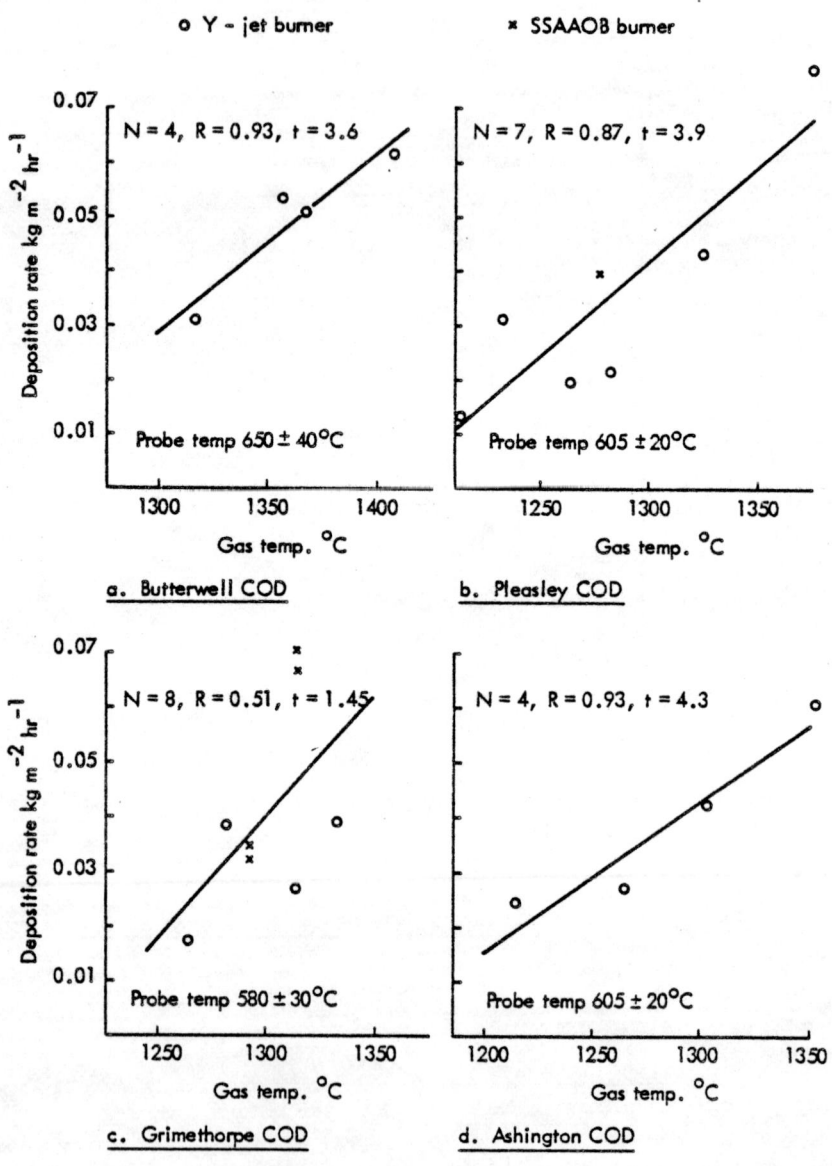

FIGURE 6. DEPOSITION RATES vs GAS TEMPERATURES
IN THE 0.5 MWt RIG

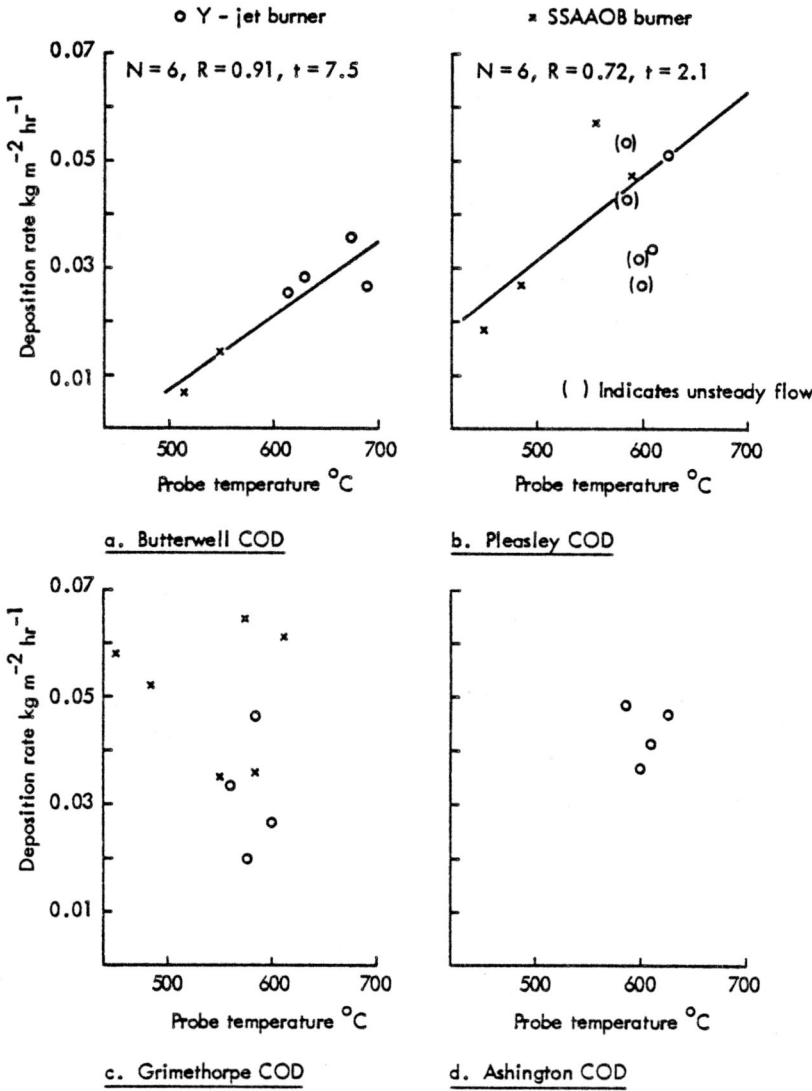

a. Butterwell COD

b. Pleasley COD

c. Grimethorpe COD

d. Ashington COD

FIGURE 7. DEPOSITION RATES vs PROBE TEMPERATURE
FOR GAS TEMPERATURE OF $1300^{\circ}C$ IN THE
0.5 MWt RIG

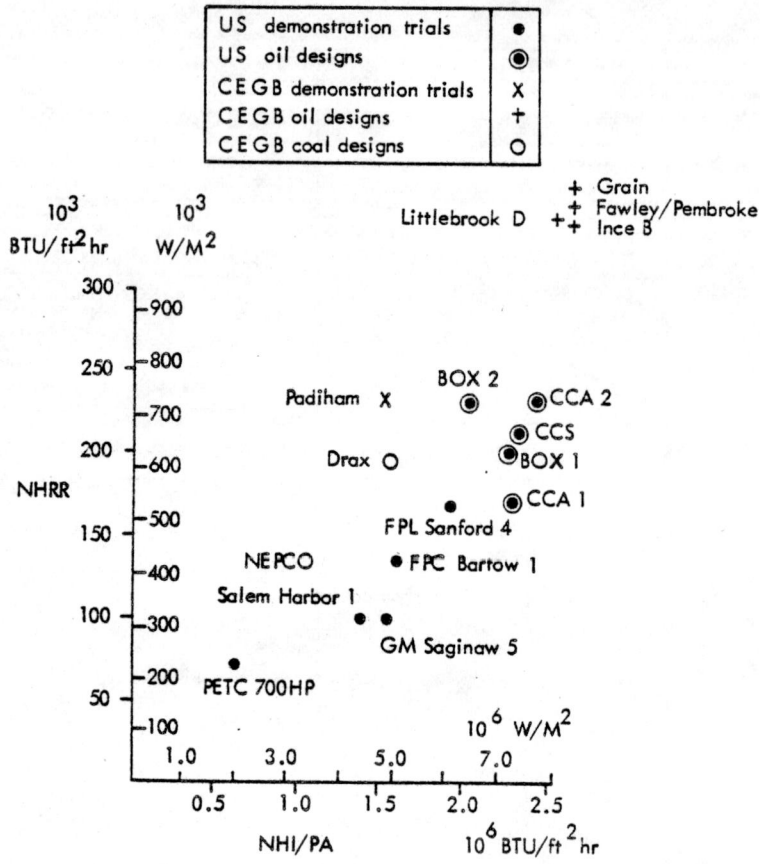

FIGURE 8. COM DEMONSTRATION TEST CONDITIONS
COMPARED WITH PERFORMANCE STUDY
DESIGN CONDITIONS[14] AND CEGB PLANT
CONDITIONS

STATUS REPORT ON THE CANADIAN COAL-LIQUID

MIXTURE PROGRAM

P.J.Read[*] and H.Whaley[**]

SYNOPSIS

A description is given of the three phases of an early
program undertaken at Chatham, New Brunswick in which
coal-oil mixtures were used in a small utility boiler.
Phase I of this program showed that burner and equipment
wear was a significant impediment to coal-oil mixture
utilization. This led to the inclusion of an oil
agglomeration coal beneficiation process into the fuel
preparation process as a means of reducing the sulphur
and abrasive ash content of the coal.

The evolution of this early program into the present
program of coal-water mixture preparation and combustion
testing for utility applications is described in detail,
together with other support programs which may enable
coal-liquid mixtures to penetrate the industrial and
transportation sectors.

INTRODUCTION

Coal-liquid mixtures could replace oil in many stationary combustors and in
some mobile uses provided that they can be burned reliably, cleanly, safely
and economically. This paper deals with Canada's ongoing development of
coal-liquid-mixture technology to meet these requirements. Canada is a net
importer of some ten percent of its oil consumption and may become more
dependent on foreign oil unless new ways are found to substitute for the
depletion of its limited conventional oil supplies. The chosen approach to
reducing reliance on imported oil is a multifaceted one which includes
conservation, upgrading of bitumens, heavy oils and residuums and
replacement by other domestic fuels, particularly natural gas and coal.
Coal-liquid mixtures offer a means of replacing oil by coal where direct
substitution of a solid fuel is impossible or uneconomic.

The present downward trend in energy prices creates economic pressures on
all potential alternatives to oil. Economics dictate that coal-liquid
mixtures, because they inherently cost more per unit of energy than coal,
must be tailored for a fuel market in which they can command a higher price

[*] Energy Policy Sector - EMR Canada
[**] Canadian Combustion Research Laboratory, CANMET - EMR Canada

than coal. To command this higher price, the coal-liquid-mixture fuel must have desirable qualities that the parent coal lacks. The primary qualities required, besides combustion reactivity, are behaviour as a liquid with appropriate viscosity for pumping, transportation and storage, minimization of ash-handling and collection requirements, and, a major environmental benefit, a decrease in sulphur content.

Utilities and other industries which might use coal-liquid mixtures are generally not in a position to switch to such a fuel or even to assess the economics of switching until there is proof that it can be burned reliably and safely. The present program is therefore demonstrating the combustion of coal-liquid mixtures at small commercial scale, has made available trial quantities of coal-liquid mixtures manufactured from Canadian coal, and will ensure that all ancillary equipment is available for the conversion of larger units. Once all these goals have been achieved it is expected that normal commercial practice will take advantage of the technology wherever it is economic to do so.

BACKGROUND

Coal-liquid mixtures have been investigated in Canada since 1972 when the Canadian Combustion Research Laboratory, CCRL, initiated the work with an in-house program to study the combustion and heat transfer characteristics of coal-oil mixtures. Results were presented in 1972[1] and 1973[2,3]. This work led to a demonstration at Chatham, N.B.[4] which included beneficiation of the coal by the National Research Council's, NRC, oil agglomeration[5] system. Although satisfactory combustion in the short term had been achieved in the laboratory[6], good combustion performance could not be maintained in the longer demonstration trials because of burner tip wear[7,8]. The conclusions drawn from operations amounting to 1500 hours using coal-oil mixtures 10 wt. % to 40 wt. % coal were:

The erosion of burner tips was the main obstacle preventing the successful utilization of coal-oil mixture technology in the small utility boiler in Chatham N.B.;

The erosion which results in progressive flame deterioration can be attributed to the use of a highly abrasive coal in the coal-oil mixture;

This problem still persists even when incorporating an in-line coal cleaning process to reduce the ash and pyrites content of the coal;

Pumps, valves and secondary grinding equipment also suffered significant wear-related damage which resulted in deteriorating performance, but it is felt that this problem can be eliminated by appropriate materials and equipment design considerations;

Pipework was relatively unaffected by wear, essentially due to the low prevailing fluid velocities.

The major problem of burner tip erosion may be solved by choice of a less abrasive coal, improved coal cleaning by ash and pyrites rejection, further reductions in coal particle size, materials selection or the use of externally atomized burners with low coal-oil mixture efflux velocities and a simple configuration.

Using the experience gained at Chatham and in consultation with experts at the NRC and at the Canada Centre for Mineral and Energy Technology (CANMET), a ceramic burner tip was designed and spray tested for 200 h on coal-oil-water mixture under simulated operating conditions. When this nozzle was compared to a conventional Y-jet tip the abrasive wear was less than one percent compared to 40 percent for the Y-jet (measured as the percentage increase in flow due to flow channel wear under standard test conditions[9]). The ceramic nozzle has now been rigorously tested under intermittent and steady combustion conditions without failure despite some extreme thermal shock procedures. There are now plans to test the nozzle in a 1000 h demonstration in an industrial boiler followed by extended testing in a kiln.

This background is expanded in some detail in the references cited and summarized in presentations to the American Flame Research Committee[10] and a more up-to-date discussion was given at the 5th International Symposium on Coal Slurry Combustion and Technology[11].

CURRENT PROBLEMS AND OPPORTUNITIES

The price range which coal-liquid mixtures can command is determined by competing fuels. In large industries and electric utilities these fuels are usually residual oil, Bunker 'C' or coal. Smaller energy consumers use No. 2 or No. 4 fuel oil or natural gas which, in Canada, may also be used in larger industries and utilities. For most of these users, the prices per unit of energy for the competing fluid fuels are approximately two-thirds that of crude oil. In order to attract customers by pricing significantly below that of the competition, coal-liquid mixtures must therefore sell for less than 60 percent of crude oil prices on a heat value basis. On this basis, the most expensive Canadian thermal coal sells for somewhat over 40 percent of the crude oil price so where direct use of coal is possible it is the most attractive fossil fuel. However, where solid fuel cannot be used, even if the starting material for a coal-liquid mixture is one of the more expensive Canadian coals, there is about a 50 percent margin above its regular price available to cover additional preparation costs and return on investment.

Feedstock costs all but preclude the use of coal-oil mixtures since, at the upper physical limit (about 50 percent) of coal concentration, the cost of coal, oil and preparation exceed the price of competing fuels. There may be an exception to this generalization in the case of proprietary fuels containing about 25 percent oil, 15 percent water and 60 percent coal, particularly for modest scale heating plant and marine use, but overall the Canadian coal-liquid-mixture program has veered away from its early interest in coal and oil mixtures only on economic grounds.

Canada has the objective of decreasing its atmospheric emissions of sulphur oxides by 50 percent between 1980 and 1990 and the substitution of low sulphur coal for residual oil can materially assist in reaching this objective. The multistage cleaning process associated with coal-liquid mixtures can reduce medium-sulphur coal to this desirable state. However, even where such mixtures might be chosen as an oil substitute preferable to coal on environmental grounds alone, such as in a furnace originally designed to burn coal but later switched to oil (to minimize particulate as well as sulphur emissions), the competitiveness of clean-burning natural gas sets an upper limit to the price.

In utilizing coal-oil mixtures in utility boilers, particularly those designed for oil-firing, there are many problems to be overcome. Usually an oil-designed boiler is smaller, the steam-raising tube banks configured differently, and the gas velocities in the banks much higher than for an equivalent capacity coal-fired unit. In addition to this, coal slurry fuels contain ash which poses problems of tube erosion and slagging and/or fouling. The ignition, flame and heat transfer characteristics of coal-liquid mixtures may be quite different from those of heavy fuel oil and therefore the heat release pattern from the flame may not be suitable for the oil-designed unit. The combination of all these factors usually means that the oil-designed unit will be derated; that is it will not be able to attain the maximum generating capacity for which it was designed when firing oil. In addition to depending on the design of the boiler, the extent of this loss of electrical output will also depend on the coal, its rank and reactivity, (volatile matter, inert macerals content, degree of oxidation) as well as its ash content and composition and the slagging/fouling propensity of the ash.

A coal-liquid mixture which burns well is not necessarily ideal for transportation or storage. The mixture should ideally have a low viscosity which does not vary with temperature and should be readily pumpable after long periods of storage. It should not freeze in anticipated cold weather conditions. It should contain a minimum of inert components to minimize transportation and storage costs. One possibility here is the use of methanol to reduce freezing tendency, improve viscosity for transportation, and improve combustion performance. In fact, the complete substitution of methanol for water has been contemplated, yielding a coal-methanol mixture which has significant advantages in combustion. At the present stage of the program, transportation of coal-liquid mixtures will be by road, rail or barge, therefore the ultimate requirement for very low viscosity, which is needed for pipeline transportation, can be waived but the need for maximum concentration of combustibles remains.

Removal of mineral matter from coal is important for boiler performance, for economy of distribution, for ash disposal, and for environmental protection. Conventional coal beneficiation removes much of the adventitious mineral matter but cannot extract minerals that are very finely interspersed or are part of the molecular structure of the coal. In the case of sulphur, the occurrence may be in pyritic, sulphatic or organic form. Very finely divided pyritic sulphur is often reported as organic because it is so difficult to remove by physical means. Where coals will be burned as a slurry and fine grinding is essential, advantage can be taken of this grinding to liberate sulphur compounds and other minerals. Therefore in the preparation of coal-liquid mixtures, conventional washing is followed by milling and separation on the basis of surface characteristics: froth flotation for coal-water mixtures and oil agglomeration for coal-oil-water mixtures. Coal from the Sydney coalfield in Nova Scotia is particularly amenable to grinding and flotation, showing promise of good yields with mineral matter in the 1.5 to 3 percent range and with about two-thirds of the original sulphur removed.

Several estimates have indicated that deposition of slag on the tubes of boilers conservatively designed for oil firing would contribute very substantially to boiler derating which could be as much as 50 percent when coal is used as fuel. The site chosen for preliminary tests was again the generating station at Chatham, New Brunswick since it has two boilers originally designed to burn coal but recently adapted to burn oil, one front-wall fired and one tangentially fired, and of 12 and 23 MW(e) capacity respectively. The results being obtained at Chatham, where coal-liquid-mixture burners have replaced oil nozzles, are yielding, at a small utility scale, virtually all the data required to assess burners and fuel without risk of damaging a bigger unit or of seriously interrupting electricity supply.

OBJECTIVES OF PRESENT PROGRAM

The ultimate objective of the coal-liquid-mixture program is to derive enough data concerning the fuels and how to burn them that potential users will be able to make decisions to replace oil, based on economics and without technical risk. An essential sub-objective is the establishment of a quality-cost-price relationship. Obviously it costs more to prepare a high quality (i.e. low sulphur, low ash) mixture than a low quality one. Research into the application of oil agglomeration to coal-oil-water mixtures has indicated the costs in terms of light oil addition for various levels of rejection of mineral matter including sulphur. Depending on the fineness of grind and mineral content needed, light oil requirements may vary from 1 to 5 percent of coal weight. For coal-water mixtures, conventional cleaning applied to the highest quality coal can reduce mineral matter to 3 percent and sulphur to 1.2 percent: grinding and multistage flotation can reduce these levels to 1.5 percent and 0.8 percent respectively: if lower quality (less expensive) coals are used, the same process is expected to attain about 3 percent minerals and 1.5 percent sulphur, the cost difference being in the starting feedstock coal rather than in the process.

Use of coal-liquid mixtures by utilities requires a delivery and storage system, including stirring vessels where necessary, and pumps which can deal with fluctuations in diurnal and seasonal demand. The program is demonstrating methods of transportation which will be applicable to industrial users and, later this year, some of the combustion tests will be scheduled in freezing weather so that any problems due to low temperature operations can be detected and solved. Addition of antifreeze such as methanol may be necessary, this will add to the cost but may improve combustion characteristics.

The performance of utility boilers designed for oil will be significantly different when using coal-water mixtures. The problem of unit derating has already been mentioned and each unit to be converted will need a detailed individual assessment to ascertain its loss in electrical generating capacity when firing a typical coal-water mixture. Again, the derating will depend strongly on the fuel and the boiler design. One of the objectives of the current program is to provide data for the determination of the inter relationship between properties and quantity of mineral matter in the coal-water fuel, the flame and unit derating. The utility company will then determine the net loss in its system generating capacity if several units are to be converted to coal-water fuel. It must be noted that a significant requirement of the coal-water mixture program is that the slurry burners be compatible with the retention of fuel oil capability to attain full generating capacity during peak demand periods.

Assuming successful demonstrations at Chatham, the next steps proposed are to conduct tests in oil-designed boilers of similar size to the larger Chatham boiler and then to design systems for burning coal-water fuel in larger utility units. In eastern Canada there are several larger front-wall and tangentially-fired oil-designed units. The current program embraces the design of coal-water systems for both configurations.

DETAILS OF PRESENT PROGRAM

The present program comprises several elements which will combine to achieve the objectives set out above. These are construction of a 7 tonne per hour pilot plant at Sydney, Nova Scotia, for preparation of a coal-water mixture containing over 70 percent coal, the design of burners suitable for reliable combustion of this fuel, the demonstration of the use of fuel and burners at Chatham and the design of coal-water burner systems for larger units. The fuel preparation pilot plant treats clean coal (- 3 mm) from an adjacent conventional dense medium coal preparation plant which reduces the mineral matter content from about 8 percent to 3 percent. The pilot plant illustrated in Fig. 1 comprises two stages of grinding, particle size control, two stages of froth flotation (further reducing the mineral matter to about 1.5 percent) and the mixer to add a stabilizer. The process is based on the proprietary CARBOGEL process. The target solids content is 75 percent with viscosity in the 800-1000 centipoise range. Use of higher quality coal is planned for the first trials to minimize problems with ash handling but different coals with higher mineral and sulphur contents and with poorer washability characteristics could save $20 per tonne (of coal). The prepared fuel is held in day-storage tanks for regular delivery by tank truck (three trucks per day for about 750 km) to Chatham: storage tanks of 500 m^3 capacity which were already in existence at Chatham form the buffer to match demand with production capacity. Fuel production costs are recovered by the producer through the price charged to the electric utility. The utility then passes on the differential between this price and normal coal-fired enerating costs, as well as the cost of burner development, to the federal government. Construction of the pilot plant began in November 1982 and should be completed by June 1983 with regular fuel production by July 1983.

Development of slurry burners for the units at Chatham has been undertaken concurrently with the construction of the coal-water pilot-plant preparation facility. The two phases of coal-water burner development are as follows:

Phase I:

This is the design, testing and evaluation of burners, rated at approximately 30 GJ/h thermal input, suitable for coal-water slurry fuel combustion in the 10 MW(e) front-wall fired unit. An evaluation program for burner and boiler performance assessment was developed for the performance trials which are being undertaken in this unit during Phase II. A parallel program for tangentially fired units is leading to performance trials in the 22 MW(e) unit.

Key elements in Phase I were a review of the state of the art of coal-liquid mixture burner technology and recommendation of the most promising burner concepts for coal-water mixture firing for each boiler configuration. Full scale burners have been designed and were tested prior to installation in the units at Chatham.

Phase II:

This phase assesses burner and boiler performance when firing coal-water mixtures in front-wall and tangentially fired boilers, with special emphasis

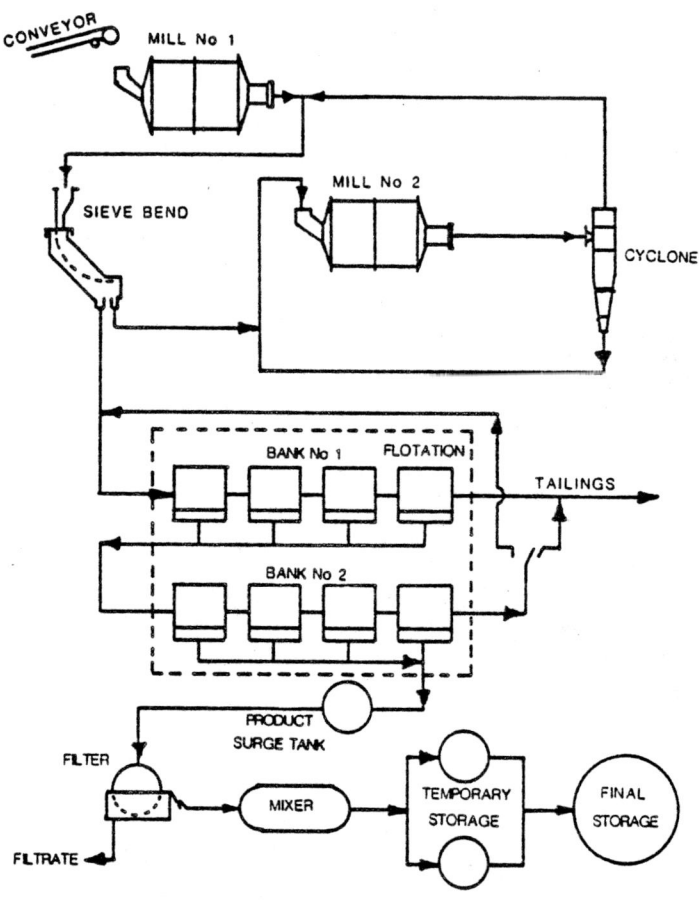

FIGURE 1

on reliability of equipment. It is anticipated that 6000 tonnes of fuel will be required for the performance trials, 2000 tonnes for Unit No. 1 and 4000 tonnes for Unit No. 2. The test fuel contains less than 2 percent ash and is similar to that used in Phase I for burner development. The Phase II performance trials are currently scheduled to begin in the summer of 1983. It is expected that these two phases should lead to the scale-up and testing of burners for demonstrations of coal-water mixture technology in oil-designed utility boilers in the 50 to 150 MW(e) capacity range and of both basic configurations typical of eastern Canada.

During the last five decades, coal has been considered as a possible fuel for diesel engines. This interest has usually been moderated by the fact that until fairly recently, the availability of relatively cheap diesel fuel together with its ease of use has made other fuels, unattractive. For the same reasons that coal-liquid mixtures are now receiving attention as industrial and utility fuels, a coal-based diesel fuel becomes more attractive. Chemically processed coal-derived fuels are very costly and some attention is now being given in Canada to mixtures of very clean coal and diesel fuel as a means of reducing the consumption of expensive refined petroleum products in diesel engines. Obviously high speed diesel engines are unsuitable for coal-liquid mixtures, but the low and medium speed diesels with longer combustion chamber residence times may be suitable for less reactive fuels such as coal-liquid mixtures. The major problem with the use of coal-liquid mixtures in diesel engines is likely to be the injector and the possibility of abrasive wear and premature failure. In order to address this problem CANMET and the NRC have been studying injector performance using a clean coal-diesel fuel mixture. The feed coal supplied was 3.3 percent ash Nova Scotia coal which was then cleaned by the oil agglomeration process to less than one percent ash. In the final mixture, the clean coal was mixed to 28 wt percent with diesel fuel and was 90 percent less than 10 micron. Some problems with stability were observed but it was concluded that with some modification and materials hardening the injector would withstand prolonged use. It is now planned to conduct stationary combustion tests in a medium-speed diesel locomotive engine.

In its role of technology support to the various coal-liquid mixture projects that are being undertaken, CANMET is involved in contract and in-house research to address the following key problem areas:

1) Burner development for coal-liquid mixtures including the study of abrasive wear of atomizer components.

2) Assessment of the potential loss of capacity (derating) when converting oil-designed boilers to coal-water mixtures.

3) Slagging and fouling assessments of coal-liquid mixtures in utility and industrial boilers and combustors.

4) Parameters for upgrading existing and designing new environmental control equipment for oil-fired boilers when converting to coal-liquid mixtures.

5) Combustion and heat transfer properties of coal-liquid mixtures in various combustion system configurations.

6) Upgrading coal quality by advanced cleaning techniques in order to minimize abrasive wear and to reduce environmental emissions of sulphur dioxide and flyash.

CURRENT PROGRESS

At the time of writing, formal contracts have been signed among the Cape Breton Development Corporation, the New Brunswick Electric Power Commission and the federal government to conduct the program, Cape Breton Development Corporation has entered a licensing agreement with Boliden - Scaniainventor to use their CARBOGEL process, the detailed design of the pilot plant has been finalized and it is being built, contracts have been issued for design and development of burners for front-wall and tangentially fired boilers. Two batches of coal-water mixture of 30 tonnes and 500 tonnes have been produced in Sweden using Nova Scotian coal. These batches met the design objectives of less than two percent ash and more than 70 percent coal with a viscosity less than 1000 centipoises; the sulphur content was reduced from about 2.5 percent in the raw coal to below one percent; the weight yield of coal to fuel was over 80 percent and the heating value yield over 90 percent.

FUTURE PROGRAM

The major emphasis of the current program is to assess whether coal-water mixtures are feasible for use in utility boilers. There will obviously be many side benefits of the program in the industrial sector, particularly in the area of burner development for coal-water mixtures. Because of the much wider variety of types of industrial boilers and process combustors it is clear that the non-utility development of coal-liquid mixture technology will be much more difficult. A start has been made in this direction with the development of the ceramic atomizer by Scotia Liquicoal, and it is anticipated that this burner will require industrial demonstration in boilers, kilns, both of which are drastically different in their burner, flame shape and heat transfer requirements. However, whilst much scale-up information will be generated as larger utility demonstrations proceed, the small Chatham units are typical of many industrial boilers which may directly utilize the operating experience gained there. Consequently, at the conclusion of the coal-water mixture program in eastern Canada, some of the industrial sector, particularly large kilns and boilers, may convert to coal-water mixtures as fuels. However, smaller units, which may not be large enough to accommodate coal-water mixtures, may be compelled to use coal-oil or coal-oil-water mixtures. There will be need for significantly more R and D support for the penetration of appropriate coal-liquid mixtures into the industrial, marine and diesel markets.

Following the Chatham demonstrations and possible demonstrations in oil-designed boilers of similar size, scale-up is the next obvious step. Design of burners for front-wall or tangentially fired boilers in the 50 to 150 MW(e) range is planned as a third phase of the coal-water mixture program. A generalized derating study which uses modelling techniques to predict boiler performance when boilers designed for oil are fired with coal-water mixtures is almost completed. A priori reasoning cannot predict specific derating effects because there is insufficient experience connecting the formation of ash from coal-water flames burning finely ground coal in an atomized spray to slagging or erosive effects on boiler tube surfaces. Also it appears that the emissivity and burning characteristics of coal-water mixtures is unlike coal and this will significantly influence derating. When more information concerning ash properties, ash formation and slurry combustion characteristics is available from the current work, the program will go on to include specific application studies to 100 and

CLM-R

150 MW(e) oil-fired boilers in Nova Scotia which will predict the minimum overall cost, by balancing the costs of boiler derating against those of fuel beneficiation.

REFERENCES

1. Canadian Combustion Research Laboratory "Proceedings coal-in-oil seminar"; Division Report FRC 72/95-CCRL: CANMET, Energy, Mines and Resources Canada; 1972.

2. Lee, G.K. and Brown, T.D. "Coal-in-oil: a substitute boiler fuel"; Presented at A.S.M.E. Winter Annual Meeting, Paper No.76-WA/Fu2, New York; Dec. 1976.

3. Brown, T.D. and Lee, G.K. "Liquid and colloidal alternatives to conventional liquid fuels"; 4th Members Conference, International Flame Research Foundation, Holland; May 1976.

4. Whalen, P.J. and Davies, F.W. "Coal-oil slurry firing of a utility boiler; Report prepared for Energy, Mines and Resources Canada under Supply and Services Canada, Contract No. 18SQ.23440-7-9033; June 1978.

5. Capes, C.E., McIlhinney, R.E., McKeever, J. and Messer, L. "Application of spherical agglomeration to coal preparation"; Proceedings 7th International Coal Preparation Congress, Sydney, Australia, Paper H2; May 1976.

6. Whalen, P.J. Davies, F.W., Lee, G.K. and Mitchel, K.A. "A study of coal agglomeration and coal-in-oil mixture combustion in a utility boiler"; Report prepared for Energy, Mines and Resources Canada under Supply and Services Canada, Contract No. 18SQ.23440-7-9055 IV; September 1979.

7. Whalen, P.J. and Davies, F.W. "Coal agglomeration and coal-oil mixture combustion in a utility boiler: Phase III"; Report prepared for Energy, Mines and Resources Canada under Supply and Services Canada, Contract No. 18SQ.23440-9125-3; January 1981.

8. Whaley, H. and Whalen, P.J. "Burner nozzle wear during coal-oil mixture combustion trials in a small utility boiler"; Proceedings Corrosion 81 International Corrosion Forum, National Association of Corrosion Engineers, Toronto; April 1981.

9. Bruno, L., Deshpande, A.S. and Whaley, H. "Coal/oil slurry combustion and tribology - a Canadian experience" 4th International Symposium on coal slurry combustion, Orlando; May 1982.

10. Read, P.J., Whaley, H. and Lee, G.K. "Evolution of Canada's coal-liquid mixture program"; American Flame Research Committee, International Symposium on conversion to solid fuels, Newport Beach, California; Oct. 26-28, 1982.

11. Whaley, H., Rankin, D.M., Nicholson, R.P. and Covill, I.D. "The development of coal-water mixture technology for utility boilers in eastern Canada"; Proceedings of the 5th International Symposium on coal slurry combustion and technology, Vol.I, pp. 809-825, Tampa, Florida; April 25-27, 1983.

COMPARISON OF COMBUSTION OF COAL-OIL AND COKE-OIL
SLURRIES IN A SIMULATED BLAST FURNACE TUYERE RACEWAY

D.P. Jenkins*

Coal-oil and coke-oil slurries have been
combusted in a laboratory simulation of the
unusual conditions obtaining in the blast furnace
tuyere raceway. The results have been used to
calibrate a one-dimensional mathematical model
of the atomisation, mixing and combustion
processes. Statistical analysis of the results
demonstrate that free coal particles are
liberated during slurry atomisation.

1. INTRODUCTION

The use of oil as a blast furnace tuyere injectant is well known,
but after the price rises of the last decade investigation into a
cheaper injectant was instigated. This paper examined the work
programme for one type of the alternative injectants currently
under investigation, namely coal-oil and coke-oil slurries.

The quantity of injectant which may be used is limited by two
conditions:
Firstly the thermochemical effect of the injectant on the Raceway
Adiabatic Flame Temperature (RAFT), which must be kept within
certain limitations. This is to avoid:
(a)   low productivity (RAFT too low) or:
(b)   distillation of alkalis within the furnace (RAFT too high)

Secondly, if the gasification of the injectant is too incomplete,
then difficulty is experienced with soot carryover into the gas
cleaning system.

It is this latter constraint which is investigated in this
paper.

* British Steel Corporation, Welsh Laboratory, Port Talbot.

2. COMBUSTION CONDITIONS IN THE TUYERE RACEWAY

The blast furnace uses preheated air at a temperature ususally greater than 1000°C, and which may be enriched with oxygen to a volumetric oxygen content of as much as 27%. This air is supplied through water-cooled copper tuyeres at high velocity, usually in excess of 200 m/s.

The momentum of the air blast clears a space amidst the furnace burden immediately in front of the tuyere and this is termed the 'raceway' and may vary in size from 0.5 to 1.5 metres in length. The semiliquid nature of the burden severely limits recirculation of gases in the raceway.

It has been empirically established that, at least in the case of oil injection, virtually complete combustion must take place within this raceway to avoid the difficulties of soot formation. This soot formation also occurs in certain conditions [1] with high volatile coals, but does not seem to occur even with extremely high injection rates, when anthracite is injected[2].

The provisional conclusion might thus be drawn that the volatile content of the injectant should be largely combusted in the raceway to avoid sooting problems. In the case of oil, the volatile content is normally over 90% of the fuel mass. It is therefore advantageous that coal-oil slurries are produced from low-volatile coals for tuyere injection.

3. EMPIRICAL INVESTIGATION

The combustion investigation apparatus was configured to give as close an approach as possible to the conditions obtaining in the blast furnace raceway, namely low recirculation, high air preheat, variable oxygen content and variable air velocity.

The preheated air conditions were achieved by means of combustion of air-oxygen-propane in a separate combustion chamber (Fig.1). The air was then fed to a refractory half-scale model tuyere where the careful computer metering and control of the flows of the air components ensured the correct velocity (Fig.2(A)).

The coal-oil mixture was pumped via a peristaltic pump and was atomised with steam (Fig.2(B)) and introduced to the tuyere via a stainless-steel injection lance.

The tuyere was situated at the end of a flame tunnel (Fig.3) which had been profiled by means of rammed refractory into a conical form of half-angle 10° to virtually eliminate recirculation. The flame tunnel had five sampling ports at 450mm intervals along its length.

Flame sampling was carried out by means of an automated sampling rig which is controlled by the microcomputer control system. Five radial gas analysis readings were taken at each port, and the area-and-velocity weighted mean of these readings were used to calculate the 'burnout' of the flame at each port. The result is a representation of a one-dimensional flame along the axis of the flame tunnel.

4. THEORETICAL INVESTIGATION

A one-dimensional iterative mathematical model of the coal-oil mixture flame was produced, consisting of the following:

(i)   Atomisation of the coal-oil mixtures.
(ii)   Mixing of the steam-borne fuel stream with the air blast.
(iii)   Combustion of the slurry droplets.
(iv)   Combustion of the larger, separate coal particles formed on atomisation.

4.1. Atomisation of coal-oil mixtures

If the slurry is atomised in a normal twin fluid atomiser then it can reasonably be expected that the resultant size distribution arises from the fact that the smaller droplets are formed at the zone of higher shear forces in the atomiser, rather than formation on a purely random basis.

The coal is also distributed in size, and assuming homogeneous distribution of the size ranges throughout the oil, then, upon the formation of oil droplets, some larger coal particles will obviously be at the position of high shear forces in the atomiser, thus coinciding with the formation of smaller droplets.

Some of the larger coal particles will obviously now appear as 'free' coal particles. In general, therefore, the smaller the droplets of slurry formed, the lower their coal content will be, and vice versa. However, since homogeneous distribution of the coal was assumed, the larger droplets of slurry cannot contain more than the original proportion of coal (indeed by the above definition it must contain somewhat less). Therefore, for distributed size ranges of the coal and distributed size ranges of the slurry, some 'free' coal must inevitably appear unless the proportion of coal is low, and the 'grind' is much finer than the slurry drop size distribution.

Having established that some free coal must necessarily appear, it now remains to ascertain what quantity of coal and at what size distribution actually is free of the slurry.

If the size ranges are numbered 0, 1,2--15 in order of increasing fineness, then the nth slurry droplet size range will only contain coal particles of the (say) $n+4^{th}$ size range or finer (n+5,N+6 etc).For a complete distribution, therefore, the coal sizes 0 to 3 will be completely free from the slurry, the 4th coal fraction will be contained only in the 0th fraction of slurry, the 5th fraction will be contained only in the 0th and 1st fractions of slurry etc. The relationship between slurry and free coal is thus 'four overlaps'.

Consider Fn to be the proportion of free coal in fraction n relative to total coal mass supplied and Sn is the proportion of slurry in the nth size range of the total of the atomised slurry (i.e. excluding free coal). If p is the relative mass of coal to oil as supplied to the atomiser and Cn is the proportion of coal in the nth size range of the total coal mass, then:

Proportion of slurry which can hold the nth fraction of coal

$$P_s = \sum_{0}^{n-4} Sn$$

It can then be assumed due to homogeneity that this proportion of the nth fraction of coal has been assimilated in the slurry, the remainder being 'free' coal, i.e.

$$F_n = (1-P_s).Cn.P \qquad \ldots\ldots\ldots(1)$$

expressed in terms of unit flow of oil.

The total atomised slurry mass flowrate relative to unit flowrate of oil is thus the mass of flurry supplied to burner - total free coal

i.e. $$S_t = 1+P-\sum_{0}^{15} Fn \qquad \ldots\ldots\ldots(2)$$

and the mass flowrate of each fraction of atomised slurry per unit oil flow is:

$$M_s = S_n.S_t \qquad \ldots\ldots\ldots(3)$$

Equations (1) to (3) above are thus used in the determination of the free coal and slurry flows.

4.2. Mixing of Fuel Stream with Air Blast

The Pieri[3] mixing correlation for annular concentric jets is used for calculating the 'mixedness' of the air and fuel streams using the Toor[4] mixing index.

For a non-swirling jet, Pieri shows that:

$$M = \frac{A'X}{1+A'X} \qquad \ldots\ldots\ldots(4)$$

where: M is the mixing index.
       A' is a constant whose value is approximately 3.
       X is the dimensionless distance along the
          flame axis which is equal to $\frac{x}{D}$ for non-swirling
          components.

where: X is the distance from the nozzles, D is the
          effective diameter found from the momentum-
          weighted mean diameters of the nozzles.

The index M is indicative of the degree of completion of a
reaction betwen a fuel gas and oxidant, the rate of which is
controlled entirely by turbulent diffusion.

In this case, however, the combustion process is not
instantaneous, and so the mixing index must be used in
conjunction with the combustion rate equations developed
later to produce a time - local oxygen concentration.

4.3. Combustion of Slurry Droplets

Long[5] proposed that for a unit surface area of an oil
droplet:

$$\frac{dm}{dt} = \frac{k}{cd} \left\{ 2 \ln \left[1+ \frac{c(Tf-Ts)}{Q} \right] + \frac{pdc}{16Xk} \left[\ln(1+ \frac{X.n}{\Omega}) \right] \right\} \dots (5)$$

where: k = mean thermal conductivity of the intra-flame
           space
       c = mean specific heat of fuel vapour in the intra-
           flame space
       d = droplet diameter
       T_f= flame temperature
       T_s= droplet surface temperature
       p = mean density of oxygen beyond the flame
       D = mean diffusivity of oxygen beyond the flame
       X = number of moles of combustion products
           derived from stoichiometric combustion of
           unit mass of fuel with oxygen
       n = mole fraction of oxygen in bulk surrounding gas
       Ω = number of moles of oxygen required for
           stoichiometric combustion of unit mass of fuel
           with oxygen

The flame temperature is calculated from a consideration of
stoichiometric combustion (by assumed model definition) of
the oil with the gases at ambient temperature and
composition for that iteration.

The equation used in the final model was thus the simplified
heat conduction version, with values as given by Long[5].

4.4. Combustion of Coal Char

A devolatilization routine was not necessary, since only low-volatile anthracites and cokes were examined. The coal burnout was thus confined to char combustion.

Work by Essenhigh[6] on coal particles combustion found that diffusion controlled combustion obtained for large particles. The diffusional differential equation of reaction rate per unit external area assuming that diffusion rate is slower than reaction rate is given by:

$$\frac{dM}{dt} = -K_d (O_g - O_s) \qquad \dots\dots\dots (6)$$

where M is particle mass at time t, O is oxygen partial pressure (g in the ambient gas and s at the surface) (for infinite kinetics and total diffusion control, O_s = zero). K_d is the diffusional rate constant given by:

$$K_d = \frac{3\phi \rho_o D_o}{4d} \cdot \left\{ \frac{T}{T_o} \right\}^{0.75} \qquad \dots\dots\dots (7)$$

where ϕ is the reaction mechanism (1 for reaction to CO_2, 2 for reaction to CO)

ρ_o is the gas density at T_o degrees absolute,
D_o is the diffusion coefficient of oxygen in nitrogen at T_o (the other gases present being ignored), d is the diameter of the particle,
T_o is the refrence temperature and
T is the mean of the gas and particle surface temperatures, (The latter has been measured practically [7],[8]).

However, the work of Essenhigh[6] was conducted with particles of 300μ and over and previous work by Hottel and Stuart[9] had suggested that for typical sizes of particles associated with pulverised fuel (<50μ) reaction kinetics could influence overall reaction rate.

If the reaction rate is slower than diffusion rate:

$$\frac{dM}{dt} = -K_s \cdot O_s \qquad \dots\dots\dots (8)$$

(where O_s = O_g if the diffusion rate is infinite)
where K_s is the surface reaction rate constant, given by the Arrhenius equation:

$$K_s = K_s \exp (-\Delta E/RT_s) \qquad \dots\dots\dots (9)$$

where   E = activation energy
K_s = pre-exponential constant

If the two rates are of the same order of magnitude, then a combined diffusion/kinetic equation can be formed by susbtituting (8) in (6) to give:

$$\frac{dM}{dt} = \frac{O_g}{\frac{1}{K_d} + \frac{1}{K_s}} \qquad \dots\dots\dots (10)$$

5. DESIGN OF THE WORK PROGRAMME

The following points must be considered:

(i) The variability of gas analysis with time is well known when sampling in flames.

(ii) The lags of the sampling system and the variation in speed of response of the gas analyses.

(iii) The pulsations of the peristaltic pump, whilst limited by an air-filled smoothing chamber, still exist.

(iv) Because of the smoothing chamber, considerable delays occur before an altered level of output is realised after a change in speed of the pump.

(v) 'Drift' in the gas measurement transducers.

As a result of these sources of inaccuracy, individual 'runs' are unlikely to be accurate enough for individual analysis.

It is desirable, therefore, to carry out the work programme on a parametric basis, investigating at the highest and lowest value of the variable to be changed. Equal numbers of 'runs' at each level of parameter are desirable.

The following values of the parameters were decided upon by actually running the rig:

Parameter	High Value	Low Value
Blast temperature	1000	750
Oxygen content of blast	26%	21%
Blast velocity	240	200
Injection level (proportion of Stoichiometric)	0.9	0.8
Atomisation pressure	50 psig	40 psig

It can be seen that the differences in the high and low value are not as large as is ideally desirable. The reason for this is the cumulative effects of the alteration of the parameters. For example, the flow of pure oxygen at the conditions 750°C, 26% oxygen and 240 m/s is three times the flow obtained at 1000°C, 21% oxygen and 200 m/s. The selection is limited on the one hand by supply limitations and on the other by orifice plate accuracy. The even larger variation of the level of injectant resulted in a constant ratio of steam to injectant being adopted to retain the steam flow within measureable limits. The result is that only the pressure of supply can be varied in the 'high' and 'low' levels of atomisation. In practice, even this small allowable variation is subject to such inaccuracies of setting that pressure variations due to mass flow corrections can span this range.

It was therefore decided to fix the steam atomisation pressure at 40 psig and a mass flow of 0.4 times the slurry mass flow. This results, for each slurry type, in four variable parameters.

Each slurry has thus a set of $2^4 = 16$ 'runs' which are to be carried out on it.

The five injectants to be investigated are:-

(i)   Heavy fuel oil (HFO)
(ii) 60% HFO + 40% coarse coke breeze
(iii) 60% HFO + 40% fine coke breeze
(iv) 60% HFO + 40% coarse anthracite
(v)   60% HFO + 40% fine anthracite

The characteristics of which are in Appendix 1.

It can thus be seen that the total work programme consists of a maximum of 80 runs.

The original intention was to investigate lower coal concentrations in addition to the runs above, but the 'coal effect' of the slurry is lessened; and the economics of the process depend upon using the maximum feasible coal concentration. The use of a concentration between 0% (HFO) and 40% is thus spurious.

6.   RESULTS FROM RIG

6.1.   By Slurry Type

The burnouts of each coal slurry type are shown in Fig.4. It can be seen that the degree of grinding of the coal has a noticeable effect on the burnout.

$$
\begin{aligned}
&\text{Coal 1 fineness} = 82.8\% <75\mu\\
&\text{Coal 2 fineness} = 91.4\% <75\mu\\
&\text{Coal 3 fineness} = 61.4\% <75\mu\\
&\text{Coal 4 fineness} = 96.1\% <75\mu
\end{aligned}
$$

6.2.   Effect of Air Blast Preheat on Burnout

It can be seen that higher preheat has higher burnout, especially in earlier phases, see Fig.5.

6.3. Effect of Air Blast Velocity on Burnout

It would seem that microturbulence is a factor in the combustion of the coal, but not a major one, see Fig.6.

6.4. Effect of Oxygen Enrichment on Burnout

Bearing in mind that the slurry flow is proportional to the oxygen loading of the blast, the small differences due to oxygen enrichment is, perhaps, unsurprising (see Fig.7).

6.5. Effect of Stoichiometric Ratio on Burnout

The mass transfer rates from oil droplets and coal particles vary markedly in relation to temperature, which is considerably higher in the high stoichiometry cases. This is in contrast to recirculatory flames, where a high stoichiometric ratio would usually mean lower burnouts. (see Fig.8).

7. CALIBRATION OF THE MATHEMATICAL MODEL

7.1. Criteria of Measurement of Model Accuracy

Any final measurement of model accuracy must contain a component not only of the overall accuracy of prediction, but also of the accuracy variation along the ports and the accuracy variation between high and low values of the preheat temperatures, blast velocity, oxygen content and stoichiometric ratio.

Given for each coal slurry species:-

$$x_n = \frac{\sum\limits_{\text{1st run}}^{\text{16th run}} \dfrac{F_p - F_a}{F_a}}{16} n \quad \ldots\ldots\ldots\ldots (11)$$

where: F is Fractional burnout and the subscripts: n refers to port number, $_p$ refers to predicted model value and $_a$ refers to actual experimental value.

$$x_m = \frac{\sum\limits_{\text{1st run}}^{\text{16th run}} \dfrac{F_p - F_a.}{F_a}}{16} m \quad \ldots\ldots\ldots\ldots (12)$$

and

where: m refers to the eight conditions of low and high levels of:
temperature, velocity, oxygen and stoichiometric ratio

Then for the entire set of results for the species:

$$\bar{x} = \frac{\overset{Port\ 4}{\underset{Port\ 1}{\Sigma}} \bar{x}_n}{4} \qquad \dots\dots\dots(13)$$

x thus given the mean error for the entire coal species. The mean deviation between the ports can be found from:-

$$\delta_n = \frac{\overset{n=4}{\underset{n=1}{\Sigma}} \left| \bar{x}_n - \bar{x} \right|}{4} \qquad \dots\dots(14a)$$

and the mean deviation between the errors of each input condition can be found from:

$$\delta_m = \frac{\overset{m=8}{\underset{m=1}{\Sigma}} \left| \bar{x}_m - \bar{x} \right|}{8} \qquad \dots\dots(14b)$$

An 'overall error' can now be defined as:

$$\delta = 1/3 \ (x + \delta_n + \delta_m) \qquad \dots\dots(15)$$

These criteria of distribution were chosen since the errors are of the same order as x. Standard deviation (root mean square error) would, relatively speaking, exaggerate the worse deviations from the mean. A short programme was written to evaluate these errors on a coal species basis using the above equations.

7.2. Slurry Variables to be Optimised

There are several input variables which must be taken into account:-

(i) Mean size of the slurry droplets formed
(ii) The distribution of the droplet size around this mean size of droplet.
(iii)The degree of internal burning of the coal char particles
(iv) The area of the char particles, given the particle size.

i.e. Area of particle $= \left(\dfrac{d}{a} \right)^2 . A$

where A, the area factor is equal to $4.\pi$ for spheres, 16 for cubes.

7.3. Optimisation Techniques

The optimisation if carried out in phases:

(i) A three-dimensional matrix is produced by the mathematical model of percentage error difference between predicted and actual values at the first port. The spans of the matrix more than cover the ranges of slurry droplet size and distribution, and slurry coal fractions overlap likely to be encountered. The mesh of this matrix is as fine as is technically feasible.

(ii) The most obvious point of low error in (i) above is taken as the aiming point of the second phase. This point is not necessarily the point of lowest error - which could be the result of a compounding of random errors - but a low error point which is surrounded by other low error points on all sides - the twenty seven nearest points. The point whose twenty seven surrounding points have the lowest standard deviation is taken as the lowest error point in the largest volume of low error, and is therefore likely to be the point of lowest actual error.

The process so far is taken as the first phase of the procedure.

(iii) The technque is repeated with a finer mesh size centering on the lowest error point found in the first phase. The analysis of this matrix is as in the first phase, this second phase is essentially a refinement of the first phase values. The extra computer time involved in producing this second phase is offset by the fact that this phase is an analysis of a two dimensional matrix due to the selection of one fixed overlap at each coal. Interpolation of overlaps is impossible, since overlap number is an integer.

(iv) Using the optimum slurry characteristics from the second phase, the model is now run to produce a two-dimensional error matrix by varying the internal burning index and the surface area factor for each coal, using the overall error concept. This is termed the third phase.

(v) The fourth to sixth phases inclusive are repeats of the first to third using the previous best values found for the various parameters, but using the overall error concept throughout.

7.4. Results of Optimisation

The techniques outlined in 7.3 were applied, and the following results were noted for each slurry type:

Parameter	Coal 1	Coal 2	Coal 3	Coal 4
No. of overlaps	6	5	6	6
Mean droplet size (μ)	85	110	90	105
Rosin-Rammler index	0.910	0.875	1.225	1.155
Coal surface area factor	16	20	22	20

The following points emerge:-

(i)   There is good agreement on the number of fractions overlap. The proposition that slurry droplet retain coal particles only below some factor of the droplet size therefore seems to have some foundation.   At six fractions overlap and a 1.38 size ratio between fractions, this factor is $1/(1.38)^6 = 0.14$

(ii) The finer grinds of coal, 2 and 4, produce larger mean droplet sizes of the slurries. This is to be expected, since finer grinds produce higher viscosity slurries, and higher viscosity liquids produce larger droplet sizes on atomisation.

(iii) The anthracitic coals, 3 and 4, produce a wider slurry droplet size distribution (higher Rosin-Rammler index) than the coke based slurries. Presumably this is either a particle shape or a coal density effect.

(iv) There is no clearly unequivocal result that can be inferred from the coal surface area factors, other than to state that coal 3 may be anomalous due to the breeze content of this nominally anthracitic coal.

8.   CONCLUSIONS ON THE COMBUSTION OF COAL-OIL MIXTURES

8.1. Atomisation and Mixing

For a coal-oil slurry of 40% coal by weight, some 10-25% appears as free coal particles on atomisation, i.e. 0.25-0.63 of the coal content.

This quantity of free coal varies according to both the mean droplet size distribution of the slurry, and the mean particle size and distribution of the coal. The criterion of six 'overlaps' means that particles of greater than 0.14 of the size of the corresponding droplet cannot stay with that droplet on atomisation.

The net result is that the smaller coal fractions are almost entirely retained within the oil droplets upon atomisation.

The Toor mixing index was adequate for this non-recirculating system.

8.2. Combustion of Slurry

It appears that the model of Braide[10] et al is
substantially correct. This is where the slurry droplets -
with fine coal particles enclosed - burns very much like
heavy fuel oil.

The more recent work of Jenkins[11] et al proposed a 'plum
pudding' model of slurry combustion for high volatile coats,
but a separate combustion for anthractie coals, paralleling
this work.

8.3. . Combustion of Coal Char Particles

This work programme has only investigated low volatile
anthracite and coke breeze, and the provision for volatile
matter evolution was unnecessary. Combustion of these
materials took place under conditions of diffusion limited
combustion and thus internal burning was not in evidence.
Clearly the diffusion limitation was due to both:

(i)   The smaller size fractions of coal combusting in the
      slurry droplets.

(ii) The very high values of gas temperature making the
      value of the surface reaction rate coefficient very
      high.

It might be expected that, with such high velocities
encountered, microturbulence might make a notable
contribution to the combustion processes. However, only
small systematic errors can be noticed between 200 and 240
ms^{-1} blast velocities. It might well be that
microturbulence does exist in the earlier stages of
combustion and is reflected in incorrect results for droplet
size distribution of the slurry. It is thus concluded that
an earlier port is needed to verify this phenomenon, and the
rate constants of the mathematical model may then be
modified.

8.4. Accuracy of the Mathematical Model

By statistical analysis of the results, it was found that
the model averaged better than 95% accuracy in 95% of cases.
It is concluded that the relatively crude model is adequate
for the predictive purposes intended. A typical set of
results from the model is shown in Fig.9.

9.   FUTURE WORK

Laboratory work has already begun on the combustion of
coal-water slurries in a similar programme to that outlined
here. This programme also compares and contrasts the
combustion obtained when the coal is air-carried in normal
pulverised fuel fashion. The whole programme is being
partially funded by the E.C.S.C.

REFERENCES

1) Dietz, J.R. Journal of Metals 1963-499.
2) Bocong, G. Metal Soc. Conf. Oct. 1982, Paper 8.
3) Peiri, G. J. Inst. F. Dec. 1973 pp 384-388.
4) Toor, H.L. J.A.I.Ch.E. March 1962 pp 70-78.
5) Long, V.D. J.Inst.F. Dec. 1964 pp 522-525.
6) Essenhigh, R.H. PhD. Thesis, Sheffield University, 195?.
7) Tu, C.M. Davis, H. and Hottel, H.C. Industr. Engineering
   Chemistry 23, 277-285 (1934).
8) Babii, C.I. and Iranova, I,P.
   Teploeneraetika 1958, 15 (12), 34-37.
9) Hottell, H.C. and Stuart, I.
   Indstr. Engineering Chemistry, 1940, 32, 719.
10) Braide, K.M.,Isles, R.G.,Jordan, J.B. and Williams,A.
    J.Inst. E (115) Sept. 1979 pp 115-124.
11) Jenkins, B.G., Alabaf, J.S., Patterson, M.C. and
    Moles, F.D.
    Proc. 64th C.I.C. Coal Symposium, October 1982.

APPENDIX 1

PROPERTIES OF FUELS USED

Slurry Component		Coal 1	Coal 2	Coal 3	Coal 4	Oil
Description		Coarsely Ground Coke Breeze	Finely Ground Coke Breeze	Coarsely Ground* Anthracite	Finely Ground Anthracite	Residual Class G
Proximate	% ASH	7.2	9.9	6.5	3.9	-
Analysis	% H_2O	8.9	1.3	7.3	4.8	-
daf.	% V.M.	1.0	1.1	2.9	4.9	-
Ultimate	% C	83.1	86.8	84.2	86.9	85.2
Analysis	% H	0.4	0.4	1.2	2.7	11.9
(As Fired)	% S	0.9	0.9	0.8	0.6	2.9
	% N	0.8	0.8	0.8	1.0	neg
	% O	0.9	1.0	0.9	0.8	neg
	% ASH	6.2	8.9	5.7	3.6	neg
	% H_2O	7.7	1.2	6.4	4.4	neg
Net C.V. (MJ/Kg)		28,950	30,250	30,100	32,900	39,700
Size	<125μ	94.5	98.6	83.7	99.6	-
Distrib-	<106μ	93.3	97.7	78.8	99.2	-
ution	< 90μ	89.4	95.7	71.5	98.5	-
	< 75μ	82.8	91.4	61.4	96.1	-
	< 63μ	76.3	86.3	52.7	92.7	-
	< 53μ	69.6	79.4	44.8	87.9	-
Rosin- Rammler Analysis	Constant (b)	77.3	68.5	329.9	42.1	-
	Exponent (n)	1.140	1.186	1.332	1.137	-
Mean particle size (μ)		45.3	35.3	77.6	27.1	-

* Found to be contaminated with coke breeze

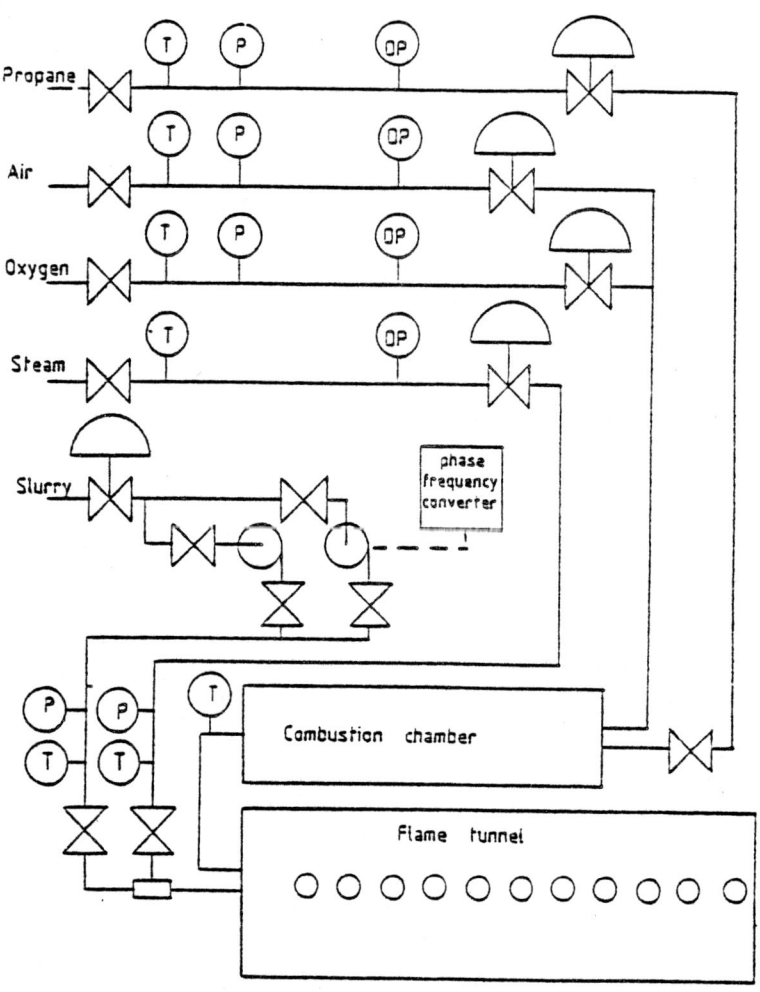

Propane

Air

Oxygen

Steam

Slurry

phase frequency converter

Combustion chamber

Flame tunnel

<u>Measurement and control of hot air blast + slurry</u>

Fig 1.

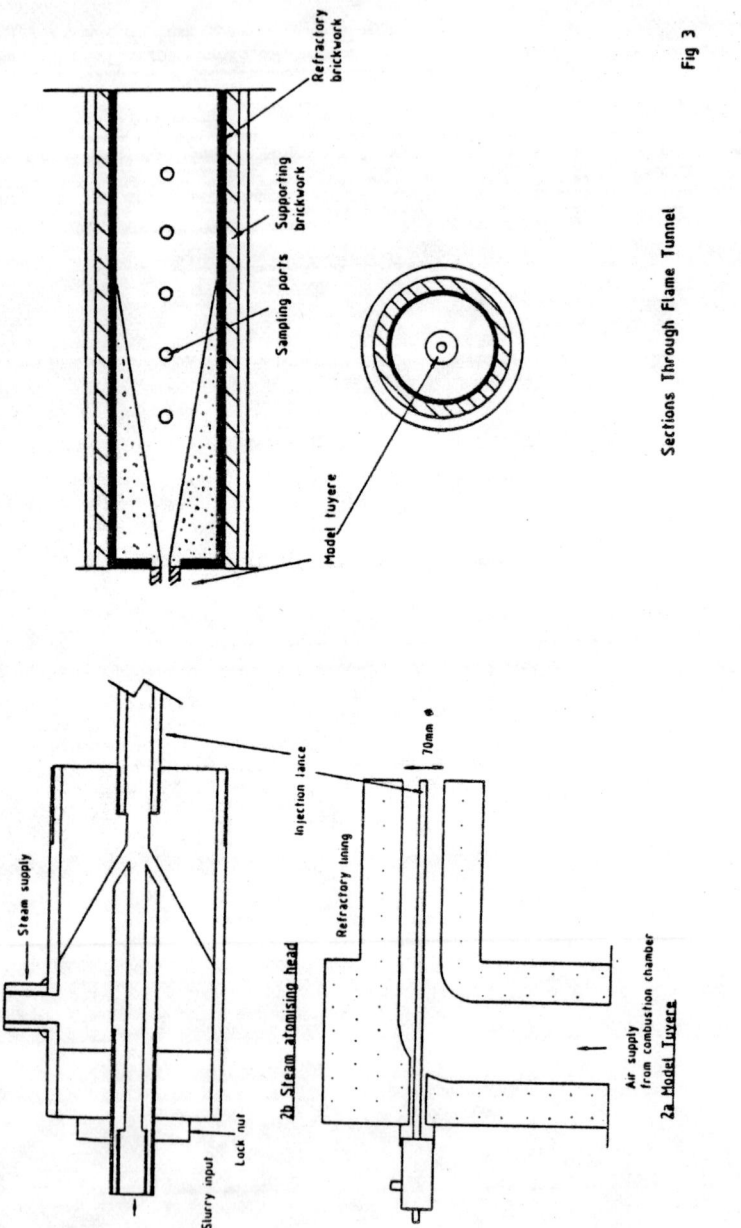

Refractory brickwork

Supporting brickwork

Sampling ports

Model tuyere

Sections Through Flame Tunnel

Fig 3

Steam supply

Injection lance

2b Steam atomising head

Slurry input

Lock nut

Refractory lining

70mm ∅

Air supply from combustion chamber

2a Model Tuyere

Fig 2

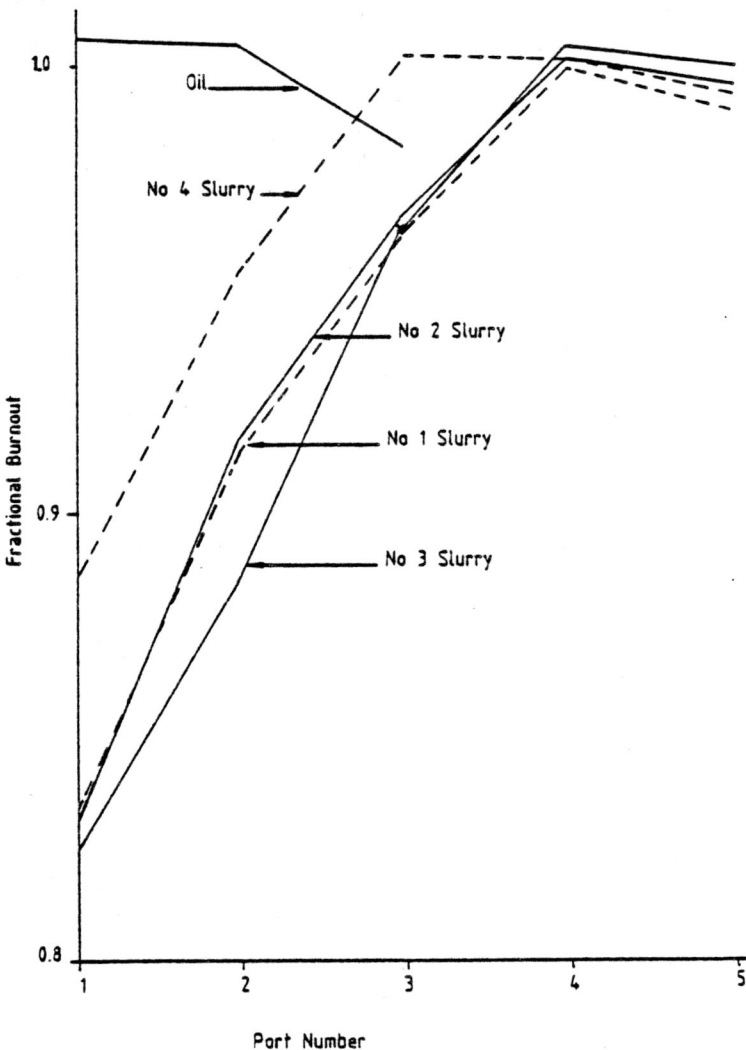

Fractional Burnout by Species of Slurry

Fig 4

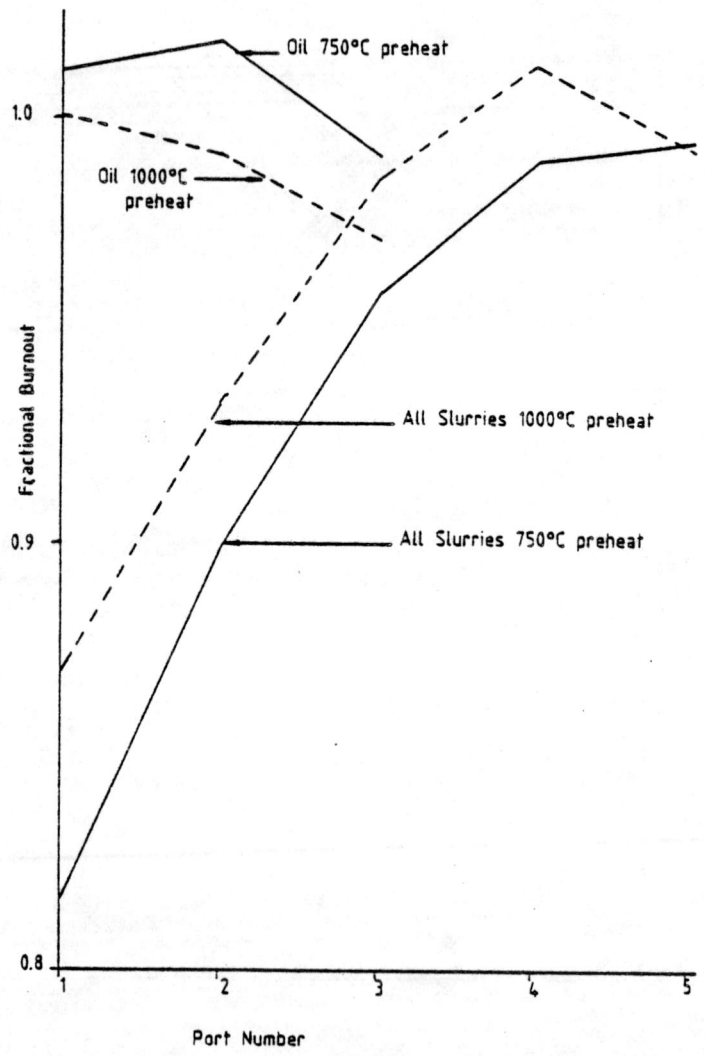

Effect Of Temperature on fractional Burnout

Fig 5

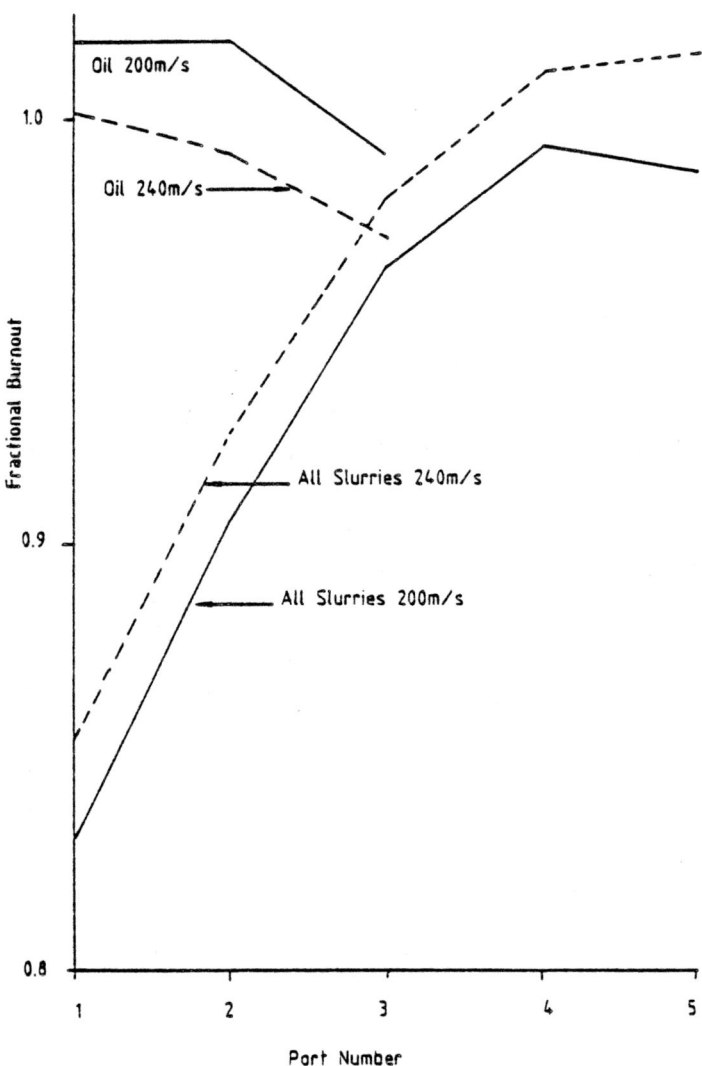

Effect Of Velocity on Fractional Burnout

Fig 6

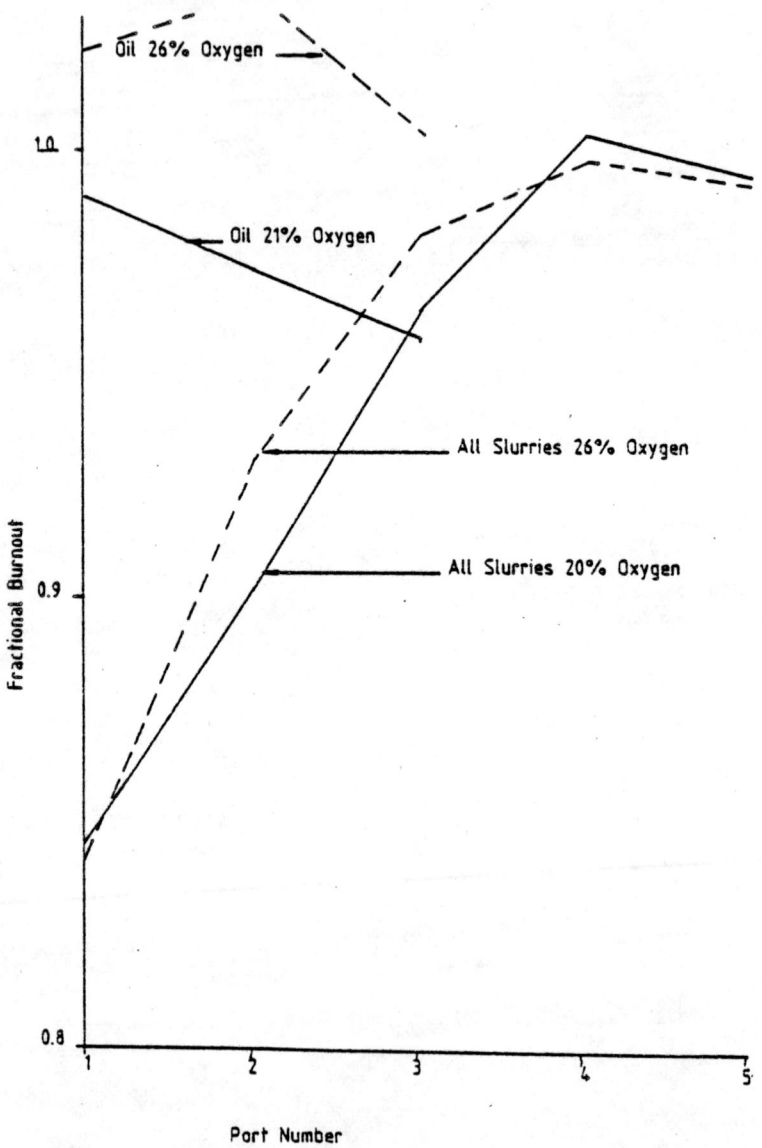

Effect Of Oxygen Enrichment on Fractional Burnout
Fig 7

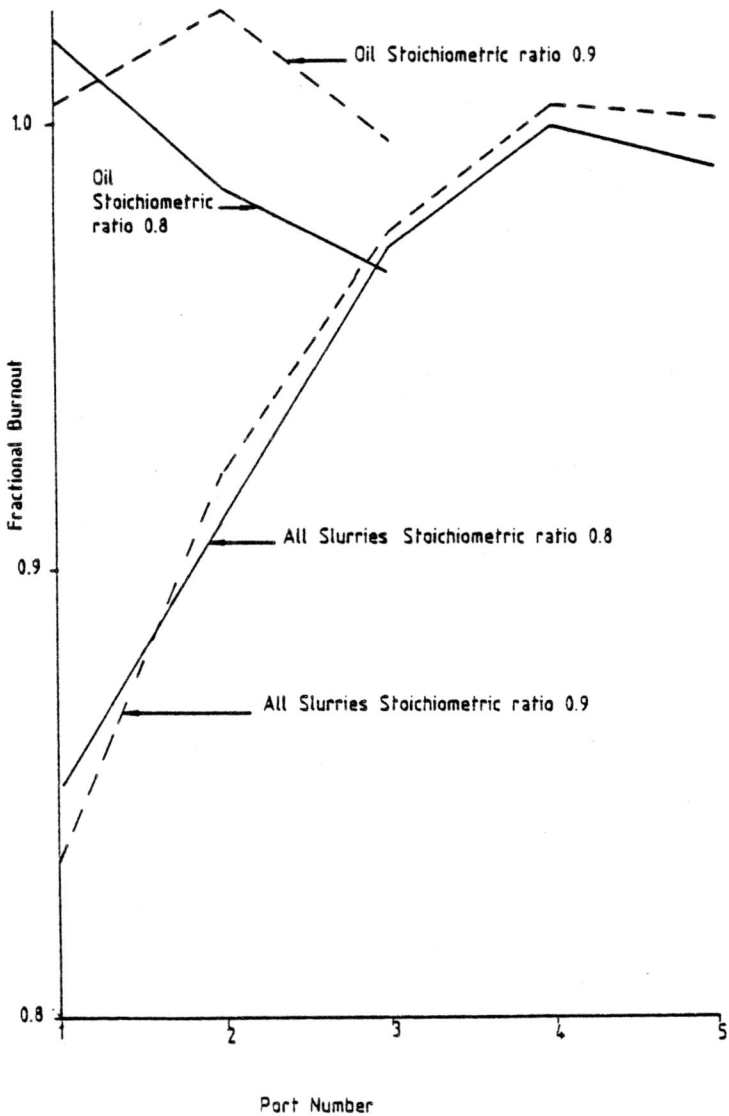

Effect Of Stoichiometric Ratio on Fractional Burnout

Fig 8

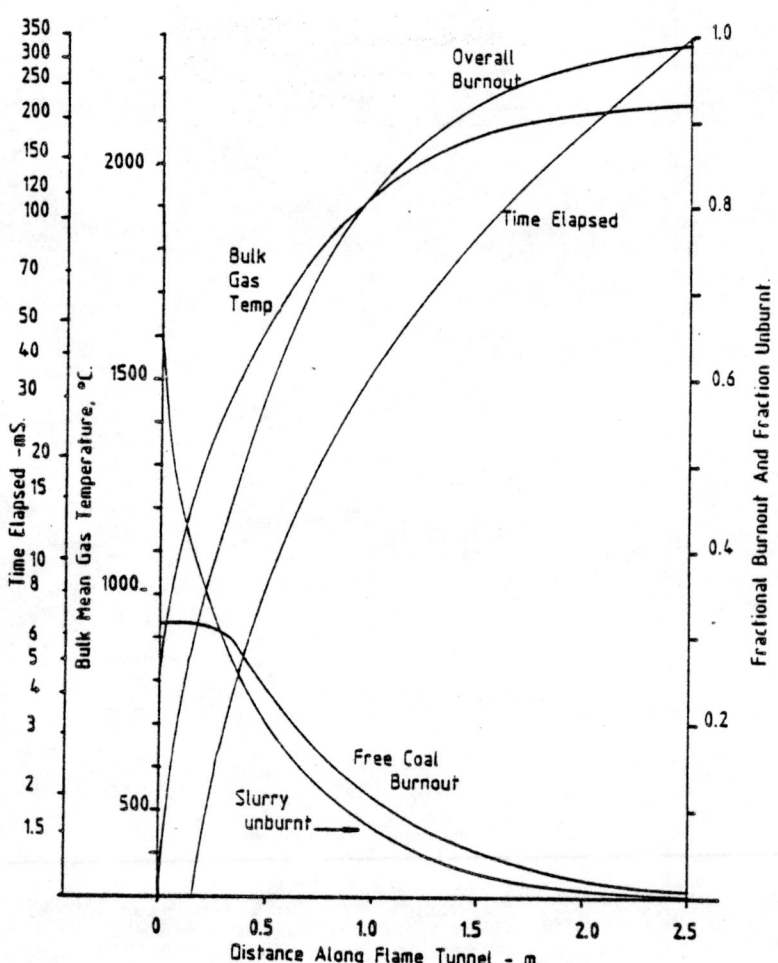

MODEL BURNOUT HISTORY OF RUN 75

FIG   9

CWM SUITABILITY FOR INJECTION INTO BLAST FURNACE

G. Malgarini, L. Palumbo and S. Palella*

* Centro Sperimentale Metallurgico, Via di Castel Romano, Roma, Italy.

The "Centro Sperimentale Metallurgico" has made an evaluation of the suitability of Coal-Water Mixture (CWM) for injection into the blast furnace. Encouraging results have
prompted investigations on laboratory and pilot scale for
CWM preparation, characterisation and handling. The feasi
bility of high coal content mixtures has been ascertained
and the influence of additives to achieve fuel products hav
ing good rheological properties and stability has been investigated. Pumping tests in a flow loop are now in progress. A 350 kg per hour pilot plant for the preparation
and injection of CWM into one tuyere of a commercial blast
furnace has been designed and is now under construction.

1. - INTRODUCTION

Much research has been performed in recent years to find new coal-based liquid
fuels that can replace fuel oil in whole or part. The work has been done main
ly by the electricity generating sector, and big energy users, such as the iron
and steel industry in particular.

In Italy, the Centro Sperimentale Metallurgico (CSM) - company of the Finsider
Group - has recently carried out laboratory research on the preparation and cha
racterisation of coal-liquid mixtures (CWM). The results obtained were then
checked on a pilot plant. At the present time the CSM is engaged on a project
financed by the European Coal and Steel Community (ECSC) for the construction
of a pilot plant for the preparation of a CWM containing 70% coal and its injection into one tuyere of a commercial blast furnace.

2. - LABORATORY STUDIES OF COAL LIQUID MIXTURES

Extensive laboratory investigations have been performed on the preparation
and rheological properties of coal-liquid mixtures to assess their feasibility.
The influence of particle size and coal content in various dispersing media
has been studied.

A summary of the laboratory characterisation results is given below.

2.1 - Preparation of coal-liquid mixtures

Coal-water mixtures (CWM), coal-oil mixtures (COM) and coal-oil-water mix
tures (COWM) have been prepared. Several types of coal have been tried, the
results reported being for a low-volatiles coal with a proximate analysis (dry

basis) of: ash $\sim$ 10%, volatile matter $\sim$ 20% and fixed carbon $\sim$ 70% (Hardgrove In
dex: 96.51). The fuel oil used was a commercial, low-sulphur type with 87.2%
C, 9.7% H, and a viscosity of 77.7 cSt at 70°C.

Fig. 1 illustrates the procedure adopted for mixture preparation, namely,
preliminary grinding with screening of the ungraded raw coal which is then mix
ed with oil or water and wet ground in a stirred laboratory ball mill without
additives.

The first fact to emerge from the mixture-preparation work was the maximum coal content attainable. In the case of coal-oil mixtures this is 60%,
which also holds good with regard to COWM when the oil and water are present in
equal proportions, namely 60-20-20%. With CWM instead, the maximum coal content is higher at 75%. It is often difficult to prepare these mixtures with a
high solids content and the use of suitable additives can sometimes be of deci
sive assistance in this regard.

2.2 - Characterisation of coal liquid mixtures

The CSM characterises coal-liquid mixtures by conventional chemical analy
sis, particle-size determination and apparent viscosity measurements. A laser
beam analyser (Microtrac) is used for size analysis of fine solids, while the
Brookfield LVT viscometer is adopted for determining apparent viscosity, the
30 rpm values being examined for the purpose of comparison.

By and large, it is confirmed that coal-liquid mixtures behave as non-New
tonian fluids, as reported by other workers, and that the apparent viscosity
increases with the suspended coal content and with diminution in mean diameter
of the solids[1]. The dispersing medium has a marked influence on viscosity, as
is evident from Fig. 2, which illustrates the apparent viscosities measured on
different mixtures and various coal contents (mean diameter 28 μm). Fig. 2 also indicates the viscosity of the oil used for mixture preparation (measured
at 80°C).

Comparison of the values shows that the apparent viscosity of the CWM at
20°C without additives is higher than that of fuel oil at 80°C, passing from
100 cP for a coal content of 50% to 1000 cP for one of 70%. The apparent viscosity of the COM at 80°C is higher than that of CWM at 20°C, for the same
coal content, passing from 350 to 1050 cP as the coal content rises from 50 to
60%.

Addition of water to COM to prepare COWM causes a further increase in apparent viscosity at 80°C.

The apparent viscosity of the mixtures can be reduced by means of additives, the value attained depending on the type and concentration involved. Fig.
3 provides an example of the results obtained on a CWM containing 70% coal. It
is evident that as the amount of additive increases the apparent viscosity first
decreases from 800 to about 350 cP, after which it rises. In the case of addi
tive A the minimum value is obtained with 0.5%, while with additive B 0.8% is
needed.

3. - FORECASTS ON THE USE OF CWM IN THE BLAST FURNACE

It is planned to use CWM in the blast furnace to replace the fuels inject ed via the tuyeres in the past, particularly fuel oil. It was thanks to the gradual perfection of the tuyere fuel injection technique that improvements we re achieved in the regularity of blast furnace operation and thus in overall efficiency. However, as a result of soaring costs in recent years, the use of fuel oil for this purpose has been abandoned[2,3]. Blast-furnace operation has been modified to permit the elimination of fuel oil or to ensure all-coke work ing[2,3], or in some cases to allow powdered coal injection (PCI)[4,5,6]. There have also been a few reports of the injection of COM[7,8] and TCM[8].

Only in the last two years has serious consideration been given to the pos sibility of using CWM as fuel[9,10]. The use of CWM in the blast furnace appears to be particularly attractive. The following increasing order of cost emerges from a recent economic analysis on the use of coal in the blast furnace[4]: all-coke, CWM, PCI, tar and oil.

The CSM has evaluated the feasible CWM injection rate in a medium-sized blast furnace under standard operating conditions[1]: Adiabatic Flame Tempera-ture 2200°C; Blast Temperature 1250°C; Blast Moisture 15 g/Nm^3. With these conditions the CWM injection rate can be as high as 63 kg/tHM and the replace-ment ratio for a 70-30 CWM injection rate of 54 kg/tHM can be evaluated at 1 kg of coal in the slurry to 1 kg of coke.

4. - PILOT SCALE CWM PREPARATION

The feasibility of continuous production of a 70-30 CWM has been tested adopting a wet grinding with a disc mill. In the pilot-plant arrangement the material to be ground is fed in above the two discs, the upper one of which is fixed (stator) while the lower one (rotor) gyrates at 3000 rpm. The material thus enters via the central opening of the upper disc and is reduced in size as it passes between the two discs.

Numerous trials were made to arrive at a flowsheet that is as simple as possible while ensuring continuous operation in the works. The trials were al-so conducted with a view to being able to use a commercially ground coal avail-able in a steelworks. As a result, a flowsheet[11] has been verified for the production of CWM with a coal content of over 70%, starting from a minus 3 mm feed.

A slurry is produced in a primary wet grinding where coal and water with an additive are fed in separately in the desired proportions (Fig. 4). The slurry produced by this primary grinding has coal content of 70-75% with a mean harmonic diameter of about 60 μm (60% < 74 μm and 20% > 250 μm). Under these con ditions, the pilot mill produces 100 kg/h slurry. Secondary grinding of this slurry, on a continuous basis at a rate of about 100 kg/h, reduces the size of the solids, producing a closer grading with a mean harmonic diameter of 50 μm and 70% minus 74 μm. Table 1 indicates the particle-size characteristics of a typical product obtained by this process.

5. - RHEOLOGICAL PROPERTIES OF CWM PRODUCED IN PILOT PLANT

The basic problems to be resolved for trouble-free commercial use of these mixtures concern pumping and storage.

Apparent viscosity determined in a Brookfield viscometer at a variable number of revolutions or the flow curve (shear stress plotted against shear rate) can be used to characterise the pumpability of the mixtures.

The apparent viscosity (Brookfield 30 rpm) of a CWM produced in the pilot mill is 230 cP. Flow curve measurements are now being made with a recently acquired Haake viscometer.

5.1 - Pumping circuit

The CSM has built a flow loop pumping circuit to acquire information on pressure losses of CWM during pipeline transport.

The circuit has been designed to simulate that which will be involved when the CWM has to be injected into the blast furnace via the tuyeres, assuming that the slurry-making unit is sited close to the BF and that the slurry is made on a continuous basis.

The circuit (Fig. 5) consists of about 60 m of $1\frac{1}{2}$" pipe with a level difference of around 10 m. The CWM is circulated by a screw pump, which draws the slurry from a barrel equipped with a mixer; the slurry is returned to the barrel after having completed the circuit. On the end of the pipe is a nozzle of the same size as the one that will be used for injection via the BF tuyeres. The constructional design of this nozzle is based on that of nozzles used to inject fuel oil, so no modifications will have to be made to the BF.

The first tests made using this rig indicate that a pressure of 7 kg/cm^2 is needed for a flow rate of 380 kg/h.

5.2 - CWM stability

Various tests have been evolved for characterising the static stability of CWM and for studying its behaviour as various relevant factors are changed. It would appear that the Rod Penetration Test[12] is the best and simplest for the purpose. This involves making weekly measurements of the time needed for a 3 mm diameter rod weighing 20 g to penetrate a 180 mm layer of unstirred material. The greater the increase in penetration time compared with that of the first measurement at the zero time, the less stable is the suspension. In some cases the rod does not penetrate to the bottom of the container; this is indicative of the tendency of the material to settle out. The measurement is repeated in various points to eliminate false results due to the formations of preferential passages. It thus ensues in some cases that there are differences in measured penetration times, which point to a lack of uniformity of the suspension as depths vary.

Tables 2-5 report the penetration times measured over four successive weeks in CWMs with different types and quantities of additives (all produced on the pilot plant). Tests have been made with steadily greater quantities of

the additive used for production of the slurry (Type A, tab. 2) and with the
addition of other substances;B (tab. 3), C (tab. 4) and T (tab. 5). For each
CWM the apparent viscosity at zero time was also measured. It can be seen that
from the practical point of view with the additions adopted there were no marked improvements in static stability compared with the reference mixture. Two
important points emerge from these initial results:

1) the CWM produced has good static stability and
2) improvement in stability (Case B 0.05%) involves a big increase in apparent
   viscosity that renders the product unusable.

In this regard it should be noted that if the CWM is produced in a plant
located close to the blast furnace, it is unnecessary for the mixture to have
very high stability. For direct use in the blast furnace it is as well for the
suspension to have an apparent viscosity that is not high, so as to limit pres
sure losses.

6. - FUTURE DEVELOPMENTS

The results presented here have been obtained in the first phase of a
project financed by the ECSC for the construction of a pilot plant for the pro
duction of 350 kg/h CWM and the injection into one tuyere of a commercial blast
furnace.

The CSM is now building the plant as per the flowsheet described. Meanwhile the study of the rheological properties of the mixtures is continuing
from the practical point of view (pumping circuit) and in the laboratory, to
learn more about the influence of additives on viscosity and static stability
to improve these parameters.

7. - CONCLUSIONS

Comparison of various types of coal-liquid mixtures (CWM, COM and COWM)
prepared in the laboratory shows that:

- for the same coal content and particle size, the apparent viscosity of CWM
  at 20°C is by far the lowest;
- with CWM it is possible to add a higher proportion of coal, 75%, against only
  60% with coal-oil mixtures;
- the addition to coal-oil mixtures of significant water amount (higher that 10%)
  further increases the apparent viscosity.

A pilot-scale process has been developed for the preparation of 100 kg/h
of 70-30 CWM with good rheological properties. The product is characterised
by an apparent viscosity of about 250 cP and good static stability after four
weeks.

A pilot plant for the production of 350 kg/h CWM and its injection into a
blast furnace tuyere is now at an advanced stage of construction.

ACKNOWLEDGEMENTS

The authors wish to thank the European Coal and Steel Community for finan
cial aid provided by Contract 7220.ED.401 for the work presented here.

BIBLIOGRAPHY

1) Malgarini G., Giuli M., Palumbo L., and Palella S.
"Evaluation of CWM use in ironmaking" Fifth International Symposium on Coal Slurry Combustion and Technology, April 1983, Tampa, Florida.

2) Kajikawa Y., et al.
"Operation of blast furnace without liquid fuel" Trans ISIJ 1982, v. 22, p. 134.

3) Palchetti M., Giuli M. and Vecchiola G.
"Marcia dell'altoforno senza iniezione di fuel-oil alle tubiere" Bolletti no Tecnico Finsider 1982, n. 393 p. 24.

4) Sakurai S., Takahashi H., Suemori A.
"Injection of coal as auxiliary fuel for blast furnace" International con ference on direct use of coal in iron and steelmaking - 19-20 Oct. 1982, London.

5) Monson J.R., Gathergood D.S.
"Review of injection of coal into blast furnace tuyeres" Ironn. & Steelm. 1981, n. 3 p. 101.

6) Hemming C.H., Carter G.C.
"Coal injection systems for blast furnaces" Ironm. & Steelm. 1981, n. 3 p. 104.

7) Miyazaki T. et al.
"Utilization of COM for injection to blast furnaces" Trans ISIJ 1982, v. 22, p. 207.

8) Kurashige I. et al.
"COM Injection into All Tuyeres of Kashima No. 3 Blast Furnace" Fifth International Symposium on Coal Slurry Combustion and Technology, April 1983, Tampa, Florida.

9) Fourth International Symposium on Coal Slurry Combustion, May 1982, Orlando, Florida.

10) Fifth International Symposium on Coal Slurry Combustion and Technology, April 1983, Tampa, Florida.

11) Palumbo L., Malgarini G.
Italian Patent pending No. 48185A83.

12) Shimamura Y., Ukigai T. and Igarashi T.
"Studies on the stabilization of COM" Ref. 9.

Composition

Coal content	(%)	:	70
Additive: type A	(%)	:	0.5

Grain size

> 250	μm	(%)	:	6.3
250-177	μm	(%)	:	8.0
177-74	μm	(%)	:	15.4
< 74	μm	(%)	:	70.3
Mean diameter	(μm)		:	48.4

"Microtrac" analysis (under 177 μm) :

90° percentile	(μm)	:	103.7
50° "	(μm)	:	23.1
10° "	(μm)	:	3.2
Mean diameter	(μm)	:	39.7
Specific surface	(m^2/cm^3)	:	0.593

Apparent viscosity (20°C)   (cP) : 230

(Brookfield 30 rpm)

TABLE 1 - Characteristics of a typical 70-30 CWM made by CSM (Sample N. 223).

Second Additive		Apparent Viscosity (Brook-30rpm) cP	Rod penetration time (s)					Static Stability Index
type	wt %	0 time	0 time	I week	II week	III week	IV week	
-		590	0.5	0.3	0.4	0.5	0.8	3
A	0.2	270	0.3	0.4	0.9	1.5	D	D
"	0.5	480	0.3	0.3	0.5	0.9	0.9	2
"	1.0	610	0.3	D	1.5	1.8	2.0	D

TABLE 2 - CWM stability (Additive A).

Second Additive		Apparent Viscosity (Brook-30rpm) cP	Rod penetration time (s)					Static Stability Index
type	wt %	0 time	0 time	I week	II week	III week	IV week	
-	-	590	0.5	0.3	0.4	0.5	0.8	3
B	0.05	3500	2.1	2.4	2.5	2.2	2.1	∞
"	0.10	8000	3.2	4.0	D	2.4	3.2	D
"	0.25	16800	———————		U	———————		U
"	0.50	85000	———————		U	———————		U

TABLE 3 - CWM stability (Additive B).

CWM characteristics:

Coal content (%) : 70

First additive A (%) : 0.5

Mean Diameter (μm) : 47

% under 74 μm (%) : 72

Temperature (°C) : 20

Note:

D : CWM not homogeneous

S : Part non penetrated shorter than 3mm

U : Part non penetrated larger than 3mm

Static Stability Index = $\dfrac{1}{t^{IV} - t^{0}}$

Second Additive		Apparent Viscosity (Brook-30rpm) cP	Rod penetration time (s)					Static Stability Index
type	wt %	0 time	0 time	I week	II week	III week	IV week	
-	-	590	0.5	0.3	0.4	0.5	0.8	3
C	0.05	490	0.5	0.3	0.5	0.7	1.0	2
"	0.10	2400	0.7	0.5	0.8	0.7	1.0	3
"	0.25	3300	0.8	0.7	1.2	1.1	1.6	1
"	0.50	3800	1.5	1.4	1.0	1.0	1.6	10

TABLE 4 - CWM stability (Additive C).

Second Additive		Apparent Viscosity (Brook-30rpm) cP	Rod penetration time (s)					Static Stability Index
type	wt %	0 time	0 time	I week	II week	III week	IV week	
-	-	590	0.5	0.3	0.4	0.5	0.8	3
T	0.05	860	0.5	0.3	0.4	0.7	0.7	5
"	0.10	840	0.5	0.3	0.4	0.7	0.5	∞
"	0.25	760	0.4	0.3	0.5	0.5	0.8	2
"	0.50	640	0.3	0.3	0.5	0.6	1.1	1

TABLE 5 - CWM stability (Additive T).

CWM characteristics:

Coal content (%)　　　: 70

First additive A (%) : 0.5

Mean Diameter (μm)　: 47

% under 74 μm (%)　: 72

Temperature (°C)　　　: 20

Note:

D : CWM not homogeneous

S : Part non penetrated shorter than 3mm

U : Part non penetrated larger than 3mm

Static Stability Index = $\dfrac{1}{t^{IV} - t^{0}}$

CLM-T

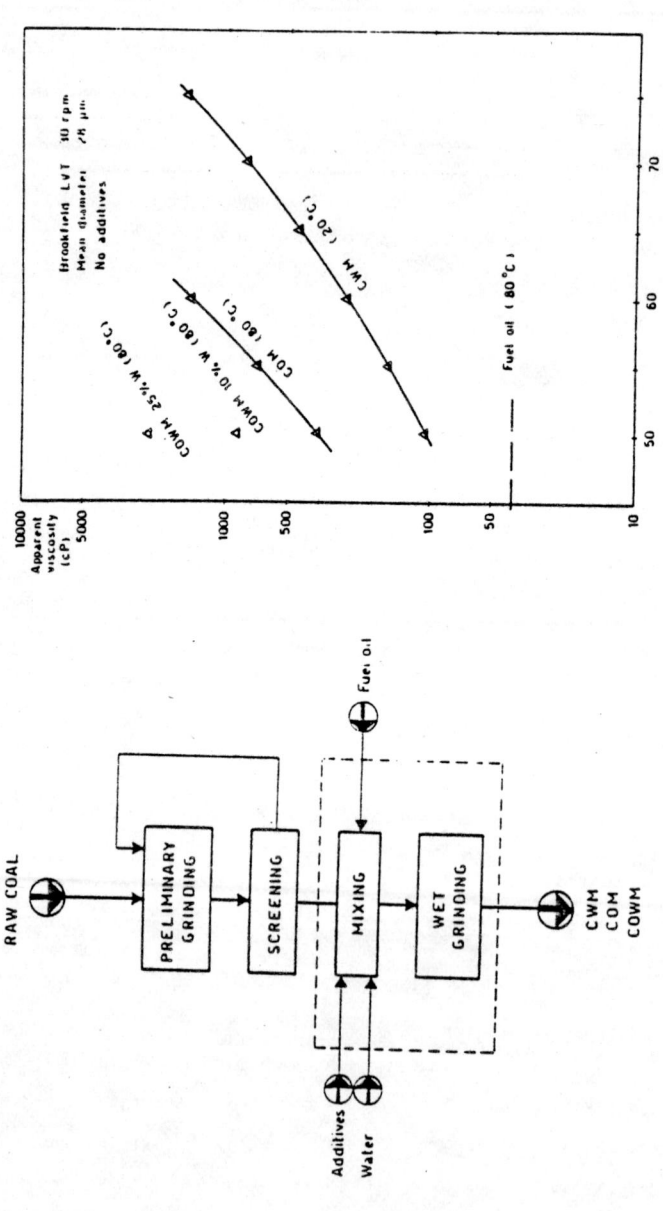

Fig 2 - Influence of the coal content on the slurry apparent viscosity of Coal - Liquid Mixtures (CWM, COM and COWM)

Fig. 1 - Block - diagram of coal liquid mixtures preparation in laboratory.

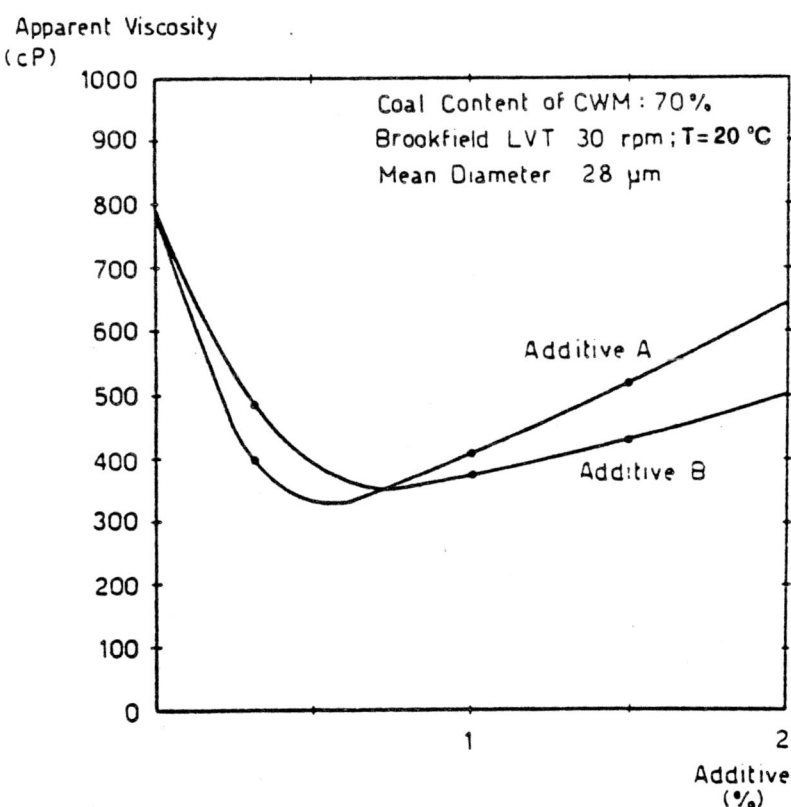

Fig.3_ Influence of additive substances on the CWM apparent
viscosity .

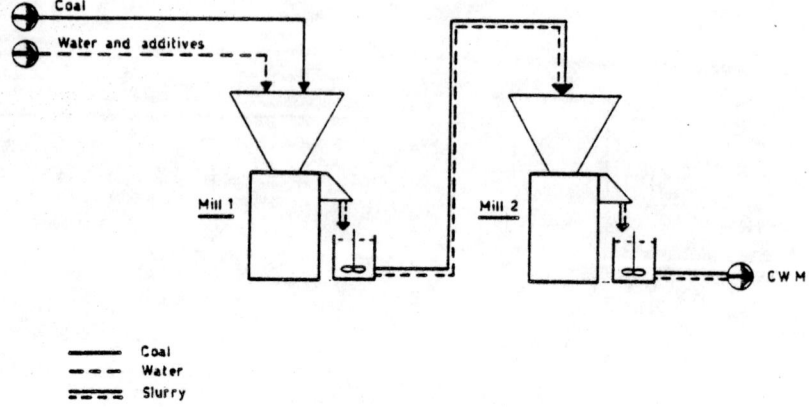

Fig. 4 - CWM preparation scheme.

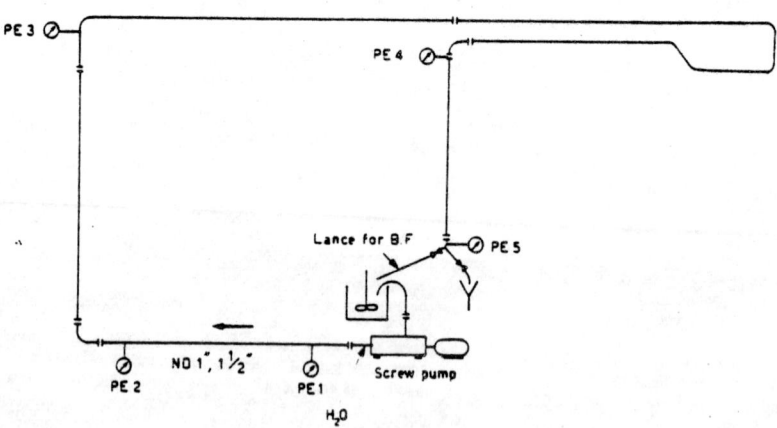

Fig. 5 - Coal-water mixture pumping test loop

A PRELIMINARY STUDY FOR THE CONVERSION OF THREE SIZES OF OIL FIRED INDUSTRIAL BOILERS TO COAL WATER FUEL

W. STOCKDALE & D.J. VINCENT

This paper presents some of the engineering considerations involved in the conversion of oil fired industrial boilers to coal water fuel (CWF). The preparation and properties of CWF have not been included as sufficient data is available elsewhere. The case for CWF firing is reviewed together with the general engineering requirements for the conversion of oil fired industrial boilers. Performance predictions have been made for three sizes of boiler and the findings discussed. In-house computer models have been used to modify existing operational data on oil firing to predict the performance of boilers when firing CWF. The experience gained from, as yet unpublished results of CWF firing trials, has also been used.

1.0 INTRODUCTION

THE CASE FOR COAL WATER FUELS

Many industrial nations are economically dependant upon imported crude oil and have, in consequence, suffered as a result of oil price fluctuation. Coal reserves far exceed those of oil and are well distributed throughout the world, resulting in a more stable and predictable price structure. This has created the incentive to develop the technology for the preparation of coal in a form in which it is compatible with oil.

Coal oil mixtures (COM) have been successfully produced (1). More recently emphasis has been placed on the use of Coal Water Fuels. These are based on the preparation of a stable suspension of beneficiated coal in water.

The purpose of this paper is to investigate the use of coal water fuels in existing industrial boilers. Three boilers have been selected to illustrate the potential for conversion in the range 14 tonnes per hour to 68 tonnes per hour of steam.

FOSTER WHEELER POWER PRODUCTS, LONDON, ENGLAND

1.0   Continued

The principle advantages of CWF are as follows:-

(a)   Predictable price structure and reliable availability
      of base coals.

(b)   Current technology needing little adaptation for the
      preparation of CWF. Minimal modification of existing
      oil storage facilities and no radical innovations in
      burner design required.

(c)   Compared with conventional coal conversions, CWF offers
      simplicity of transport, handling, storage and eliminates
      dust and associated hazards at transfer and storage points.

(d)   The ready adaptation of existing liquid fuel distribution
      networks to suit the smaller user.

(e)   Potential for pipeline transportation which may be attract-
      ive for the larger user.

(f)   Beneficiated coals used in CWF result in lower SO_x emiss-
      ions and reduced ash related problems.

(g)   The possibility of extending the economic life of suitable
      existing oil boilers by converting to CWF.

(h)   Where high turndown is required, the advantages of CWF
      over dry PF result from the substitution of water for
      transport air providing stable flames at lower loads
      eliminating support fuel requirements for high volatile
      coals.

2.0 ENGINEERING CONSIDERATIONS

2.1 Fuel Storage

It is anticipated that CWF could well be produced at
central preparation plants with distribution by road
or rail tankers to local storage depots, or direct to
customers.

Existing oil storage tanks may well be appropriate to
receive the pumped slurry, providing they are adequately
cleaned. The energy density of CWF, is typically
30.4 GJ/m^3, intermediate between that of bulk coal
(22 GJ/m^3) and oil (41 GJ/m^3). Hence additional storage capacity may be required particularly if the oil
fired option is retained to enhance fuel flexibility.

2.1 Continued

The stability of CWF during storage is generally accepted
as satisfactory. However, as a precaution, a low speed
impeller could be used in tanks to prevent settling over
long periods.

In contrast to oil CWF is pumpable at ambient temperature
and tank or trace heating is not normally required except
when sub-zero temperatures are expected over a long period.

In the designs of CWF transfer lines precautions must be
taken to avoid local areas of high temperature, which
could result in cake formation on the pipe walls.

2.2 Pump Selection

Positive displacement pumps are the most satisfactory for
slurry handling, since they operate at low shear rates
and it has been demonstrated that high shear rates adversely affect the properties of the slurry (2).

To define the required pump specification, it is necessary
to predict the friction losses for the pipe diameter and
flow rates in question. For this purpose the average
viscosity of the mixture along the length of the pipeline
must be evaluated. Details have been described elsewhere
(3 & 4).

In practice it is advisable to size the pump for a pressure loss twice that determined from these procedures
(4).

Coal water slurries are generally classified as moderately
abrasive and appropriate measures should be taken to maximise the service life of rotating parts. Positive displacement pump manufacturers have specified either 18/2
stainless steel or chrome plating for the rotor and shaft
with a natural or synthetic rubber for the stator. Pump
speeds should be restricted to minimise abrasion and to
limit the shear rate.

2.3 Pipe Sizing

To reduce the shear stress and abrasion, pipe velocities
should be in the range 0.2 to 1.0ms^{-1}. Long radius bends
should be installed instead of elbows to reduce erosion
and pressure losses.

2.4 Flow Control System

The control system should be designed in accordance with
appropriate codes (5).

2.4 Continued

Two pumps with variable speed motors should be specified
for boiler conversion, each being capable of providing
the total flow for MCR output. Relief valves would be
installed at pump outlets. Control of flow under normal
operation maybe achieved by means of the variable speed
motors with actuator operated safety shut-off valves at
each burner.

A manually operated ball valve should be provided at each
burner for commissioning and light up purposes.

CWF has been shown to be non-corrosive (4) and carbon steel
can be specified for pipework.

2.5 Flow Measurement

Flow rate is directly proportional to the speed of rotation of the progressive cavity pump and calibration of
the pumps during commissioning, will enable flow rates
to be monitored during operation.

2.6 Burners

To produce stable CWF flames with no support fuel requirement, purpose designed burner systems are required which
atomise the slurry and promote rapid mixing with combustion air. Sufficient radiant heat must be provided at
the burner throat in order to evaporate the water and
release and ignite the volatile matter. In practice this
is provided by a refractory quarl.

Several burners have been developed to achieve this and an
example of a commercially available system (made by Forney
International) is shown in Figure 1. Relatively low
pressure atomising air is used in order to avoid erosion
of the burner tip. Pre-heated air (100-150°C) is introduced with a high degree of swirl to provide a stable
flame. Factors such as reactivity and volatile content
of the coal may have a significant effect upon preheat
requirements and in some cases stable flames have been
achieved with no pre-heat.

Following proving trials in a flame tunnel this burner
was subsequently fitted to a utility boiler in order
to demonstrate CWF firing.

2.6 Continued

If a conventional air preheater is not available, pre-heated combustion air can be conveniently provided by direct gas firing at a suitable point in the air duct between the forced draught fan and burner windbox. Air temperatures of 100-150°C result in only a small decrease in oxygen concentration which does not appear to adversely affect combustion.

The burners specified are capable of dual fuel firing; oil firing will normally be used for warming the boiler through and heating the refractory quarl in order to provide a stable flame when CWF is introduced.

Conventional ignition and flame monitoring equipment has been specified for each burner. Turndown ratios in excess of 3:1 have been shown to be practical possibilities without using support fuel.

2.7 Furnace Design

Furnace design is of considerable importance in determining the suitability for conversion. The furnace must be large enough to enable complete combustion of coal particles to occur and provide sufficient cooling of combustion gases to minimise slagging. The principle furnace parameters controlling the effectiveness of conversion to CWF are:

          a) Furnace configuration
          b) Furnace heat release rates (volumetric)
          c) Furnace liberation (surface)
          d) Furnace exit gas temperature
          e) Furnace exit gas velocities
          f) Flame path available and residence time
          g) Ash removal

The listed parameters have been evaluated and compared with those appropriate to conventional pulverised fuel (PF) firing and on this basis some de-rating of the boiler may be required either as a result of high exit temperature (leading to slagging) or high exit gas velocities (leading to erosion of screen tubes). However, as a result of benefication, the problems typically encountered with PF firing should be considerably reduced. More accurate design data will accrue for CWF firing as boilers are converted which may change the design basis. Flame characteristics have been observed to be more akin to oil flames than PF flames and it may transpire that design data nearer that for heavy fuel oil firing may be more appropriate than PF design data.

2.8 Boiler Passes

Acceptable gas velocities for oil firing are considerably higher than for coal firing due to the low solids loading. Conventional coal firing design data has been used for the CWF conversions to date. In some cases this limits the maximum output especially with modern designs of oil fired package boilers which have close pitched super-heaters and generally high velocities throughout.

The use of beneficiated coal (as in CWF) may raise the acceptable gas velocities. However, flyash collecting in boiler passes is likely to require attention.

The installation of sootblowers for the periodic removal of flyash is preferential on economic grounds to modification of pressure parts for continuous collection of ash in hoppers.

2.9 Boiler Emissions

In order to specify the most practical and cost effective gas clean-up equipment, data has been obtained from CWF firing trials regarding dust loading, particle size distribution and particle density. Electrostatic precipitators, bag filters and combinations of high efficiency cyclone and bag-filter, have been considered in each case and selection for any boiler conversion will depend on overall factors rather than the technical performance for a particular piece of equipment.

As a result of coal benefication sulphur emission levels can be reduced compared with direct coal firing.

Levels of nitrogen oxides have also been monitored during CWF firing trials and available data indicates that levels of NO_x vary between 50% and 100% of those typically produced in the combustion of pulverised coal (6 & 7).

The majority of ash will be collected from cyclone bins and bag-filter or electrostatic precipitator hoppers. This will be transported pneumatically into suitable sealed containers for disposal in a conventional manner.

2.10 Forced Draught & Induced Draught Fans

Existing fans should be utilised whenever practical. In most cases the additional fan duty resulting from higher air and gas weights is offset by the boiler derating. In cases when only forced draught fans are installed, resulting in a pressurised furnace, the problems of gas leakage are exacerbated by the fly ash particles. However, the three boilers considered for the study were all balanced draught units.

3.0 CASE STUDIES OF BOILER CONVERSIONS

FROM OIL TO C.W.M.

3.1 General Descriptions

Three different balanced draught liquid fuel fired
boilers were selected for study. Deliberately, both
a range of evaporation rates and of commissioning dates
were considered. Specific details of design and the
effects of conversion are provided in the sections
which follow, but, in general terms, the study embraced:-

| | | | | | Furnace Details | |
Boiler	Age (Yrs)	Evaporation (t.p.h.)	Pressure (Bar)	Temp. (°C)	m^3/MW	% Output
1.	25	14	10	210	2.5	Basis
2.	20	34	17	238	1.9	76
3.	15	68	34	316	1.4	56

Boiler No. 1 (Fig. 2) is typical of small, site assembled
bi-drum, "D" type boilers of its era. The furnace is
constructed with front, side walls and roof of open tube
construction with refractory backing and it is fitted
with an uncooled refractory floor. Four oil burners
are fitted in a rectangular array and they are disposed
to fire directly across the furnace towards an exit
screen of staggered tubes behind which is located a
single loop pendant superheater. The convection bank
is also typical of older bi-drum boiler designs in that
it comprises two gas passes separated by a central wall.
The installation is completed by a separate economiser,
containing rectangular gilled cast iron tubes on in-line
square pitch arranged in two banks

Boiler No. 2 (Fig. 3) is an early module built unit of
skin case construction utilising tangent tube construction for the major portion of the furnace. Floor tubes
are protected by loose super fire brick, which material
is used also for the construction of the boiler front
wall, which is uncooled. Two oil burners are fitted
on vertical centres and they are disposed to fire along
the furnace. A four row square pitch screen is provided
at the furnace exit aperture i.e. the entry to the reverse pass, behind which is located a cavity with a twin
vertical tube drainable superheater. The square pitch
is extended partially into the convection bank, the
majority of which is arranged on conventional triangular
pitch without baffles. A separate single tier economiser

3.1 Continued

arranged with two gas passes is also provided external to the boiler.

Boiler No. 3 (Fig. 4) is a relatively modern works assembled unit of mono wall construction including the baffle wall between the furnace and the convection bank. Floor tubes are protected by castable refractory which is also used in conjunction with super fire brick as the construction material of the boiler front wall which is also uncooled. Two oil burners are fitted on vertical centres and as with boiler No. 2 they are disposed to fire along the furnace. Convection bank tubes on triangular pitch are provided in the way of the furnace exit aperture and a super-heater of inverted drainable type, arranged for ease of withdrawal and repair, is interleaved within the convection bank. This unit also is fitted with an economiser which is of welded steel fin type.

3.2 **Boiler Performance**

Performance data for the original fuel and heat transfer characteristics were modified to account for the new characteristics appropriate to firing coal water fuel. Furnace, superheater and convection bank performance results are detailed below but it is of interest to note that overall the conservatism inherent in the design of boiler No. 1 enables us to predict that no deterioration in performance should occur when fuelled by C.W.F. In contrast, the later boilers are dimensioned such that de-rating is essential if C.W.F. is to be adopted. The following tabulation will be of interest as it illustrates that the reduction in heat input required to meet furnace design limitation (furnace exit temperature) coincides with the de-rating required in order to meet convection pass limitation (gas velocities).

Boiler	Age	Furnace	De-rating
1	25 yrs	100%	Nil
2	20 yrs	76%	26%
3	15 yrs	56%	40%

Modern designs are marginally more highly rated than boiler No. 3 and conversion to C.W.F. may necessitate de-rating by up to 50%.

3.3 Furnace Performance

Furnace performances were reviewed from the standpoints
of flame length, flame clearance, furnace exit temperature and inter-tube velocity at the furnace exit
aperture.

For boiler No. 1 the predicted F.E.G.T. of 1066°C
was considered to be within acceptable limits as
was the maximum gas velocity at 5 metres per second.
However, the furnace depth to the exit screen is
approximately 3 metres and it is considered prudent
to protect the screen tubes from direct flame impingement by the installation of a refractory wall to deflect
the flames upwards. This proposal is illustrated in
Figure 2. This wall will increase the residence time
and will also provide another radiant source to assist
combustion of coal particles. An initial setting phase
during re-commissioning will be required to attain the
optimum flame shape to avoid direct impingement on to
the refractory and reduce any tendency for slag accumulation. In view of the very low gas velocities predicted, a proportion of the furnace exit aperture may be
occluded by the wall proposed without detriment.

For boiler No. 2 the F.E.G.T. predicted was in excess of
1200°C which was regarded as inappropriate. Therefore,
it is proposed to de-rate the furnace to 74% of the original M.C.R. when fired with C.W.F., at which rating the
F.E.G.T. approximates to 1000°C. The flame path available is approximately 6 metres long which is adequate
to provide residence time at above 1000°C sufficient
to ensure high carbon utilisation. Under the de-rated
condition, the square pitch of the screen is such that the
maximum flue gas velocity is approximately 15 metres per
second, which is regarded as suitable.

For boiler No. 3 the F.E.G.T. predicted was also considered to be excessive without de-rating. In this case,
however, in order to limit the temperature to 1000°C
the furnace must be de-rated to only 60% of the original
value. Even in this case, the maximum velocity over
the screening convection bank in the way of the furnace
exit aperture is 21 metres per second; close to the
limit to which most boiler manufacturers normally recomm-
end.

3.4 Burner Position

Four burner positions are available in boiler No. 1
of which two will be utilised for the conversion to
CWF. The clearances between burner centres and furnace
side walls are in excess of 1 metre but the clearance
between the lower pair of burners and the refractory
floor is just over half a metre. It is felt that this
precludes their use due to the possibility of flame
contact and slag accumulation. Consequently, it is
proposed to install CWF burners in the upper row and
to fit the refractory wall mentioned in section 3.3.

The minimum clearances in boilers 2 and 3 are 1 metre
and 1.6 metres respectively and no alteration to the
burner locations is considered necessary.

3.5 Superheater Performance

On all the boilers considered the superheater performance
will depend upon the condition of the tubes in addition
to the flue gas heat transfer properties. However,
each boiler is fitted with soot blowers located in
cavities between the superheaters and convection banks.
Assuming that the blowers are effective, it is predicted
that the superheater in boiler No. 1 will over perform
by approximately 5°C, despite the influence of the refractory wall mentioned in 3.3 above. The superheater
of boiler No. 2 is predicted to over perform by approximately 3°C whereas that of boiler No. 3 is expected to
manifest no alteration in performance.

Since these values are well within the range of practical
operating tolerances, the substitution of C.W.F. for the
original fuel will necessitate no modification to any of
the three superheaters considered.

3.6 Convection Bank Performance

The performance of the unbaffled convection banks of
boilers 2 and 3 are comparatively simple to analyse.
However, the baffles fitted to boiler No. 1 will create
complex flow patterns which could affect the accuracy
of performance prediction. As with the superheater
performance, the convection bank of boiler No. 1 is
predicted to over perform slightly and it is concluded
that C.W.F. may be utilised in this boiler with no
significant alteration in overall performance. For
boilers Nos. 2 and 3 the convection bank calculations
indicate that performance will correspond to the derating proposed from furnace considerations.

3.7 Economiser' Performance

The performances of the three economisers have been
calculated to assess the feed water and flue gas
exit temperatures. All figures predicted are in
close agreement with existing values.

3.8 Fly Ash

The problems of deposition of fly ash in furnace and
boiler passes have been considered in the light of
experience gained in the course of firing trials of
C.W.F. In no case considered is ash deposition on
to the floor of the furnace anticipated to be a significant problem.

Both ash and unburnt carbon will accumulate when gases
change direction significantly, however, for boilers
2 and 3 the use of the soot blowers is considered
likely to eject such material into the economiser
from which some fall-out may have to be removed.
For boiler No. 1, however, the convection pass design
is such that material deposited may fall through the
tubes and accumulate in the brick built chamber which
encloses the water drum. Manual raking via the access
doors is not an acceptable method of removal and it is
proposed to effect minor modifications to enable a solids handling system to be installed.

3.9 Fans

C.W.F. firing requires higher air weight and results in
higher gas weight than does oil firing for equivalent
boiler output.

However, the increase can be accommodated in normal fan
margins when design output is obtained and if downrating
is essential, this factor is of no consequence unless
the pressure drop across flue gas cleaning equipment is
excessive.

4.0 DISCUSSION

It is apparent from this study of three different oil
fired boilers of differing capacities, steam conditions
and age that the primary considerations in any assessment of suitability for conversion to CWF are furnace
design, gas velocity and ash removal. Other aspects,
eg. superheaters, convection banks, fans etc. have
been shown to be of lesser importance in comparison.
Moreover, one overriding factor influencing suitability
for conversion would appear to be the age of the boiler.

4.0   Continued

Over recent years, the effectiveness of heat transfer
surfaces has been improved as refractory materials
have been replaced by Mono-walls. Developments in
design data, predictive methods, materials and manufacturing technology resulting from the increased
competition in the boiler market has also resulted
in a reduction in design margins. Modern boilers
are tailored with precision to the requirements of
the fuel to be fired and offer, therefore, less flexibility if conversion to CWF firing is considered.

To overcome problems of modern furnace heat release
rates and gas velocities higher than acceptable for
CWF firing, boiler down-rating is essential. Sophisticated "Energy Management" on individual sites frequently result in much reduced steam demand and over-
capacity of boilers installed. Under such circumstances,
de-rating to accommodate CWF may still be acceptable
and worthwhile.

Firing trials have demonstrated the acceptable combustion
characteristics of CWF. The market for conversion of
industrial boilers to coal firing has been reported by
others and this report confirms that there is an area
of this market where conversion to coal via coal/water
firing is well worthy of consideration.

It is anticipated that based on current demonstration
plants and initial conversions that the available design
data will be rapidly refined to allow CWF to be adopted
with confidence backed by commercial guarantees for
purpose designed back-fits in the future.

REFERENCES

1. General   eg 2nd Int. Symp. on Coal Oil Mixture
                            Combustion. Nov. 1979 Danvers Mass.

2. Mathiesen M. et al   "Some Rheological Data & Atomisation
                            Behaviour of CWM's Containing 68/83%
                            Coal".
                            Fifth International Symposium on Coal
                            Slurry Combustion & Technology.
                            April 1983 Tampa Florida.

3. Dillan M. et al   "Prepare To Evaulate & Design Coal
                            Slurry Pumping Systems".
                            Power December 1982.

4. Dillan M. et al   Factors Affecting Pump Selection For
                            Slurry Type Fuels.
                            Proc. Fourth Int. Symp. Coal Slurry
                            Combustion.
                            Orlando Florida May 1982.

5. Codes of Practice   eg NFPA 85, BS799 pt 4, BS806 etc.

6. Maier G.A. et al   Coal Slurry Fuels: Moving Toward
                            Commercial Utilisation. Fifth
                            International Symposium on Coal
                            Slurry Combustion & Technology.
                            April 1983 Tampa Florida.

7. Farthing G.A. et al   'Combustion Tests of Coal Water Slurry'

                            EPRI CS2286 March 1982.

CLM-U

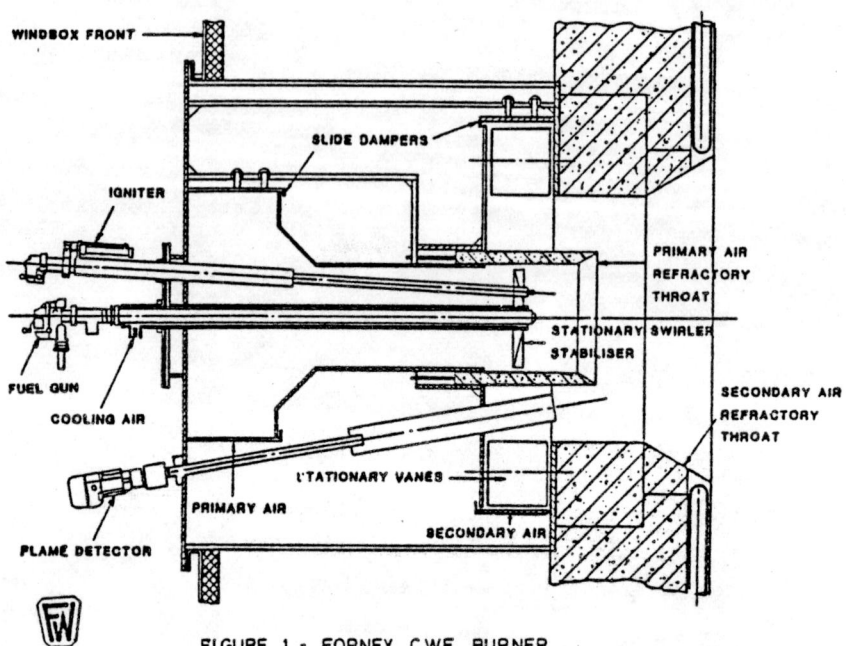

FIGURE 1 - FORNEY CWF BURNER

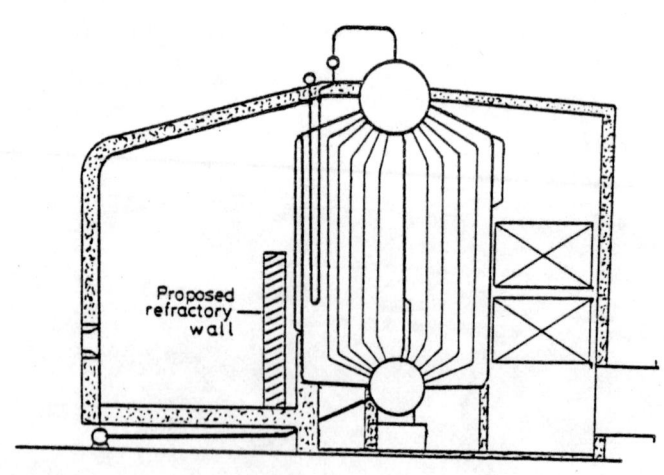

FIGURE 2 - 14 TONNE/H BOILER

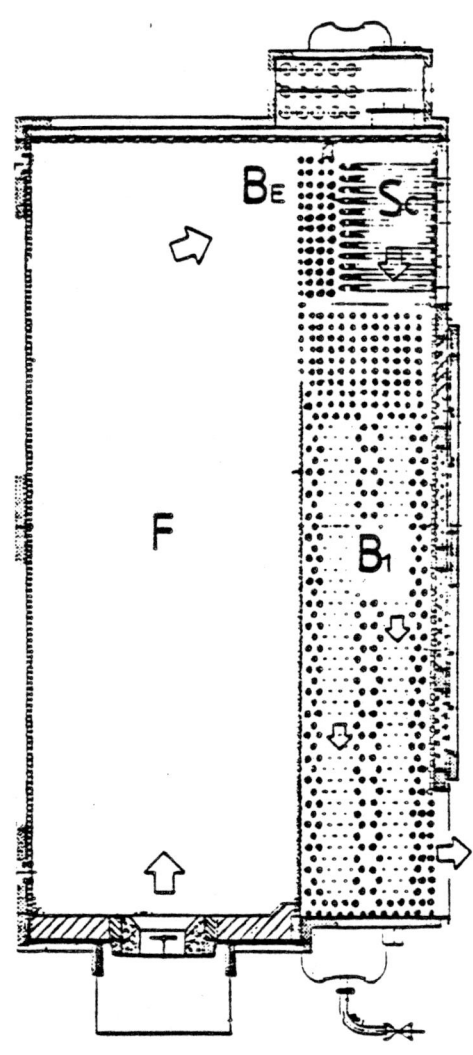

FIGURE 3 - 34 Tonne / hour Boiler

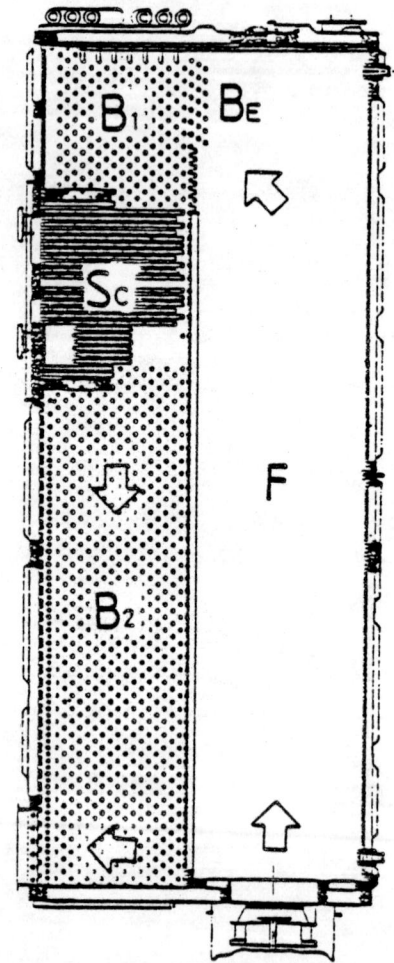

FIGURE 4   68 Tonne/hour Boiler

THE ASSESSMENT OF COAL WATER SLURRIES IN COAL CONVERSION PROCESSES USING A

FLOWSHEETING COMPUTER PROGRAM

N.W.J. Ford[*], J. Holmes[*], K.K. Pillai[**] and P.J. Read[***]

The reduction of water content and variation in ash content of coal-water mixtures as feedstocks for pressurised fluidised bed combustion and Texaco gasification have been studied using the NCB's ARACHNE process flowsheeting program. In both these applications, best efficiencies were predicted for use of the minimum fraction of water at which a practically pumpable slurry can be made.

INTRODUCTION

Coal handling, i.e. pre-processing and feeding, accounts for a significant fraction of the total capital cost in both gasification and fluidized bed combustion systems operating at elevated pressure (refs. 7 and 8). Furthermore, process efficiency is penalised by the energy consumption involved in drying and crushing the coal and in the provision of inert gas for pressurising lock hoppers. Clearly, options which either reduce the use of capital-intensive plant or increase the process efficiency must be investigated.

The use of a pumpable coal-water slurry is one technique which has the potential of reducing some of the costs. The following benefits are expected to accrue:-

1. Coal drying is eliminated; in consequence, capital costs are reduced and the associated heat requirements of this operation are avoided.

2. Some coal crushing is still required but may be performed in the wet state with some reduction in process energy requirements.

3. Inert gas usage is eliminated and the necessity for pressurised lock hoppers is obviated.

* Coal Research Establishment, National Coal Board, Stoke Orchard, Cheltenham, England.
** Coal Utilisation Research Laboratory, National Coal Board, Leatherhead, England.
***Energy, Mines and Resources Canada, Ottawa, Ontario, Canada.

4.   Control systems are simplified.

5.   Response to changes in firing rate are expected to be faster than
with pneumatic feeding and the feed rate will be less sensitive to
pressure fluctuations in the gasifier or combustor.

The penalties associated with feeding a slurry are caused by the losses of
evaporative heat that may prevail.   These losses are different in gasification and combustion systems and need to be offset against the potential
benefits.   The benefits must also be offset against the cost involved in
manufacturing the slurry.

This paper investigates the effects of feeding coal-water slurries to the
following two systems:

(1)   A pressurised fluidised bed combustion (PFBC) power generation cycle.

(2)   A Texaco entrained phase gasifier producing medium heating value fuel
gas.

In both cases, the sensitivities of the process efficiencies to the ash and
water contents of the slurries are presented.

These assessments of the system effects of coal-water slurries have been
greatly facilitated by the use of the NCB's process flowsheeting computer
program ARACHNE.   This program is described briefly in the next section.
The results of the studies on the PFBC combined cycle and the Texaco gasifier
are then given in the following two sections.

THE ARACHNE FLOWSHEETING PACKAGE

The process flowsheeting computer program is a generalised program which can
predict the mass and energy flows for a range of chemical processes.   A
flowsheeting program generally consists of a set of calculation routines
representing individual processes (e.g. heat exchangers and mixers),
calculation routines to predict the thermodynamic properties of the chemical
components taking part in the process, and a means of translating the
specification of the process flowsheet into a computer model.   The program
therefore enables the engineer to simulate chemical processes with relatively
little effort and without knowledge of computer programming.

While process flowsheeting programs are widely used in the chemical and
petrochemical industries, existing programs are not generally suitable for
investigating coal-based processes.   In particular, most of them do not
have the ability to handle solids.   The NCB has therefore developed the
ARACHNE ('A Reactor And Cycle Handling Network Evaluation') package, which
embodies the following features:

(1)   The ability to handle solids; those most commonly occurring in coal
processing are coal, char, coke, ash, dolomite and limestone.

(2)   The rigorous solution of chemical reactions, for example combustion
and gasification reactions.

(3)   The accurate representation of power generation cycles.

(4)   Information flows and control loops may be incorporated in process models.

(5)   The program is flexible, operates conversationally and is cheap to run.

The program has been applied to the modelling of coal gasification, power generation and synthesis processes (see for example references 3, 4 and 5). Developments are currently in progress to permit the modelling of coal liquefaction processes and the assessment of the capital costs of coal conversion systems.

An example of the type of model which can be generated by ARACHNE is given in figure 1.   This shows a coal-to-SNG process based on a capability for using sub-networks and, in this example, these are employed for the gasifier (FIXEDBG) the clean-up system (CLEANUP), the Claus unit (CLAUS), the Rectisol units (RECT 1 and RECT 2), the shift reactor (SHIFTER3) and the methanator (METHAN8).   The total numbers of blocks and streams (including those within the sub-networks) are 100 and 130 respectively.

PFBC POWER GENERATION CYCLES

The form of the PFBC power generation cycle considered is the supercharged boiler cycle shown schematically in figure 2.   The compressed, fluidising air reacts with the coal in the bed and is heated to the bed temperature. It then enters a cyclone train in which the concentration of particulates is reduced to a level acceptable for the gas turbine.   The hot gas is expanded through the turbine and the residual sensible heat is recovered by feedwater heating.   High pressure steam for the steam turbine cycle is raised, superheated and reheated by tubes in the bed.

The principal effects of feeding water with the coal to a PFBC combined-cycle scheme are:

(1)   To increase the mass flow through, and hence the power output from, the gas turbine, provided that the gas turbine can be modified to achieve the additional swallowing capacity required.

(2)   To reduce the heat available to the steam cycle from the combustor by the amount needed to evaporate and superheat the water fed with the fuel.

(3)   To increase the heat available in the exhaust gas stream for abstraction into the steam cycle due to the increased mass flow and heat capacity of the exhaust gas.

(4)   To increase the stack loss as steam.

The design conventions employed in the study of the PFBC combined cycle are summarised in table 1, which also gives the analysis of the coal used. Gas turbine and compressor efficiencies, and steam turbine operating conditions are typical of commercial practice, and allowances are made for gas-side pressure losses, shaft losses, leakage losses, pumping power, generator and transformer losses.   No credit was taken for heat recovery from the solids.

The energy consumptions for coal preparation and feeding were not included. The predicted changes in cycle efficiency arising from the use of coal-water slurries are therefore those arising directly from the increased water

content of the coal-water slurry over a 'dry' fuel fed to the combustor.
In practice, the efficiency penalty might be reduced if coal drying were
required in order to feed the coal pneumatically from lock hoppers.

The results of the computations on the effects of coal-water mixture water
content on the performance of the cycle are summarised in table 2 and
figure 3. Table 2 gives the results for a combustor excess air of 30 %,
gas turbine pressure ratio of 10:1 and a 10 % ash coal. Figure 3
summarises the cycle efficiency penalties for a range of excess air levels.
Results for a wide range of fuel water contents are presented: feasible
coal-water mixtures would have total water contents of around 22 % upwards.
Taking coal with zero surface moisture (corresponding to 5 % inherent
moisture) as the reference case, a coal-water mixture of 30 % water content
results in a loss of about 0.8 percentage points on cycle efficiency.
This contrasts with the performance of a conventional steam cycle, in which
a coal-water mixture of 30 % water content would result in an efficiency
loss of about 1.5 percentage points.

The difference between a combined cycle and a conventional steam cycle is
that some of the energy used to convert the water fed with the CWM into
steam is recovered in a combined cycle. Since a gas cycle operates at a
higher temperature than a steam cycle, conversion of the recovered energy
to shaft power can be accomplished at a higher efficiency in the gas turbine
and combined cycle than in the conventional steam cycle. The energy
associated with the steam escaping via the stack is irretrievable.

When utilising CWM, the total plant output decreases slightly as the stack
loss increases. The gas turbine power output increases as the mass flow
increases, but not sufficiently to compensate for the reduction in steam
turbine power. CWMs of high water content might cause problems in
increasing the swallowing capacity of the gas turbine, but this is not
likely with moisture contents of 30 % or less. Figure 4 summarises the
effects of fuel water content and combustor excess air on the proportion
of the total power contributed by the gas turbine.

In practice, the coal ash content may vary and hence the effects of this
have been investigated. Since the solid matter consists of coal and its
associated ash content, the effect of increasing coal-ash content is to
require a larger ratio of water to dry ash-free coal. Thus, for the same
thermal input, a larger quantity of water is fed with a higher ash coal.
The ratio of water to dry ash-free coal is given

$$RW = \frac{f_w}{(1 - f_w)(1 - f_a)}$$

where f_w = fraction water to be maintained in the CWM as fed
       f_a = ash content of dry coal

Table 3 illustrates the effect of increasing ash content for a CWM
requiring 30 % water as fed.

THE TEXACO GASIFIER

The system investigated employs a Texaco coal gasifier with a gas cooler,
as shown in Figure 5. Pulverised coal is slurried with water and is
injected with an oxidant into the top of a refractory-lined pressure vessel.

The coal is rapidly devolatilised and the volatile products then burn with some of the fixed carbon to produce the heat required to evaporate water in the feed and sustain the endothermic gasification reactions. All reactions occur within a few seconds and up to 99 % of the carbon in the coal is gasified.

Hot gases from the gasifier, together with ungasified solids and molten slag, are directed downwards through a radiant, water-jacketed cooler where high pressure steam is generated as the gases are cooled. The cooler is designed to cool gases and slag in the gas stream to a temperature below the softening point of the ash. Most of the solids are collected in a water-filled chamber at the bottom of the radiation section. Cooled gases containing small amounts of particulates are further cooled in a convective section where additional steam is produced. At the exit of the convective section, the gases are scrubbed with water to remove any remaining particulates. The product synthesis gas is then ready for further processing.

In the ARACHNE representation of the gasifier the following reactions are at equilibrium at the gasifier outlet temperature:

(1) $CO + H_2O \rightarrow CO_2 + H_2$

(2) $CO + 2H_2 \rightarrow CH_4 + H_2O$

(3) $H_2S + CO_2 \rightarrow COS + H_2O$

(4) $N_2 + 3H_2 \rightarrow 2 NH_3$

In addition, the radiant heat loss from the gasifier is specified as 0.25 % of the coal input energy (ref.1), the carbon conversion is specified as 99 % (ref.2) and the operating pressure is 30 bar. The oxygen flow is adjusted to give a gas temperature at the outlet of the gasification zone of approximately 1400 $^{\circ}$C. The analysis of the coal used in the study of the Texaco gasifier is given in table 4.

The following two sensitivities have been investigated:

(1) Texaco have indicated that it should generally be possible to feed slurries with adventitious water contents (defined as total water content less inherent coal moisture) of 26 % to 36 % (by mass). As a base for the analyses of the Texaco gasifier, the operation of the gasifier has been evaluated for an adventitious water content of 31 % (total water content of 36 %). It may be possible to pump slurries having lower water contents than this, and hence a range of adventitious water contents of the slurry between 17 % and 24 % (corresponding to total water contents between 22 % and 29 %) have been investigated. In addition, calculations have been performed for a 10 % adventitious water content slurry, although with present technology it is unlikely that such a slurry would be pumpable. In each case, the ash content of the dry coal is 10.6 %.

(2) It may be economic to make a high ash content slurry from washery rejects. The effects on gasifier operation of ash contents of dry coal between 27 % and 40 % were therefore investigated. In each case the adventitious water content of the slurry was maintained at 17 %

The predicted effects of varying the water content of the slurry are summarised in table 5.

The main effects of reducing the water content of the slurry may be
summarised as follows:

(1)   The gasifier efficiency (defined as the ratio of the chemical
        energies in the product gas and the coal feed) increases.

(2)   The hydrogen to carbon monoxide ratio decreases.

(3)   The gas calorific value increases.

(4)   The oxygen consumption, and consequently the power requirements,
        decrease.

The increase in gasifier efficiency arises because with less water present,
the latent heat required to vapourise the water decreases and less coal is
needed for the combustion reactions.   Thus more coal is available for
gasification, resulting in a higher chemical energy content of the gaseous
product.

The possibility of being able to process low water content coal slurries
has advantages, chiefly in the increased gasifier efficiency and reduced
power requirements of the system.   The increase in gas calorific value is
partially offset by the decrease in product gas yield.   Applications for
this type of system would appear to be in fuel gas and as synthesis gas for
low H_2:CO ratio products such as acetic acid and isocyanates.   The suitability
of the gas as a feedstock for methanol synthesis is less certain because of
the lower H_2:CO ratio obtained compared with that under normal operation.
However, the extra water-gas shifting necessary (in addition to that required
under normal operation) could be offset by the increased gasifier efficiency
and the reduction in power requirements.   Evaluation of these effects would
require a study of the methanol synthesis process in full.

The predicted effects of varying the ash content of the coal are summarised
in table 6.   The gasifier efficiency lies between 81.5 % and 79 %, for ash
contents of the coal of 26 % and 40 % respectively.

Although it is inert, the ash must be heated to the reaction temperature,
and therefore higher ash content requires the combustion of more coal,
leaving less for the gasification reactions.   Also, the sensible heat of
the product decreases with increasing ash content, reflecting the increased
proportion of input energy contained as sensible heat in the ash rather than
in the gaseous product.   With an ash content of the coal of 33 % and a
slurry containing 17 % adventitious water, the system has a gasifier
efficiency higher than that for the base case with a slurry adventitious
water content of 31 % and coal ash content of 10 %.

Carbogel AB has conducted a practical evaluation of using low-water content
slurries as input into a Texaco gasifier designed for residual oil feedstock.
The results of Carbogel's tests agree well with the predictions of the
ARACHNE model, i.e. that the process efficiency increases as water content
of the feed decreases to below 25 per cent by weight.

CONCLUSIONS

(1)   The present analysis of the use of coal-water mixtures as a feedstock
for combustion and gasification processes has been facilitated by the
ARACHNE flowsheeting package.

(2)   In PFBC combined cycles, the adoption of coal-water mixture feeding techniques leads to a small cycle efficiency penalty. This, however, may be more than offset by reductions in coal preparation costs and by the simplification of feeding and control systems.   The efficiency penalty is kept small because of the additional mass flow through the gas turbine. Coal-water mixtures show greater potential benefit, therefore, in combined cycle systems than in conventional steam cycles.

(3)   For the Texaco entrained phase gasifier, low water content slurries give an increased gasifier efficiency and reduced power requirements compared with operation with normal slurries; this has been confirmed by experiment.   The hydrogen to carbon monoxide ratio is, however, reduced. The efficiency advantage of low water content slurries can more than offset the energy penalty associated with gasification of high ash content feedstocks.

REFERENCES

(1)   'Economic studies of coal gasification combined cycle systems for electric power generation', Fluor Engineers and Constructors, EPRI AF-642.

(2)   'The Texaco coal gasification process-synthesis gas for chemical feedstocks,' W.B. Branch, presented at the International Coal Conversion Conference, Pretoria, South Africa, August 1982.

(3)   'Coal gasification for combined cycle power generation: a general simulation method', J. Holmes, M. Smith, D. Merrick and J.S. Harrison. Proceedings of 'Future Energy Concepts' conference, Institution of Electrical Engineers, London, 27-30 January 1981.

(4)   'The development and application of the ARACHNE process flowsheeting package to synfuels production processes using hydrogen', D. Merrick, A. Bogle, J.E. Davison, J. Holmes, P.F.M. Paul, Proceedings of '3rd International Seminar on Hydrogen as an Energy Carrier', Lyon, 25-27 May, 1983.

(5)   'Design-point and part-load analyses of combined cycle plant incorporating a pressurised fluidised bed coal combustor (PFBC)', K.K. Pillai, J. Holmes and D. Merrick.   To be published in the Journal of the Institute of Energy.

(6)   'Evaluation of intermediate-btu coal gasification systems for retrofitting power plants', D.A. Waitzman et al. EPRI AF-531, August 1977.

(7)   'Engineer, design, construct, test and evaluate a pressurised fluidised bed pilot plant using high sulphur coal for production of electric power', Phase 1: preliminary engineering, commercial plant conceptual design'.   Curtiss-Wright Corporation, US/DOE/FE-1726-20A, March 15, 1977.

(8)   'Economics of Texaco gasification - combined cycle systems', Fluor Engineers and Constructors, EPRI AF-753, April 1978.

ACKNOWLEDGEMENTS

The authors wish to thank Dr. D. Merrick for his many useful contributions to the formulation of the paper, and the National Coal Board and Energy, Mines and Resources Canada for permission to publish.   The views expressed are those of the authors and not necessarily those of their respective organisations.

TABLE 1

PFBC study design conventions

(1) Combustor

				mass %
(1)	Coal analysis (daf)	C		78.86
		H		5.14
		O		8.06
		N		1.38
		S		5.56

Ash content : 10 % of dry coal

(2) Dolomite feed rate : Ca/S molar ratio = 2.0

(3) Bed temperature = 850 ^{o}C

(4) Pressure drops : fluidised bed   0.25 bar
            gas clean-up   4.8 %
            valves and
            ducting   3 %

(2) Gas turbine

(1) Compressor : outlet pressure   10 bar
                 isentropic efficiency   86 %

(2) Turbine   : isentropic efficiency   87 %

(3) Steam turbine

HP turbine : inlet temperature   538 ^{o}C
               "   pressure   159 bar (abs)
               outlet pressure   42 bar (abs)
               isentropic efficiency   84 %

IP turbine : inlet temperature   538 ^{o}C
               "   pressure   40 bar (abs)
               outlet pressure   2 bar (abs)
               isentropic efficiency   89 %

LP turbine : 1st stage : outlet pressure   0.21 bar (abs)
                       isentropic efficiency 86.6 %

              2nd stage : outlet pressure   0.05 bar (abs)
                       isentropic efficiency 82.4 %

Boiler feedpump : efficiency   90 %
                 outlet pressure   181 bar (abs)

Deaeration at 2 bar : bleed steam for IP exhaust

(4) Miscellaneous

Generator efficiency	98.7 %
Transformer efficiency	99.5 %
Gas and steam turbine mechanical loss	1 %
Steam turbine auxiliaries & leakage loss	1 %
Pinch point temperature limitation	25 ^{o}C
Stack temperature	140 ^{o}C

TABLE 2

Effect of fuel-water content

Turbine entry temperature   850 °C
Compressor outlet   10 bar
Excess air   30 %
Stack temperature   140 °C

Fuel water content %	0	5 'Dry' coal	10	20	30	40
Cycle efficiency %	39.9	39.8	39.6	39.3	39.0	38.5
Proportion gas turbine power %	16.5	16.8	17.3	18.3	19.9	21.8
Gas turbine power MWe *	69	70*	72	76	81	87
Steam turbine power MWe	349	347	344	339	326	312
Total power MWe	418	417	416	415	407	400

*For comparison of the effects on power output and swallowing
capacity a gas turbine output of 70 MWe is taken as the datum
condition with 'dry' coal at 30 % excess air

TABLE 3

Effect of varying coal-ash content

Ash content of dry coal	10	15	20	25
Kg of water fed per kg of dry ash-free coal	.48	.50	.54	.57
Cycle efficiency	39.0	38.8	38.6	38.3

Coal water mixture containing 30 % water as fed

TABLE 4

Coal analysis for Texaco study

		mass %
Ultimate analysis	C	81.5
	H	5.3
	O	7.9
	N	1.4
	S	3.9

TABLE 5

Output stream properties with varying slurry water content
(Ash content of dry coal = 10.6 %)

Adventitious water content of slurry (% mass basis)	10 %	17 %	24 %	31 % Base case
Total water content	15 %	22 %	29 %	36 %
Product Gas Composition (% volume basis)				
N_2	0.62	0.59	0.56	0.53
H_2O	2.68	7.68	12.99	18.73
H_2	31.86	30.71	29.19	27.33
CH_4	0.04	0.01	-	-
CO_2	1.69	4.25	6.72	9.08
CO	61.77	55.73	49.51	43.38
H_2S	1.08	1.02	0.97	0.90
COS	0.08	0.07	0.06	0.05
Exit temperature (^oC)	1400	1401	1402	1400
Heat loss (% coal HHV)	0.25	0.25	0.25	0.25
Yield (kg wet gas/kg daf coal)	2.48	2.66	2.87	2.62
Coal Energy Input (MJ/kg daf coal)	34.1	34.1	34.1	34.1
Chemical Energy in Gas (MJ/kg daf coal)	28.9	28.4	28.0	27.4
Sensible Heat in Gas before cooling (MJ/kg daf coal)	4.33	4.74	5.18	5.71
Gasifier Efficiency (%)	84.6	83.4	82.0	80.3
H_2/CO (molar)	0.52	0.55	0.59	0.63
System Power Req. (MJ/kg daf coal)	1.28	1.44	1.52	1.58
Calorific Value of Gas (MJ/m^3) Dry	11.62	11.30	10.91	10.49
Wet	11.31	10.43	9.49	8.53
Oxygen consumption (kg/kg daf coal)	0.89	0.93	0.98	1.03

TABLE 6

Output stream properties with varying coal ash content
(Slurry water content: adventitious = 17 %, total = 22 %)

Ash content of coal (% mass basis)	26	33	40
Product Gas Composition (% volume basis)			
N_2	0.58	0.58	0.57
H_2O	11.24	13.24	15.58
H_2	29.11	28.18	27.13
CH_4	-	-	-
CO	6.08	7.11	8.29
CO_2	51.95	49.86	47.42
H_2S	0.99	0.97	0.95
COS	0.06	0.06	0.06
Exit Temperature (^{o}C)	1406	1403	1401
Heat Loss (% coal HHV)	0.25	0.25	0.25
Yield (kg wet gas/ kg daf coal)	2.35	2.42	2.50
Coal Input Energy (MJ/kg daf coal)	34.1	34.1	34.1
Chemical Energy in Gas (MJ/kg daf coal)	27.8	27.4	27.0
Sensible Heat in Gas before cooling (MJ/kg daf coal)	3.71	3.45	3.24
Gasifier Efficiency (%)	81.5	80.4	79.0
H_2/CO (molar)	0.56	0.57	0.57
System Power Req. (MJ/kg daf coal)	1.54	1.59	1.66
Calorific Value of Gas (MJ/m^3) Dry	11.02	10.85	10.65
Wet	9.78	9.42	8.99
Oxygen consumption (kg/kg daf coal)	0.98	1.01	1.04

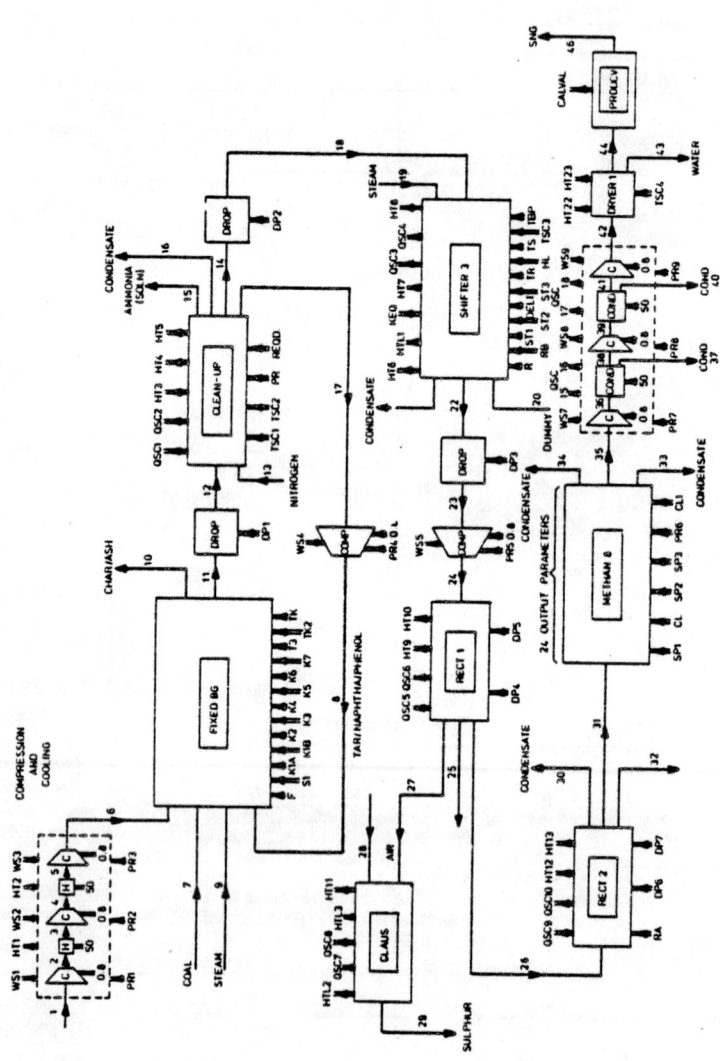

Figure 1. ARACHNE flowsheet for SNG production from fixed-bed gasification

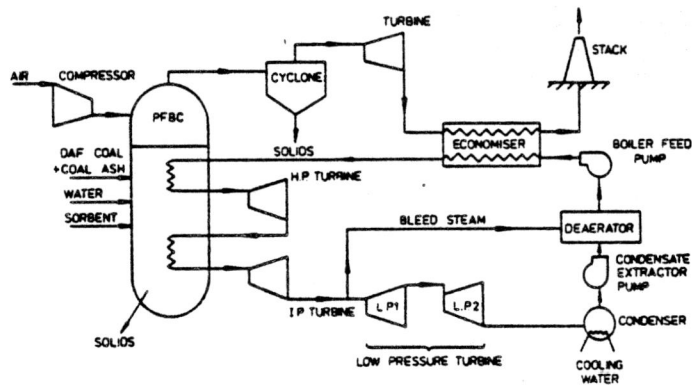

Figure 2. PFBC combined cycle (supercharged boiler)

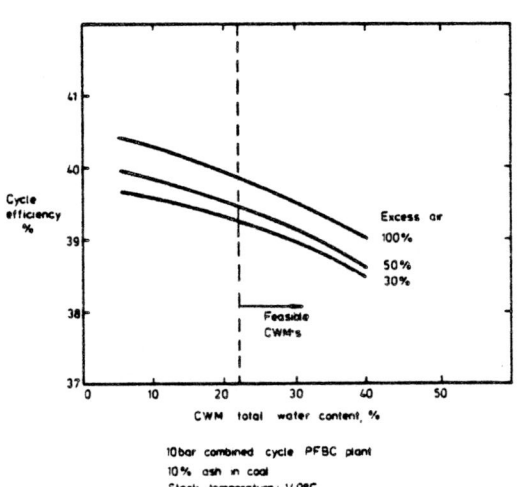

10 bar combined cycle PFBC plant
10% ash in coal
Stack temperature : 140°C

Figure 3. Effect of fuel water content on cycle efficiency

331

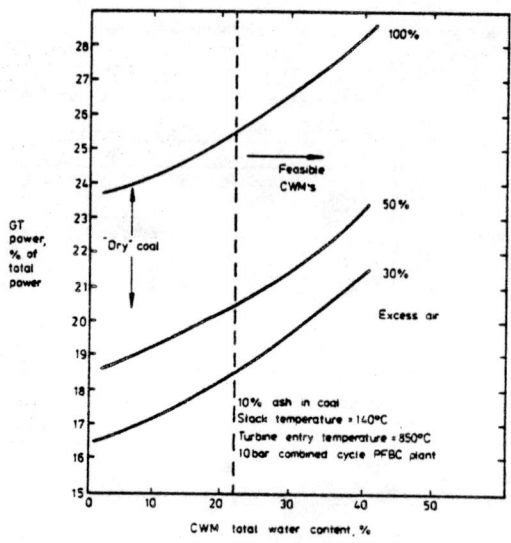

Figure 4. Effect of fuel water content on GT power

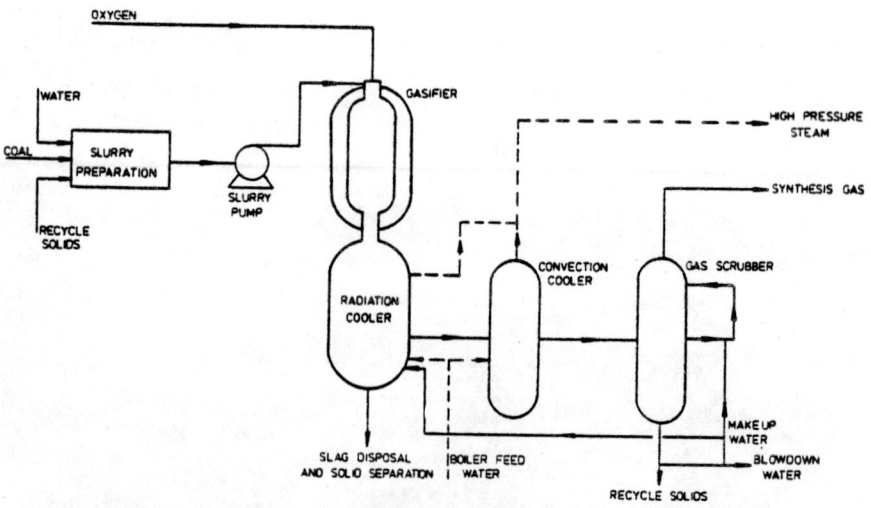

Figure 5. The Texaco coal gasification process operating in the gas cooler mode

PRIMARY ENERGY DEVELOPMENTS IN EUROPE
- A CHANCE FOR COAL

Ray Long and Anthony Baker *

The first part of this paper summarizes energy trends
since the 1973 oil price increases. The second part
surveys general energy prospects to the year 2000, and
the third part looks more specifically at the role of
coal in industry.

INTRODUCTION

At a conference which is concerned primarily with technical and
marketing developments in coal-liquid mixtures, it is helpful to
consider potential markets for coal in industry (and thus the
opportunity for CWM to play a role) against the background of likely
energy developments. The first part of this paper summarizes energy
trends since the 1973 oil price increases. The second part surveys
general energy prospects to the year 2000, and the third part looks more
specifically at the role of coal in industry.

ENERGY DEMAND SINCE 1973

The first table in this part of the presentation dealing with recent
trends in Western Europe shows the growth of energy demand relative to
economic growth on a year-to-year basis (tables with references are at
the end of the text). That last point is important. There has been a
tendency to see the post-1973 woes in monolithic terms; but, as these
figures show, the period breaks into three distinct phases: the
immediate impact of slower growth and energy cutbacks; the period from
1975 to 1979 when the West European economies grew by more than 3% per
annum with more than matching energy growth; and the sharp recession and
sharper fall in energy demand induced by the 1979/80 oil price increases
which continued into 1982.

*   Anthony Baker is Head of the Economic Assessment Service of IEA
      Coal Research and Ray Long works in the Service on coal supply,
      transport and trade.

Over the period 1973-81 it appears that energy requirements per unit of output fell by about 9%. It is suggested in IEA and other analyses that this decline is due about half to increasing energy efficiency and half to structural moves away from energy-intensive industries, but the trends are very uncertain.

Table 2 shows energy growth by source. The major impact of the energy cutback was on oil, but while coal demand did grow over the period, faster growth occurred in all other sources, particularly in nuclear power rising from a small base.

Table 3 shows a breakdown of demand by sector. The interesting features are the above-average growth in energy demand in the transportation and service sectors and in power generation. There is nothing here to suggest that moves towards a service-based economy will inevitably cause a drop in energy demand. And if the trend towards services is accompanied by increasing electricity consumption this in itself increases primary energy requirements more than proportionally because of the higher energy losses in power generation.

Table 4 indicates some quite marked differences in energy demand growth and also in the energy/GDP growth coefficient from one part of Europe to another. It may be no coincidence that the 'other Mediterranean' group of countries - Portugal, Spain, Greece and Turkey - recorded both the fastest rate of economic growth and also a very high energy/GDP coefficient in the 1973-81 period. As the least advanced region of Europe its performance can be likened to that of the 'Newly Industrialising Countries' in the less developed world - an emphasis on basic industries with overall growth having a priority over energy efficiency.

The final table in this statistical overview shows the major advantage which coal has as an internationally traded fuel - its cheapness. In response to an 11-fold increase in nominal oil prices since 1973, gas prices have been pushing their way up towards comparable levels on an energy-equivalent basis. The disadvantage with gas as a major component of internationally traded fuel is that the high cost of piping it over long distances or liquefying it leave net-back prices which offer little incentive to producers (though exported methanol made from natural gas would give the producers greater downstream benefit). Thermal coal, on the other hand, has shown relatively low real price increases since the early seventies and can be delivered into Europe currently at prices about one third those of the equivalent energy in oil.

ENERGY PROSPECTS TO 2000

The hazards of making energy projections are at least as great as those associated with any form of forecasting; but views have to be taken of the future, if a presentation of future prospects for coal is to be made. One set of energy projections, the best estimate or reference case, is used to illustrate the arguments; but this is not to suggest any certainty of what will happen, and a lower demand case is also given.

The three critical estimates in making a forecast of coal demand are:
(a) economic growth; (b) energy demand relative to economic growth; and
(c) changes in the different fuels demanded. The view taken in this
paper and in the detailed work which EAS has carried out on energy
markets as background to its thinking on coal prospects, is that
economic growth must recover from the current depression. This view
corresponds with most long-term forecasts that are being made. Opinions
differ considerably, however, on the intensity of energy use Many
people now argue that because of structural changes in economic activity
and consumer response to price changes, the growth in energy demand will
continue to fall even with strong economic growth. Our view is that
conservation and structural changes will have a continuing effect, but
that sustained economic recovery accompanied by rising real incomes will
eventually pull energy demand up again towards the rates of increase in
economic growth. On the third point of energy substitution, the earlier
figures on energy costs indicate that there is considerable room for
coal to penetrate other markets, and there is the additional "political"
impetus that greater use of coal within Europe contributes to security
of energy supplies.

Table 6 sets out some recent forecasts of economic growth and energy
demand for comparison with EAS's own views. The general point is that
there is consensus about a degree of economic recovery over the next two
decades and a growing role for coal. The major differences are in the
rate of energy demand relative to economic growth (the ratio varying
from 0.54 up to 0.88), and in the level of coal imports required to fit
the other energy parameters (a variation of 65% in coal imports with an
economic growth variation of only 20%).

Two EAS views are presented in Table 6 and a fuller analysis of energy
use consistent with these two scenarios is presented in Table 7. It
should be pointed out that the "reference case" is the "best estimate"
case which we used in our report on coal trading constraints published
early in 1983. The reference case reflects a considerable amount of
work which went into projecting complex energy scenarios whose parts
were consistent with each other and with the coal trade projections
which were a main focus of the constraints report referred to. The "low
demand case" in Table 7, by contrast, presents some broad numbers which
have not been worked out in the same level of detail. They are put in
here as what we would argue is a 'minimum' scenario for examining coal's
potential.

EAS's reference case analysis, like many other analyses of fuel mix,
suggests a number of important trends which will affect the competitive
postion of coal in European energy markets.

It seems clear that the policy of individual countries, and concerted
action at the IEA and EEC levels, will aim at a continued moderation in
oil demand. North Sea oil production is likely to fall away before the
end of this decade unless substantial increases in world prices induce a
big push to develop marginal fields, whereas world prices would have to
fall back perhaps to pre-1979 levels to reverse the policy of reducing
reliance on Middle East supplies. This suggests that oil will
increasingly be used in the premium markets - transportation fuels and
chemical feedstocks - which account for about 240 Mt or 40% of its use

at present. The next largest user of oil is the residential market
followed by industrial use and electricity generation. Residential use
of oil has fallen in the past decade both as a proportion of total oil
use and as a proportion of residential fuel requirements and this is
likely to continue. The building of new coal-fired stations and the
conversion of oil-fired stations are beginning to have an impact; 23 GW
of coal-fired capacity is due to come on stream in EEC countries during
the period 1982-87 compared with 115 GW of coal-fired capacity available
at the end of 1981; in the same period 6½ GW of oil and gas-fired
capacity will be phased out (13).

Use of oil in industry other than iron and steel and chemicals accounts
for about half of this sector's total energy requirements. Prospects
for coal depend primarily on the development of new technologies for
burning coal in small-scale plants and on the availability of natural
gas. The CIAB's report on the use of coal in industry last year (14)
took the view that we are on the beginning of a learning curve with new
technologies for industrial coal use, and although developments might be
slow for the next decade the market could boom in the nineties. Gas may
find its premium market increasingly as a chemical feedstock and in the
replacement of household use of oil; but gas could pose the major
competitive threat to coal in the industrial market as well, depending
on how much is available.

Despite the huge availability of natural gas resources in the world it
hardly makes strategic sense for Europe to replace reliance on distant
oil supplies with reliance on distant gas supplies, generally in the
same fragile areas of the world. Imports of piped or liquefied gas,
with the exception perhaps of Mediterranean countries close to North
Africa, are unlikely to change the supply situation fundamentally. Net
imports currently, i.e. allowing for Dutch and Norwegian exports to
other European countries, are only 10% of total natural gas
requirements; and the controversial USSR pipeline is scheduled to supply
40 Gm^3/a by the mid-eighties, the equivalent of less than 20% of today's
gas requirements.(7) The major uncertainty is with gas available within
Europe itself, particularly in the Norwegian sector of the North Sea.
If the extent of the gas fields there and the cost of their development
can be established, the Norwegians may still exact a high price for
maximum development which would be in Europe's interest but not in the
interest of Norway's tiny economy. Yet this could be seen as a
strategically better alternative than accelerated coal use reliant on
imported coal.

The final area of major uncertainty for coal is in power generation.
European studies (15) continue to show nuclear as more competitive, the
cost difference varying between countries and according to different
discount rate assumptions. Coal's advantage may lie in less quantified
directions such as the long lead-times involved in enquiries over
nuclear power development; lower output ratings arising from faults
reported in some European stations; the economic concern about
wholesale plant shutdowns if serious faults were discovered in plants
which are in widespread use; and the uncertainty over costs of fuel
reprocessing and disposal and plant decommissioning. Basic decisions
about nuclear power expansion will be made this decade by a number of
European countries and this will clearly affect coal's prospects in the

nineties. The energy projections presented earlier, however, suggest that there is room for both nuclear and coal-fired power generation in the long term.

The effect of these trends on prices is likely to be renewed upward pressure on oil prices eventually, particularly at the margin of expanded North Sea production and diminishing imports from outside Europe. Gas prices will move up closer to the oil equivalent price, with the marginal price set either by LNG imports or the cost of SNG from coal. Coal prices also will rise both from demand-pull and cost-push reasons; but although the gap between coal and oil prices could narrow, coal is unlikely to lose much of its competitive strength.

Our. views on likely price trends to the end of the century, again adopting the format of 'reference case' and 'lower demand case' are set out in Table 8. As in the earlier analysis of demand volumes, the reference case figures are based on detailed considerations of feasible trends between the prices and costs of the competitive fuels, whereas the low demand figures are offered only as a suggestion that, with subdued energy demands, real prices of energy are unlikely to rise much in real terms between now and the year 2000.

THE CHANCE FOR COAL IN INDUSTRY

The major growth point for coal use in W. Europe for the next 20 years is still of course in electricity generation. Despite assumed increasing use of nuclear power, coal-fired power stations are projected to provide about half of the extra electricity needed. By contrast the market for metallurgical coal is expected to grow considerably less than the total economy and barely grows at all in the subdued energy demand scenario - the result possibly of incremental steel production being installed largely outside Europe. The trends are set out in Table 9.

The 'other' coal market in this table is largely for steam raising and direct firing in industry. This is the market where the contrast between 'political' coal use referred to earlier and projected coal use could be greatest.

The IEA/CIAB study on the Use of Coal in Industry suggested that for the OECD region as a whole (not just Europe) the potential industrial market for coal at the end of the century is over 1000 Mtce, whereas projected coal use in industry based on a sectoral analysis of demand would be no more than 400 Mtce. Pro-rata for W. Europe these figures would be about 400 Mtce and 160 Mtce respectively, though these levels are probably too high since the pace of conversion from oil and gas to coal in Europe could be slower than in the USA where the price incentive is greater. In comparison with the projections mentioned above, Table 9 shows coal in industry (including iron and steel which are included in the CIAB's projections) between 255 Mtce and 304 Mtce by 2000 (179 and 213 Mtoe).

By comparison the amount of oil products used in the industrial sector in OECD Europe is 1981 was 123 Mtoe (roughly 80% fuel oil and 20% gas oil), and gas use in industry was 66 Mtoe. Thus, even without allowing for industrial growth over the next 20 years, the figures for potential industrial coal use suggested in Table 9 do not appear unduly high if the need to phase oil and gas out of these non-premium markets is accepted.

The constraints on coal in industry reaching its full potential were detailed in the CIAB report mentioned earlier. They are:

(a) the high cost of investment in coal handling and combustion equipment;
(b) the requirement of a very high rate of return on such investments (1-2 year paybacks are typical);
(c) environmental considerations; and
(d) physical difficulties at particular sites (e.g. no space for coal stocks).

In addition, the price of coal to industry is likely to be higher than the price of power station coal, reflecting the requirements for cleaner, ash-free, larger-sized coals. Whereas in Table 8 the price of thermal coal in the reference case for the year 2000 is given at $100/tce, the price of coal to industry would probably be 10-20$/tce higher - still leaving coal at less than half the price of gas or oil.

Let us look finally at the potential for coal/water mixtures. If the need can be obviated for completely new boilers and for major expenditure on environmental control (e.g. dust collection equipment), then CWM could make a substantial contribution to bridging the gap between modest projections of coal use in industry and the much larger potential use. EAS has begun to examine how economic CWM may be in comparison with the direct use of coal in different situations. At this stage there are two clear hurdles for CWM to overcome:

(a) the extra cost of preparing the CWM blend, both in operational cost of a plant and through any extra loss in total thermal value of the coal in the deep cleaning necessary; this may show the need for preparing at least two different products from one raw coal, and raises questions of plant location and transfer prices; in general this must pose seemingly greater problems in Europe than the USA, but it is interesting that some European organisations are expressing interest in the possibilities;
(b) the extent of boiler derating with CWM use; clearly this depends on the boiler design, but the range of trials being carried out in the USA and elsewhere should provide some guidance.

Both these topics are receiving attention in other papers at the conference. At this stage it is perhaps sufficient to say that the published costs suggest that the extra cost of a CWM may be 30-60$/tce more than the original coal cost; this, compared with a coal/oil price difference of more than $100/tce, gives some scope for savings if the boiler is not heavily derated. But much will depend on whether practical experience proves the economics of preparation and use.

Table 1   Economic Growth and Energy Demand in OECD Europe
           (percent per annum increases)

	Real GDP	Primary Energy Requirements
1974	+2.1	-1.5
1975	-0.8	-3.5
1976	+4.6	+5.8
1977	+2.4	+1.1
1978	+3.0	+2.6
1979	+3.2	+4.8
1980	+1.4	-2.8
1981	-0.2	-2.6
Annual average (1973-81)	+1.9	+0.4

Source: (1)

Table 2   Primary Energy Growth by Source in OECD Europe

	Quantities (Mtoe)		Percentage change
	1973	1981	1973-81
Solid fuels	258.9	285.7	+10.4
Oil	703.8	590.5	-16.1
Gas	122.2	175.2	+43.4
Nuclear Power	17.9	71.7	+300.6
Hydro & Geothermal	74.9	94.9	+26.7
TOTAL	1177.7	1218.0	+3.4

Source: (1)

Table 3   Energy Demand by Sector in OECD Europe

	Quantities (Mtoe)		Percentage change
	1973	1981	1973-81
Iron & Steel	87.9	63.9	-27.3
Other industry	245.5	254.3	+3.6
Transport	167.9	197.9	+17.9
Housing	249.0	204.1	-18.0
Agriculture	16.7	17.2	+3.0
Services	40.1	65.2	+62.6
Losses in electricity	203.8	264.2	+29.6
Other	166.8	151.2	-9.4
TOTAL PRIMARY ENERGY REQUIREMENT	1177.7	1218.0	+3.4
incl. electricity consumption	101.3	127.2	+25.6

Source: (1)

Table 4   Regional Trends in OECD Europe

	GDP (1975 $ billions)		Primary Energy Requirements (Mtoe)	
	1981	Percentage change 1973-81	1981	Percentage change 1973-81
Nordic countries	190	+19.4	119.2	+7.5
France, Germany, Italy, UK	1373	+17.5	793.7	-1.4
Benelux & Ireland	181	+16.0	117.4	-3.0
Switzerland & Austria	105	+10.8	51.1	+10.2
Other Mediterranean	204	+23.5	136.6	+45.0
TOTAL OECD EUROPE	2053	+17.8	1218.0	+3.4

Source: (1) and (2)

Table 5   Prices of fossil fuels in international trade

	Quantum change in current price (1973-82)	Price per toe early 1983
Crude oil	x10.7	220
European domestic thermal coal	x2.6	90
European imported thermal coal	x2.9	68
N. Sea gas to UK	x5.5	150
N. Africa gas to S. Europe	x5.5	220

Source: (3), (4), (5), (6) and (7)

Table 6   Energy Scenarios for Western Europe

	Percent per annum growth to 2000			Coal in 2000 (Mtoe)	
	Economic Activity	Energy Demand	Coal Demand	Total Demand	Imports
WOCOL (1980)	+3.0	+2.4	+3.6	813	402
Exxon (1980)	+2.8	+1.5	+2.8	694	300
DRI Europe (1982)	+2.7	+2.0	+2.6	692	362
IEA (1982)	+3.0	+2.2	+3.5	789	426
EAS (1982/83):					
(i) reference case	+2.5	+2.2	+3.7	896	495
(ii) low demand case	+2.5	+1.5	+2.4	700	334

Source: (8), (9), (10), (11) and (12)

Table 7   Fuel mix in EAS energy scenarios for Western Europe
(Mtoe)

	1981	2000	
		Reference Case	Low demand Case
Oil	591	626	554
Gas	175	302	267
Coal	286	597	467
Nuclear	72	274	242
Other	95	125	110
TOTAL	1218	1924	1640

Source: (12)

Table 8   Trends in W. Europe's delivered energy prices to 2000

	1982/83	2000 (marginal price)#	
		Reference Case	Low demand case
Oil: $/barrel	30	50	35
$/tonne	220	365	255
$/GJ	4.8	8.0	5.6
Gas: $/million BTU	5	8	6
$/toe	220	350	250
$/GJ	4.8	7.7	5.5
Thermal Coal:			
$/tce	45	100	45
$/toe	64	143	64
$/GJ	1.4	3.1	1.4

in 1982/83 dollars

Source: (12)

Table 9   Coal uses in Western Europe to 2000

		Mtoe		Percent p.a. growth	
	1981	2000		1981-2000	
		Reference Case	Low Demand	Reference Case	Low Demand
Economic growth				+2.5	+2.5
Electricity demand	429	838	686	+3.6	+2.5
incl: coal	169	384	288	+4.4	+2.8
oil/gas	87	55	55	-2.4	-2.4
nuclear	72	274	242	+7.3	+6.6
other	101	125	101	+1.1	-
Coal demand	286	597	467	+3.9	+2.6
incl: electricity	169	384	288	+4.4	+2.8
iron & steel	39	53	45	+1.6	+0.8
other	78	160	134	+3.9	+2.9

Source: (12)

Sources:

(1) Energy Balances of OECD Countries, 1981
     IEA, Paris. April 1982 and earlier annual editions

(2) Main Economic Indicators
     OECD, Paris. December 1982 and earlier monthly editions

(3) The Petroleum Economist. Various editions

(4) Monthly Bulletin of Statistics
     United Nations. Dec 1982 and earlier editions

(5) Yearbook of World Energy Statistics
     United Nations. 1981 and earlier editions

(6) Digest of UK Energy Statistics
     HMSO. 1982 and earlier editions

(7) World Gas Report (Monthly)
     Noroil Publishing House. Various editions

(8) Coal: Bridge to the Future (WOCOL)
     Ballinger Publishing. June 1980

(9) World Energy Outlook, 1980-2000
     Exxon. December 1980

(10) European Coal demand to 2000
     DRI Europe. June 1982

(11) World Energy Outlook
     IEA. October 1982

(12) Constraints on International Trade in Coal.
     EAS. December 1982

(13) Programmes and prospects for the electricity sector
     UNIPEDE. September 1982

(14) The Use of Coal in Industry
     IEA/CIAB. August 1982

(15) Nuclear Power in Europe
     UNIPEDE. June 1982

COAL-LIQUID MIXTURES IN THE UK INDUSTRIAL SECTOR

Alan J. Bogle[*] and Martin Smith[**]

This paper reviews the potential for coal-liquid mixtures
in the UK industrial sector. The growth of a market for
coal-liquid mixtures depends not only on their technical
viability, but also on their economics. It is necessary
that coal-liquid mixtures offer cost savings both over
oil and conventional coal-burning plant. These savings
are estimated in terms of the payback period for the
investment required. The paper concludes that there
exist significant markets for coal-water mixtures in
shell boilers and in site constructed water-tube boilers.
The market for coal-oil mixtures exists for only a
relatively narrow range of conditions.

INTRODUCTION

Background

Mixtures of coal with liquids have been considered for over 100 years as a
means of burning and transporting coal. During this period coal-oil
mixtures (COM) and coal-water mixtures (CWM) have received attention as a
substitute for liquid heating fuels, particularly during the two World Wars
when the security of oil supplies was in doubt.

The use of coal-liquid mixtures as a fuel has two main attractions:

(1)   they are cheaper than fuel oil on an energy content basis (because
      coal is cheaper than fuel oil).

(2)   they are more convenient to use than coal because some of the
      advantages of liquid fuels (for example, ease of handling and
      combustion) are retained.

Interest in coal-liquid mixtures revived during the 1970s following the oil
crisis and a number of R&D programmes were initiated. At first, these were
concerned with coal-oil mixtures, but more recently considerable interest
has been shown in coal-water mixtures, particularly as a means of converting

[*] Operational Research Executive, National Coal Board, Harrow, England
[**] Coal Research Establishment, National Coal Board, Stoke Orchard,
Cheltenham, England

existing boilers and furnaces from fuel oil to coal-firing. However, there may also be applications for coal-liquid mixtures in new, purpose-designed boilers, although this possibility has not been included in the present study.

Raw material sources

In order for coal-liquid mixtures to substitute for fuel oil, it is desirable to use coal with as low an ash content as is possible in order to minimise the disadvantages arising from the presence of ash. The most economic ash content will depend on the application. Present methods of coal cleaning allow the ash to be reduced to the level of inherent ash (a minimum of about 5 per cent).

The principal grade of coal to meet a demand for coal-oil mixtures would be washed smalls, although washed singles could also be used. For coal-water mixtures, there is the additional possibility of using froth flotation fines, although this option has not been considered in the present paper.

Preparation of coal-liquid mixtures

The aim of coal-liquid mixture preparation technology is to produce a mixture which contains a high percentage of solids, which has a sufficiently low viscosity that it can be handled by conventional pumps, valves and burners and which does not settle during transportation and storage.

The main aspects of the preparation of coal-liquid mixtures are:

(1)   grinding of the coal

(2)   mixing with the liquid

(3)   stabilisation of the slurry and viscosity
       adjustment by additives

Grinding to a maximum particle size considerably less than 0.1 mm assists in obtaining a suitable particle size distribution. This is necessary in order to achieve a high proportion of solids in the mixture and to avoid blockages in the burner atomiser. When grinding is carried out on dry coal, then simple mixing with fuel oil or water is required. However, in most preparation plants, grinding in the presence of oil or water is preferred. Chemical additives, usually of a proprietary nature, are often added during or after the grinding stage. These reduce the viscosity of the mixture and act as stabilisers.

Combustion of coal-liquid mixtures

The combustion of coal-liquid mixtures may require the development of burner designs which differ from those used for the combustion of heavy fuel oil, although this does not appear to be a major obstacle to their use.

A major consideration in the conversion of oil-fired boilers to coal-liquid mixtures is that some derating of the boiler output may be necessary. The economic impact of derating may be represented by a capacity charge that is the cost of increasing effective capacity to meet demands. This charge is zero if the revised output is still adequate to meet maximum steam demand of the site. Analyses of the industrial sector in the UK suggest that considerable spare capacity does exist. Alternatively, if maximum steam demand occurs only for brief periods, the ability to burn oil for these periods may be retained. This would reduce the fuel cost savings resulting from conversion, but would avoid the need for more boiler capacity. The effect of derating is of more significance at sites with high load factors, where the boiler operates at a level close to maximum output for long periods. The effect of derating therefore varies considerably from site to site.

ECONOMICS

The economics of coal-liquid mixtures have been considered in three sections: the cost of these new fuels, the costs of converting existing oil-fired boilers to use them, and the consequent payback periods which can be expected from these costs.

Fuel and preparation costs

Current UK fuel and preparation costs for coal-water and coal-oil mixtures are presented together with industrial coal and heavy fuel oil costs in table 1. Capital and operating costs have been estimated for producing coal-water and coal-oil mixtures from wet grinding of washed smalls on a 500,000 t/a scale.

In order to simplify the comparison of the fuels, the costs per GJ steam output are also given. These take into account the reduction in boiler efficiency associated with coal-water mixture combustion, arising from the increase in latent heat losses. On this basis, coal-water mixture offers a fuel cost saving of £0.4/GJ compared with heavy fuel oil. For coal-oil mixture, however, the fuel cost saving compared with heavy fuel oil is only £0.1/GJ.

Capital costs of conversion

The conversion of current oil-fired boiler stock to coal-liquid mixture firing requires several modifications to the plant, namely:

- alterations to the storage and handling equipment
- alterations to the burners
- provision of sootblowers
- provision for furnace bottom ash removal
- provision of grit arresting equipment
- modification of fan capacity

Of these alterations, grit arresting equipment represents the largest cost item.

Based on the published literature[2-7], estimates of the total capital cost for conversion from oil-firing to coal-liquid mixtures have been derived and are presented in table 2. The costs are given both in absolute terms and as a percentage of the cost of the corresponding new coal-fired boiler systems.

Both coal-water mixture and coal-oil mixture require similar modifications, and therefore only one set of conversion costs have been used in the analysis. The alterations required for coal-water mixture conversion could be somewhat more extensive than those required for coal-oil mixture conversion, although the full extent of the modifications will depend on the original boiler design.

Payback periods

The potential attraction of coal-liquid mixtures is that they offer lower
fuel costs than oil-firing, although additional capital investment is
required to convert an existing oil-fired boiler for coal-liquid mixture use.
Whether such conversions are economic can be assessed from the fuel cost
savings and the investment levels in terms of the payback period.

The results of the economic analysis are illustrated in figure 1 for the
three cases of site-constructed water-tube boilers, packaged water-tube
boilers and shell boilers.

In each case the payback period depends on the load factor at which the
boiler operates. Since the economics are dominated by capital costs at
low load factors and fuel costs at high load factors, oil-firing is
generally preferred for low load factor applications with coal-liquid
mixtures and direct coal-firing being preferred at higher load factors.
In each figure, three cases are considered:

(1) The continued oil-firing of an existing boiler in competition
     with direct coal-firing in a new boiler.

(2) The continued use of oil in an existing boiler in competition
     with the use of direct coal-firing in a new boiler and the
     conversion of the existing boiler to coal-water mixtures.

(3) The continued use of oil in an existing boiler compared both
     with the use of direct coal-firing in a new boiler and conversion
     of the existing boiler to coal-oil mixtures.

For each of these cases the results are illustrated in terms of the range
of load-factors for which the three options are economic at the specified
payback period.

For site-constructed water-tube boilers, the construction of a new coal-
fired boiler does not meet a two-year payback criterion compared with the
continued use of fuel oil in an existing boiler, although longer payback
periods can be met. However, conversion to coal-water mixtures allows
coal to penetrate further into the market, reducing the number of consumers
who would continue to use oil. The range of conditions under which
conversion to direct coal-firing is attractive is also reduced in this case.
The smaller fuel cost savings associated with conversion from oil to coal-
oil mixtures are responsible for the comparatively narrow range of load
factors for which this fuel is economic.

The range of load factors for which conversion from oil-firing to coal-water
mixtures is economic for packaged water-tube boilers is much smaller than
for site-constructed water-tube boilers. This reflects the lower capital
costs associated with the option of direct coal-firing using a new boiler
in this case. For similar reasons, together with the low fuel cost savings
from conversion to coal-oil mixtures, there is no range of load factors for
which conversion from oil-firing to coal-oil mixtures is economic.

For shell boilers, the range of load factors for which conversion from oil-
firing to coal-water mixtures is economic are similar in size to those for
packaged water-tube boilers but at a lower level and slightly smaller.
This range reflects the capital cost associated with conversion to coal-water
mixtures. For similar reasons as for packaged water-tube boilers, there is

no range of load factors for which conversion to coal-oil mixtures is
economic and the case is not illustrated in figure 1.

THE MARKET FOR COAL-LIQUID MIXTURES

Using the results of the economic analyses presented in the previous section,
estimates have been made of the maximum potential UK market demand for coal
in replacing fuel oil by the various options considered. The estimates are
made assuming that all of the existing oil-fired capacity within the range
of load factors for which coal-firing or coal-liquid mixture firing is
economic with a two-year payback criterion undergo conversion. The estimates
do not include capacity which would prefer to switch to direct coal-firing
even where coal-liquid mixtures are available. The market estimates in the
cases of conversion to coal-liquid mixtures therefore represent the maximum
potential market for these fuels. In practice, the actual size of the
market for coal using these technologies is likely to be much less.

The market estimates are summarised in table 3. As shown in the table,
conversion from oil-firing using coal-oil mixtures is unattractive in the UK.
Premature retirement of existing oil-fired capacity in favour of new coal-
fired capacity has a significant potential in the shell boiler market, and
there is also a smaller potential for packaged water-tube boilers. If,
however, coal-water mixtures become available, there exists a significant
market potential for an increase in coal sales.

Although shell boilers appear to offer the best prospects both for conversion
to coal and coal-water mixtures, several factors are likely to limit the
market to well below the maximum potential. In particular, it may be
necessary in many cases to construct a new boilerhouse (or modify the
existing one) when converting to coal or coal-liquid mixtures. This
represents a significant additional cost item.

In the case of conversion to coal-water mixture the expected remaining life
of the oil boilers is important. The expected lifespan of shell boilers
(20-25 years) is much shorter than for water-tube boilers (30-50 years)[6].
Over half the non-coal fuel burned in shell boilers is used in units with
less than half their lifespan remaining. In these cases, the industrialist
may prefer to defer an investment decision until he is forced to replace his
boilers.

CONCLUSIONS

(1)   Based on a coal price of £1.7/GJ, coal-water mixtures would cost
      £2.1/GJ and coal-oil mixtures £2.4/GJ. These estimates imply
      fuel cost savings compared with heavy fuel oil of £0.4/GJ and
      £0.1/GJ for the same output of steam, respectively.

(2)   The introduction of coal-oil mixtures increases the ability of
      coal to penetrate the industrial market compared with direct coal-
      firing. Coal-water mixtures are economic for a wide range of load
      factors and payback periods. Coal-oil mixtures are economic only
      for a relatively narrow range of conditions.

(3)   The most attractive markets for coal-liquid mixtures as a replacement
      for heavy fuel oil in existing boilers are for coal-water mixtures in
      site constructed water-tube boilers and shell boilers. The maximum
      potential UK market size in these two cases are 4.6 and 5.6 million

tonnes coal equivalent per annum respectively. However, demands of this magnitude are unlikely to be realised in practice.

REFERENCES

1. Economics and Statistical Division of the Department of Energy, Energy Trends, March 1983.

2. National Industrial Fuel Efficiency Service, Fuel Economy Handbook, 1979, p.47.

3. Bergman, P.D., Joubert, J.I., Bienstock, D. and George, T.J., Proceedings of First International Symposium on Coal-Oil Mixture Combustion, Florida, 1978, p.351.

4. Koda, F., Proceedings of Third International Symposium on Coal-Oil Mixture Combustion, Florida, 1981, p.529.

5. Blake, J.C., Foo, O.K. and Jamgochian, E.M., Proceedings of Third International Symposium on Coal-Oil Mixture Combustion, Florida, 1981, p.861.

6. Coal Industry Advisory Board of the IEA, The Use of Coal in Industry, 1982.

7. Morrison, G.F., Conversion to Coal and Coal-Oil Firing, IEA Coal Research Report No. ICTIS/TR07, 1979.

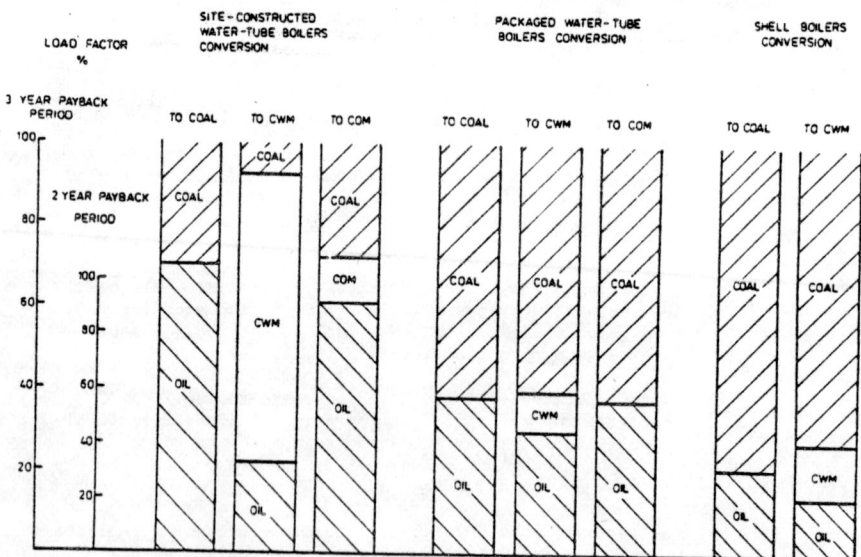

Figure 1. Load factors for conversion to coal, CWM and COM

Table 1. Fuel and preparation costs (£/GJ output (1982 £))

	Coal-water mixture (70% coal)	Coal-oil mixture (50% coal)	Delivered to large industrial consumers	
			Coal	Heavy fuel oil
Preparation costs				
Capital charges	0.1	0.05		
Operating and maintenance cost	0.2	0.1		
Surfactant cost	0.1	0.05		
Total preparation cost	0.4	0.2		
Input fuel costs				
Coal	1.7	0.7	1.7[1]	
Oil		1.5		2.6[1]
Total delivered fuel cost	2.1	2.4	1.7	2.6
Cost/GJ steam output	2.8	3.1	2.1	3.2
Cost difference from oil/GJ steam output	0.4	0.1	1.1	

Assumptions
(i)   The capital cost of plant was estimated at £13,000,000 and discounted at 5 % over 20 years to provide the annual capital charge.
(ii)   The annual operating and maintenance cost was estimated at £2,000,000.
(iii)   Coal for CWM and COM production had the same cost per GJ as coal to industrial consumers.
(iv)   An allowance for delivery cost of CWM and COM was included in coal cost.
(v)   The cost per GJ steam output, was calculated from delivered fuel cost taking the different conversion efficiencies into account, i.e. coal and oil 80 %, CWM 70 %, COM 78 %.
(vi)   Gross calorific values:
        Coal for CWM/COM production   29.9 GJ/t
        Coal to industrial consumers   27.5 GJ/t
        Heavy fuel oil to industrial
        consumers   42.6 GJ/t

CLM-X

Table 2. Capital cost assumptions for the conversion of oil boilers to CLM

Type of system	New coal-fired boiler cost (£/kW)	Conversion cost from existing oil-fired boiler (£/kW)	Conversion cost* as a percentage of a new coal-fired boiler (%)
Site-constructed water-tube boiler	95	14	15
Packaged water-tube boiler	55	16	30
Shell boiler	30	7.5	25

Table 3. Maximum potential UK markets for conversion from fuel oil assuming a 2-year payback period (Million tonnes of coal equivalent)

Boiler type	Technology available		
	Direct coal-firing in a new boiler	Conversion to CWM	Conversion to COM
Site constructed water-tube	0	4.6	0
Packaged water-tube	0.3	0.7	0
Shell	2.0	5.6	0
Total	2.3	10.9	0

ACKNOWLEDGEMENT

The authors wish to acknowledge the many useful discussions with Dr. D. Merrick. The authors also wish to thank the National Coal Board for permission to publish the paper. However, the views expressed are those of the authors and not necessarily those of the National Coal Board.

AN ECONOMIC ANALYSIS OF THE FUTURE FOR
COAL-LIQUID FUELS

Frank D. Moles*, Kevin J. Lapwood* & Barrie G. Jenkins

Over the last 6 years the Fuels and Energy Research Group
of the University of Surrey has conducted research into
the combustion, handling and emissions of coal-oil fuels.
The implications of this work are considered in the light
of current economic conditions and trends thus enabling an
analysis to be made of the potential market for coal-
liquid fuels.
The move away from oil-firing appears inevitable, although
it will not occur immediately but the role of coal-liquid
fuels as an intermediate between oil and coal-firing is
far from certain. The authors conclude that, on both
technical and economic grounds the market for coal-liquid
fuels is confined to small/medium sized industrials users
and cite steam raising for the marine industry as the
most likely potential application.

1. Coal-liquid Fuels

The Industrial Revolution was fuelled by coal, burnt in lumps on
grates. Unfortunately, the very nature of the coal surface deter-
mines that this process is technically difficult and horrifically
inefficient. If, however, coal is finely powdered and then mixed,
either with air (PF) or a liquid fuel (CLF), both its handling
and combustion characteristics are thereby significantly improved.
Both processes date back over 100 years but, whereas pulverised
fuel firing is a long established technology, interest in coal-
liquid fuel firing has only really arisen since the U.S.Depart-
ment of Energy began publicising it by means of their Symposia,
which took place in 1978, 1979, 1981, 1982 and 1983.

Our research, specifically on coal-oil dispersions, began at the
University of Surrey in 1977 in response to a contract awarded
by BP. During the last six years the Fuels and Energy Research
Group of the University of Surrey (FERGUS) have dealt with five
aspects of the subject: handling, combustion and emissions[1];
the effect of coal rank on the heat release patterns[2]; the
future technical feasibility of such fuels[3]; an examination
of the combustion behaviour of these fuel droplets[4]; and the
technical problems associated with substituting such fuels in
heavy fuel oil-fired installations[5].

* Fuels and Energy Research Group, Department of Chemical
  Engineering, University of Surrey, Guildford, Surrey GU2 5XH

† Aeroil-Flaregas Ltd., Horton Road, West Drayton, Middx. UB7 8BG.

In the current paper the authors have attempted to round this
work off by analysing the by now voluminous literature on the
subject in an attempt to answer the question "Have coal-liquid
fuels a significant economic future and if so where does it lie?"

2. The Switch from Oil to Coal

The current interest in coal slurry fuels has been re-stimulated
both by the rise in oil prices over the last decade and the
long term doubts about the availability of this valuable resource.
From a technological viewpoint oil may have to be reserved, in
the future, for transport   fuels and chemical feedstocks.
Current coal reserves are estimated at between 30 and 50 times
greater than recoverable oil reserves[6], and despite the present
glut of oil resulting in a drop in price, it has been forecast
that oil prices could rise by 10% per annum until the year 2000[7].
It is against this background of inevitable change in the world's
energy mix that coal-liquid fuels have emerged as possible alternatives to heavy fuel oil. However, the idea of mixing pulverised coal with oil owes little to the 1973 Oil Crisis. The
original patent dates back to 1879[8], and there were several useful studies prior to World War II. The economic prospects for
coal-oil mixtures (COMs), or the more stable counterparts, coal-
oil dispersions (CODs), have, however, never been better than
in the last 10 years. Demand for oil reached an all time high
in 1979, whilst the spot price for a barrel of oil peaked at
$37.00 in 1981. The economic incentives for a rapid conversion
to coal firing were well reported at the time[9] and coal-oil
fuels were proposed as an intermediate solution, requiring less
capital outlay and technological innovation than complete
conversion to coal.

The switch from oil, however, has not occurred as rapidly as was
first anticipated. Oil product prices have doubled in real terms
in the past ten years, but oil's share of the U.K. market has
only dropped from 48% to 42%[10]. The effect of the current oil
surplus can only be to reduce the incentive to convert to COM
or COD. It is perhaps for this reason that the fashionable
intermediate solution now being proposed is conversion to coal-
water mixtures (CWM), which represent a further step towards
the ultimate goal of complete coal firing.

3. What are the Problems?

When considering the conversion of any oil-fired installation
to coal liquid fuel the following technical problems have to
be taken into account.

3.1 Manufacture and Handling

Most potential users would prefer to purchase a manufactured
coal liquid fuel as opposed to investing in the required slurry
preparation plant. Commercial experience has shown this to be
the case even for large users like base-load power stations

which have both the resources and the man-power to operate such equipment. Which organisation will therefore supply such fuels? Much of the initiative in the development and manufacture of COM's and COD's has come from the major oil companies[11,12]. The attraction for these companies to be involved in the marketing of oil-based fuel is obvious, but it is not so obvious who will supply the CWM's. In the United States the interest in CWM has only been supported to a small extent by the private coal companies[13] and some research organisations[14] but potential suppliers to the U.K. market have yet to emerge.

An advantage which coal liquid fuels have over dry, pulverised coal is their potential to use the existing oil transport and delivery network. A considerable cost advantage exists for CWM over dry p.f. if the slurry can be delivered by pipeline. However, this is not feasible for anyone except the very large users situated close to a manufacturing plant. The majority of coal slurries would have to be transported by marine or road tankers resulting in higher transport costs per unit calorific value than for dry p.f. or straight fuel oil.

The plant modifications required for storing and pumping the delivered fuel have been well documented[1,15], and will not present difficulties provided the slurry is stable and the coal concentration is not so high as to unacceptably increase the viscosity.

The design of the burner is crucial. It is necessary to achieve good atomisation with minimum burner wear.

3.2 Boiler Derating

Derating of conventional oil-fired equipment may be necessary if excessive fouling and slagging occurs on the heat transfer surfaces. This will be dependent on the amount of mineral matter present in the fuel and the behaviour of the ash formed. The problem is particularly serious in more modern equipment which is designed to much tighter specifications. Insufficient fan power may also lead to derating. Higher levels of excessive air are necessary to prevent partial combustion and excessive carbon in ash levels. Even if the increased fan capacity is available the increased velocity could cause erosion problems on the heat exchange surfaces. Derating effects are frequently ignored in economic evaluations of coal-liquid fuels by assuming that the oil-fired equipment is not operating at its full load before conversion. However, the effect of derating on payback time was shown to be significant by Borio and is illustrated in Figure 1, the data for which was obtained from a study of the possible conversion of Alamitos power station, California[17].

3.3 Emissions

In addition to the provision of bottom ash handling equipment and soot blowers to clean heat exchange surfaces it will also be necessary to remove particulates from the flue gases in order

to meet environmental constraints. This will necessitate the installation of electrostatic precipitators or fabric filters which represents a major portion of the conversion cost. Figure 2 shows that plant which does not require flue gas cleaning equipment, marine boilers for example, is considerably cheaper to convert than plant which does require such equipment.

4. What are the Solutions?

The technical feasibility of burning coal slurry fuels is not in doubt, in spite of the plant conversion and operational problems. Fuel suppliers will obviously emerge if a large enough market exists; however, the size of that market will depend on there being sufficient incentives to burn coal liquid fuels as opposed to heavy fuel oil, or indeed coal.

Research into the combustion of COM's and COD's has shown that suitable burners already exist for use with these fuels. For small installations a low pressure air burner has been recommended[1], whilst other burners have been developed or modified for larger applications[18]. More development is necessary before CWM atomizers are available but suitable designs have already been proposed[19].

(a) Initial heating and devola-   (b) formation of swollen char.
      tilisation

Figure 4.   The combustion of a single droplet of CWM made with a mid-rank bituminous coal.

The derating and particulate emissions problem associated with
ash deposition and incomplete combustion of char residues can
be reduced by careful selection of the constituent coals in
coal liquid fuels. Research into the combustion of coal-oil
fuels has shown that the use of low swelling, non-caking coals
(CRC 100-202 and CRC 700-902) ensures a burnout mechanism which
favours the complete combustion of carbonaceous char residues,
especially with the more reactive, low rank coals[2,4]. The same
burnout mechanism has been observed in petroleum coke dispersions,
which have an added advantage of a very low ash content. There
is evidence to suggest that the coal rank is equally important
in determining the burnout characteristics of coal-water mixtures. Figure 4 shows a single droplet of CWM, made with a high
swelling, bituminous coal which forms a large well-fused char
after devolatilisation. This means that very fine grinding
of such coals in coal slurries has no useful effect on combustion.

Problems due to ash deposition could be solved by coal cleaning,
although there is currently no commercial process which achieves
the necessary reduction in ash content. However, recent studies
have indicated that it would be feasible to spend up to 1½ times
the raw coal price on coal cleaning[20].

5. Where is the Market for Coal-Liquid Fuels?

It is a fact that many of the applications currently being
proposed for coal liquid fuels would be better served by straight
pulverised coal firing. Several COM and COD trials have been
conducted on power station boilers originally designed to burn
coal, and it has been clearly demonstrated that very little
modification is required for successful operation. Coal-oil
fuels do provide a cheaper alternative to fuel oil, for less
capital outlay than complete conversion to coal, in such boilers.
It can be seen in Figure 2 that estimated conversion costs vary
enormously for plants of similar capacity. This is an indication
of the degree to which "suitability for conversion" can vary.
Plant which was originally designed to burn coal has much lower
heat release rates than oil-fired plant, and hence is less likely
to require derating. It is also usually easier to incorporate
ash handling equipment on such plant. However, these coal-fired
units are the smaller and older boilers which are currently
closed in th U.K. by the Central Electricity Generating Board.
Advances in burner design, coal cleaning and a better understanding of combustion mechanisms will probably widen the
market for coal slurry fuels to include more highly rated oil-
fired equipment; however, the technology which will enable this
expansion to occur can be equally well applied to the combustion
of pulverised coal.

In short, large users have the technology and resources to
convert directly to coal where such a conversion is feasible.
The current fall in oil prices can only make this option more
attractive than a switch to coal liquid fuels.

Smaller users rarely have the option of converting existing

equipment to coal-firing. Most economic studies compare the
cost of converting existing equipment to coal slurry fuels
with the cost of replacement with new coal-fired equipment.
On this basis Figure 3 shows the former case to be more attractive, especially since operating coal-fired plant requires
rather more expertise than operating oil-fired plant. However,
the demand for coal-fired equipment will inevitably grow in this
sector over the next decade, being actively encouraged by
Department of Industry and EEC grants. In the meantime there is
a potential market for coal slurry fuels among small users.
Coal-water mixtures have been successfully employed in small
rotary driers[21], and petroleum coke-oil mixtures have found
an application in marine boilers[22], The latter being particularly
suitable for conversion despite the fact that they are designed
to have high heat release rates, since they will not require
flue gas cleaning equipment.

6. The Future for Coal-liquid Fuels

In the future, when coal has to be used as a fuel, because oil
and natural gas fields have been drastically depleted, then
coal should not be used raw but ought to be gasified, or failing
this, pulverised. Both processes can be carried out in large
scale units where costs can be reduced and cleanliness closely
controlled. For small and medium sized installations, coal-
liquid fuels become a feasible and cheaper alternative to straight
oil firing, provided such installations can be suitably modified.

Coal-liquid fuels must be marketed as a package, which includes
storage tank, preheaters, valves and burners, all specifically
designed to suit the fuel. Provision must also be made to
remove the resulting ash from the system; periodically from
the combustion chamber and continuously from the flue gases.

Provided that the ash has been dealt with then the actual
combustion of coal-liquid fuels presents no difficulties with
regard to other emissions. However, it is widely accepted
that more modern, highly rated units will have to be de-rated
by around 10%.

Running through the early literature on the subject is the
thread of marine application, and, in the authors' opinion,
this is the most likely economic application in the near future.
The fly ash eminating from the stack can simply be allowed to
blow into the sea.

References

(1) ALABAF, J.S., PATTERSON, M.C. and MOLES, F.D.
    "The Handling and Combustion Emissions from Coal-Oil
    Dispersion when Fired through an LPA Burner", pp 1031-1100.
    3rd International Symposium of Coal Mixture Combustion,
    Orlando, U.S.A., 1-3 April (1981).

(2) JENKINS, B.G., ALABAF, J.S., PATTERSON, M.C. and MOLES, F.D.
    "The Effect of Coal Rank on Heat Release Rates from Coal-Oil

Dispersions"
Proceedings 64th Canadian Inst. of Chemists Coal Symposium,
Ottawa, October (1981).

(3) MOLES, F.D., JENKINS, B.G., PATTERSON, M.C. and ALABAF, J.S.
"Coal-Oil Dispersions"
Future Trends in Oil Firing Symposium, pp 74-87,
Inst. of Energy, Portsmouth, November (1981).

(4) LAPWOOD, K.J., STREET, P.J. and MOLES, F.D.
"An Examination of the Behaviour and Structure of Single
Droplets of Coal-Oil Fuels during Combustion"
5th Int. Symp. on Coal Slurry Combustion and Technology,
Tampa, USA, 25-27 April (1983).

(5) JENKINS, B.G., MOLES, F.D. and PATTERSON, M.C.
"Technical Expectations for Coal-Oil Fuels as a Replacement
in Fuel on Fired Equipment"
1st European Conference on Coal-Liquid Mixtures,
Inst. of Chemical Engineers, Cheltenham, 5-6 October (1983).

(6) World Energy Resources 1985-2000. World Energy Conference,
IPC Sci. and Tech. Press, Guildford (1978).

(7) Anon.
"Demand, Supply and Prices Summarised to the Year 2000",
pp 2-6, Energy World, 101, March(1983).

(8) SMITH, H.R. and MUNSELL, H.M.
Liquid Fuel.
U.S. Patent 219, 181, 24 February (1879).

(9) HAWTHORNE, W.
48th Melchett Lecture, 20 November 1980
also:
Energy World, 81, pp 2-12, May (1981).

(10) Anon.
"Oil Glut will Halt Switch to Other Fuels",
Energy World, 97, pp 8-9, November (1982).

(11) CLAYFIELD, E.J. and VAN KLINKE, J.
"Colloil: The Shell Approach to Stable Coal-Oil Mixtures"
4th Int. Symp. on Coal-Slurry Combustion, Orlando, U.S.A.
May (1982).

(12) WALL, D.R.
"The Handling and Combustion of Stable Coal/Fuel Oil Dispersions"
3rd Int. Symp. on Coal-Mixture Combustion, Orlando, U.S.A.
1-3 April (1981).

(13) ADAMS, D.R.
"Commercial Demonstration of a Coal-Oil Mixture in Industrial
Applications"
5th Int. Symp. on Coal Slurry Combustion and Technology,
Tampa, U.S.A., 25-27 April (1983).

(14) FORD, F.W., MADGAVKAR, A.M., MOORE, R.E., McCORMICK, R.J.
"Coal-Water Mixtures Commercial Development Evaluation"
5th Int. Symp. on Coal Slurry Comb. and Tech., Tampa, U.S.A.
25-27 April (1983).

(15) CASSASA, E.Z. et al.
"Rheology of Coal/Water Slurries"
5th Int. Symp. on Coal Slurry Comb. and Tech., Tampa, U.S.A.
25-27 April (1983).

(16) BORIO, R.W. and HARGROVE, M.J.
"COM and CWM Fuels for Steam Generators"
4th Int. Symp. on Coal Slurry Combustion, Orlando, U.S.A.
May (1982).

(17) LIPS, H. and DERBRIDGE, C.
"Evaluation of Technical, Environmental and Economic
Feasibility of using Coal/Oil Mixtures in California Power
Plants"
3rd Int. Symp. on Coal-Oil/Mixture Comb., Orlando, U.S.A.
1-3 April (1981).

(18) DUDLEY, W.E., HIGGINS, M.E. and MEYER, D.X.
"COM Burner Development, Paul L. Bartow Unit"
5th Int. Symp. on Coal Slurry Comb. and Tech., Tampa, U.S.A.
25-27 April (1983).

(19) HICKMAN, R.H. and BUCKINGHAM, F.P.
"Burner Development for Coal and Water Slurry Fuel"
5th Int. Symp. on Coal Slurry Comb. and Tech., Tampa, U.S.A.
25-27 April (1983).

(20) CUTTING, J.C., STROM, S.S. and WARYACZ, R.E.
"Design and Economics of Utility Power Plants Utilising
Deep Cleaned Coal Water Slurries"
5th Int. Symp. on Coal Slurry Comb. and Tech., Tampa, U.S.A.
25-27 April (1983).

(21) FOX, W.K. et al.
"Economics and Technical Feasibility of CWM on Rotary Rocks
Dryers and Other Industrial Applications"
5th Int. Symp. on Coal Slurry Comb. and Tech., Tampa, U.S.A.
25-27 April (1983).

(22) BESHORE, D.G. and YAGIELA, A.S.
"The Development and Commercialization of a Low Cost
Petroleum Coke-Oil Slurry Fuel for the Marine Industry"
5th Int. Symp. on Coal Slurry Comb. and Tech., Tampa, U.S.A.
25-27 April (1983).

(23) BLAKE, J.C., FOO, O.K. and Jamgochian, E.M.
"Conversion of Oil-Fired Industrial Boilers to Coal-Liquid
Mixtures"
3rd Int. Symp. on Coal-Oil Mixture Combustion, Orlando,
U.S.A., 1-3 April (1983).

(24) THOMPSON, J.F. et al.
"US Military Coal-Oil Mixture Conversion Study"
5th Int. Symp. on Coal Slurry Comb. and Tech., Tampa, U.S.A.
25-27 April (1983).

(25) HAWKINS, G.T., CAPES, T. and COPPLE, M.
"Coaliquid Coal Slurry Fuel Program 1982"
4th Int. Symp. on Coal Slurry Combustion, Orlando, U.S.A.
May (1982).

(26) FURMAN, R.C.
"Power Plant Conversions to Coal, Coal-Oil Mixtures and
Coal-Water Slurries"
4th Int. Symp. on Coal Slurry Combustion, Orlando, U.S.A.
10-12 May (1982).

(27) De LESDERNIER, D.L., JOHNSON, S.A. and ENGLEMAN, V.S.
"Conceptual Design and Economic Analysis for Coal-Water
Mixture Utilisation in an Industrial Utility Boiler"
4th Int.Symp. on Coal Slurry Combustion, Orlando, U.S.A.
10-12 May (1982).

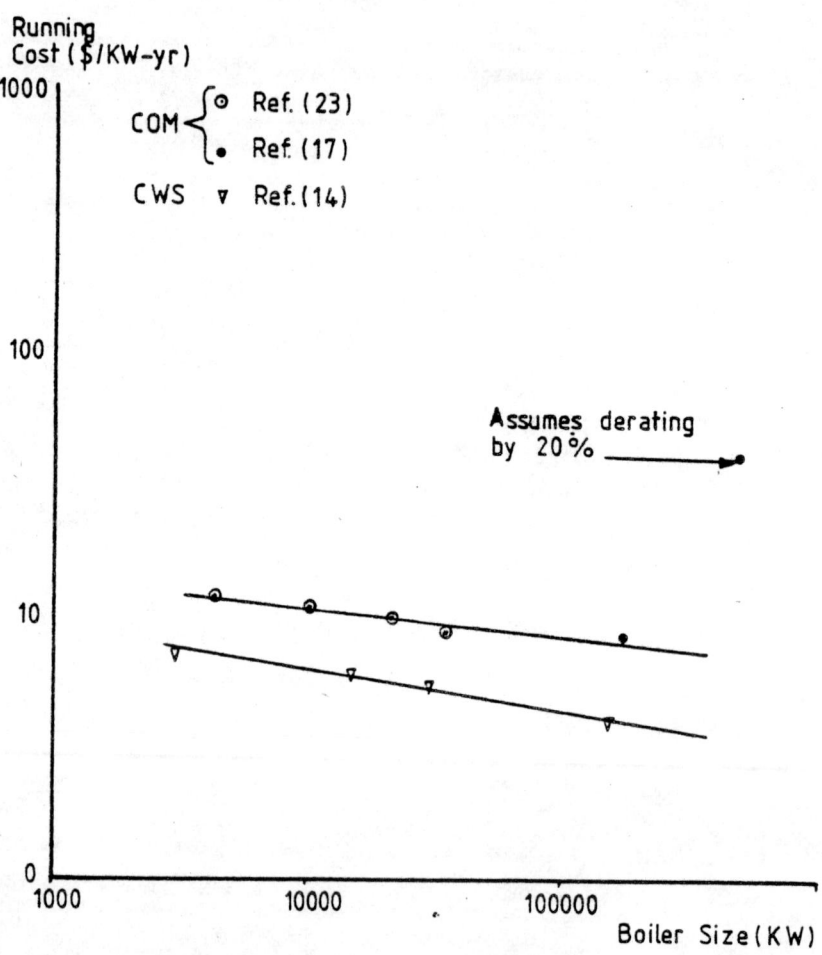

Figure (1) Running Costs of Converted Plant (Fuel, Operation
& Maintenance etc.)

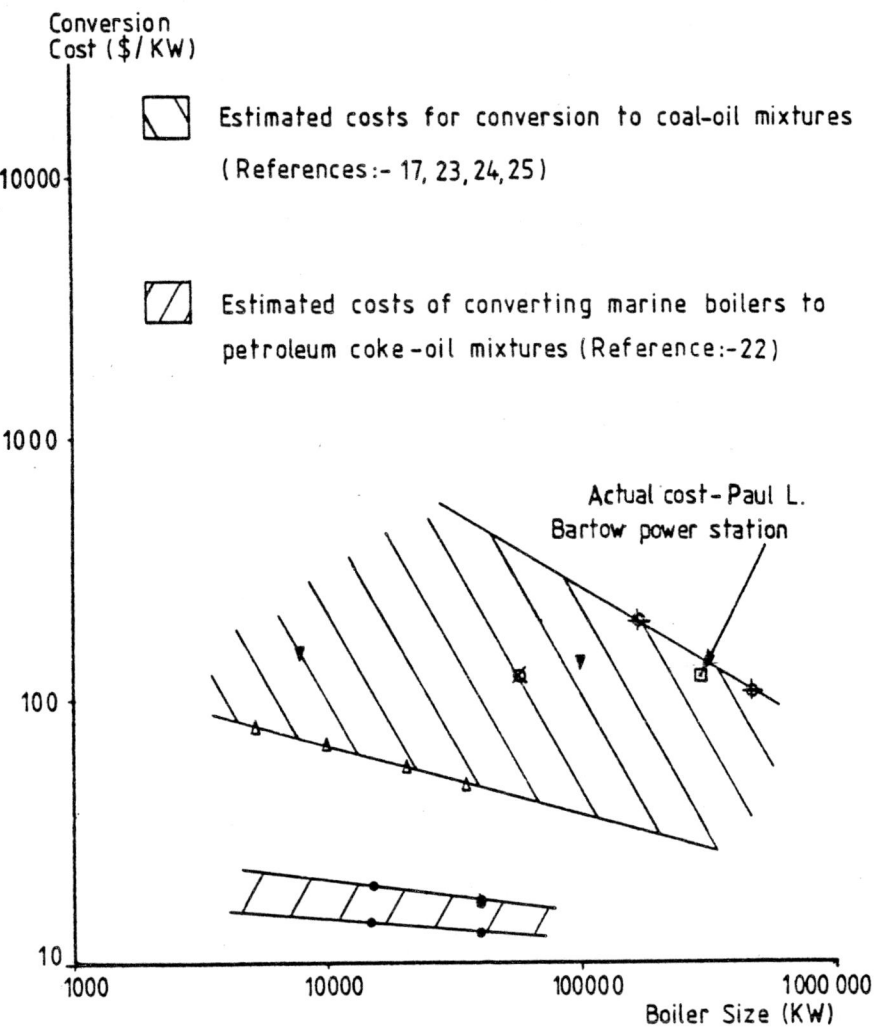

Figure (2)   Capital Costs For Conversion of Existing Equipment
To Coal-Oil Fuels

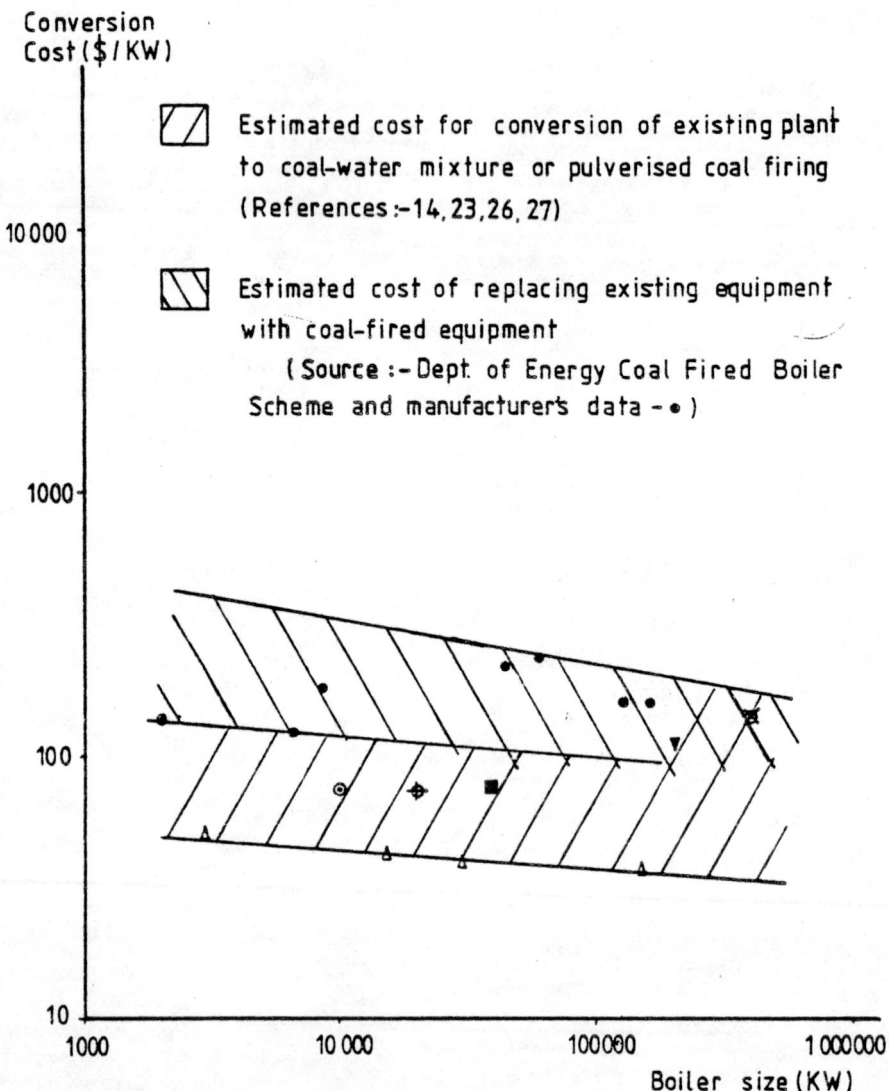

Figure (3)   Capital Costs For Conversion to Coal Water Slurry
or Coal Firing

BURNER DEVELCFMENTS FOR COAL-WATER MIXTURES

George F. Lomax * and John Harris **

This paper discusses the problems that confront the burner designer when using coal-water slurry as the fuel. Although coal-water slurry may be categorised as an independent class of fuel, certain information from previous testing with coal-oil fuels can be directly applicable in understanding the combustion process. It is significant to note that due to the unusual combustion characteristics of coal-water slurry, the designer must conceive and develop out-of-the-ordinary burners that will accomodate coal-water fuels in general, and specific coal-water formulations in particular.

INTRODUCTION

Howe-Baker Engineers, Inc., is a company primarily engaged in modular petrochemical process equipment and became involved in burner design and manufacture in the 1960's.

Since this time the company has designed many burners for unusual fuels ranging from refinery offgas to the heaviest of liquid fuels including coal slurries. The company's first experience with coal-oil fuels was in 1975 with tests in a small furnace at its Tyler facility in Texas. It was during these early tests that the handling problems with this type of fuel became apparent. However, this paper deals specially with the development of the fuel burner and it is assumed that the problems with slurry fuel handling during preparation and delivery to the burner are considered by other authors.

SLURRY FUELS

The early tests with a 1 MM BTUH (.293 MW) furnace at Tyler culminated in 1978 with the testing of two coal-oil slurries using a Howe-Baker burner in a 40,000 lbs/hr boiler at HumKo Champaign. This burner shown in figure 1 was designed with the capability to burn natural gas, No.2 through No.6 fuel oil and coal-oil slurries. Direct performance comparisons were made with an identical boiler utilising its original natural gas/fuel oil burner. The first coal-oil slurry formulation consisted of 35% wt. coal in No. 6 oil, and the second of 50% wt. coal in No. 4 oil.

* Howe-Baker Engineers Inc, Tyler, Texas, U.S.A.
** Howe-Baker Engineers Inc, European Division, Elstree, Herts

Nominal coal particle size was 200 mesh. Compared to firing No. 6 oil, the coal-oil slurries appeared to have a slightly brighter flame, but the firebox and stack discharge were slightly hazy due to the high ash content. As might be expected, the tests were conducted at a derated boiler condition (26,900 lbs/hr vs. rated 40,000 lbs/hr) due to the longer burn time requirements of the coal particles. It may be noted that these tests were conducted with the existing fuel oil valves and piping on an uninterrupted firing basis. After a flame was established with No.6 oil, the coal slurry was introduced as the No. 6 oil flow was decreased to zero; then No. 2 oil was introduced for system flush as the coal slurry flow decreased to zero, after which No. 6 oil was again introduced as the No. 2 oil flow was shut off.

In late 1981 a 2 MMBTUH (.586 MW) burner was supplied to Louisiana State University in Baton Rouge to support their investigations of heavy fuels, including slurry fuels. One such series of tests involved lignite-No. 6 slurry and these are described in a Master's Thesis by Dixit (1). The burner used in these experiments is illustrated in fig. 2. An important characteristic of the fuel in burner design is its viscosity and the relationship between fuel viscosity - lignite concentration and temperature, obtained by Dixit is shown in fig. 3. Combustion data obtained from these tests are given in fig. 4 and fig. 5. Tests with coal-water slurries in this burner are currently being carried out.

In addition to the above, several demonstration firings have been made at Tyler, on request by companies interested in coal-methanol and coal-water slurries. These demonstrations were not instrumented for test data, since the primary interest was to observe flame characteristics, using burner hardware on hand and not specifically designed for the fuels.

COAL-WATER SLURRY BURNER DESIGN CONSIDERATIONS

The burner designer is faced with a variety of problems when considering slurry fuels in general, and some of these problems are magnified in particular by coal-water slurry fuels. For sake of brevity, the 'shared' problems such as the selection of erosion-resistant materials, temperature vs. viscosity relationships, etc., have been ignored and our discussions are strictly limited to coal-water combustion.

By its very nature, coal-water slurry has a built-in fire extinguisher. Whereas coal-oil for example, exhibits ready ignition and combustion support for the coal particles, water acts as a retardent and a heat thief. On this basis alone, overall combustion efficiency dictates that a high percentage of coal must be used.

Atomisers

Perhaps in the particular case of coal-water slurry fuels, the term 'atomiser' should be changed to 'coal particle disperser'. The coal particle size does not change during the burner phase, and the coal particles are directed through a trajectory that ensures individual particle ignition, followed by sustained burning of the particles. This population of burning particles should be evenly dispersed to present an even and stable flame envelope for heat transfer purposes.

The 'atomiser' must impart a certain amount of energy to the fuel to achieve the desired condition of the fuel for ignition and burning. The rheological characteristics of the fuel can greatly influence the specific energy level required of the 'atomiser'. The manner in which this energy is imparted to the fuel, via momentum exchange becomes important. For instance, high velocity of the 'atomising' medium can introduce desirable shear forces to decrease droplet sizes in straight liquid fuels; however, a fuel having properties similar to wet sand would suggest a relatively low velocity 'atomising' medium to prevent compaction of the solid particles in the fuel mix.

Some degree of success has been achieved in burning coal slurry fuels with the two basic atomiser designs shown in Figures 6 and 7. In both of these designs, the atomising medium is introduced to the fuel at sonic velocities. By discharging into a sudden expansion zone, a network of shockwaves is created, through which the fuel must pass. So that, in addition to high shear forces, the fuel is subjected to the high pressure and temperature 'spikes' associated with shock waves. The Hyperbolic Atomiser produces a relatively long longitudinal spray pattern, whilst the Dual-Phase Atomiser essentially produces a radial spray pattern. As shown by Figure 2, an air diffuser plate was incorporated during forced draught operation with the Hyperbolic Atomiser, to provide an intermediate spray pattern between longitudinal and radial.

The primary advantages of the Hyperbolic design are:-

> Ease of fabrication and hence lower cost.
> A straight-through fuel tube that permits ease of cleanout.
> The extremely wide variety of sizes available.
> Low fuel pressure requirements.

Primary disadvantages are:-

> The fuel is encapsulated by the atomising medium.
> Relatively lower turndown capabilities and flow control
> result from lower fuel pressures

The primary advantages of the Dual-Phase design are:-

> Finer fuel atomisation.
> The atomised fuel is exposed directly to the combustion air.
> The highest intensity flame is achieved.
> High turndown capabilities and flow control result from higher
> fuel pressures.

Primary disadvantages are:-

> Machining complexities and associated higher cost.
> The minimum size is limited to approximately 3 MW duty.
> The application is generally limited to forced draught type
> operation.

In demonstration tests with coal-water slurry, a dual-Phase Atomiser that had been originally designed for straight fuel oil operation was used as it has been previously used successfully in coal-oil testing. However, the slurry available consisted of coal particles in the 325 mesh and below size range, and this together with water as the liquid base promoted rapid plugging of the individual fuel ports emanating from the central fuel tube. Consequently, a Hyperbolic design in conjuction with an air diffuser plate was tried. Some tip fouling was experienced since the water tended to flash as it approached the hot tip, leaving some of the coal particles as a deposit on the tip surface. New designs are emerging for a modified Dual-Phase atomiser that will accomodate coal-water fuels and this concept will be tested shortly. Again, one of our design problems is to define the specific fuel to be used, i.e., coal particle size coal percentage etc since these factors vary from one supplier to another.

Ignition and Burning

It is generally concluded that solid fuels exhibit surface burning characteristics, and that this is promoted by subsurface heating of the solid. Therefore both particle size and type of coal will influence both ignition and burning characteristics of the fuel. The tendency of the coal particles to agglomerate affects the completeness of combustion, hence it is not too surprising that lignite appears to burn better than higher rank coals because it does not agglomerate so readily.

There is a considerable amount of on-going research being made to pin down the combustion kinetics of coal-water slurry. Yao and Liu (2) in their work on "Behaviour of Suspended Coal-Water Slurry Droplets in a Combustion Environment, " demonstrated the effect of coal type and fuel droplet size on particle temperature history and burning characteristics. This type of information is valuable,but the single droplet data must be correlated with burner performance to permit a burner designer to take into account the influence of inter-particle reactions, air/fuel mixing effects, etc.

At the other end of the spectrum, Mapels, Cundy and Buch (3) of Louisiana State University are concurrently with this conference presenting a paper on "Combustion of a Fuel Mixture Composed of Coal and Light Fuel Oil" at the American Flame Research Committe International Symposium on Combustion Diagnostics in Akron, Ohio.

These tests were also conducted in the Howe-Baker burner depicted in Figure 2. In this paper, the coal carbon content is shown to be proportional to the amount of CO in the combustion products at a fixed O_2 level. This is described in Figure 8.

In a coal slurry fuel, each coal particle must first receive sufficient heat energy to initiate ignition, and must subsequently be exposed to the correct environment to sustain a flame envelope for complete burnout, as opposed to a glowing body or 'sparkler'. Ideally, there should be sufficient energy interchange between the particles to ignite the incoming raw fuel and to promote a homogeneous flame envelope, all without an external fuel heat source. In contrast to fuel oil based slurries, this energy interchange for coal-water slurries must be predominantly of the radiant type. This suggests that combustion air as well as water content in the fuel should be minimised to promote maximum flame temperatures of the individual burning particles.

Here we consider what occurs during the life history of coal-water slurry in the burner and beyond. First, somewhere in the process the water must be evaporated and then superheated to the combustion products temperature. On the negative side, this is a system heat loss; the coal particles are cooled by water evaporation, thereby retarding ignition and subseqent burning; and the coal particles are at least initially, encapsulated by the inert water vapour, again causing delays in ignition and burning. However, on the positive side, if the water evaporation occurs like a microexplosion during the 'atomising' process, particle agglomerations should be more readily broken up and dispersed. The bulk combustion products temperature can be maintained below the ash fusion temperature and water vapour at the combustion products temperature will not only absorb but will emit radiant energy (this is in contrast to the diatomic gases such as air, oxygen and nitrogen, which also absorb heat but are very poor radiators). Figure 9 is a simple model of this phase.

The water vapour must also be dispersed to permit exposure of the coal particles directly to combustion air. If we are to maximise system heat transfer efficiency, the amount of combustion air must be at the minimum level possible. Ideally, we should create the maximum possible velocity differential between the coal particles and the combustion air to maximise the diffusion of the air to the coal particles, thereby increasing the probability of association of the fuel with oxygen. In existing burners, the 'atomiser' provides high velocity for the fuel so that the combustion air velocity can be fairly low, requiring less energy from the combustion air fan.

After the coal particle is released from its water envelope, its specific size and shape will determine its burning and aerodynamic characteristics. Roughly, a 325 mesh particle will burn in about half the time as a 200 mesh particle. This is due primarily to the increase in the burning suface area with the smaller particles. If the particles are examined microscopically, it is seen that, in addition to the difference in mean diameters, there is a distinct difference in surface irregularities, which augment the surface area for the smaller particles. Figure 10 is a photomicrograph showing this difference. The difference in surface irregularities promote differences in, for example, aerodynamic drag, so that in addition to burning times, individual particle trajectories are also affected. This relation between burn time and overall particle trajectory is a most important consideration in a practical system.

Finally, various research investigations have shown that during burning, the metamorphosis characteristics can vary widely between the different coal particles, depending on initial size, shape, rank, etc. Because of this, some particles will more likely burn to completion than others. Since this is largely uncontrollable and unpredictable for the entire mass of solid particles, the brink widens between predicted vs. actual performance, especially as to unburned carbon in the residual ash. At best, for practical burner applications, statistical combustion test data on a given coal-water slurry fuel will be necessary to evolve an optimum burner design.

Coal-Water Slurry Burner Design Features

It is concluded from the foregoing, that the general fuel category, "coal-water slurry" covers a very broad spectrum of variables of concern to the burner designer. However, there are certain design features that stem from both logic and experience which may be applied for a baseline design as follows:-

i) The 'atomiser' must first ensure dissipation of the water from the coal, and directly expose the coal to the combustion air in a pattern that will ensure continous, stable ignition.

ii) The burning coal particles should be dispersed through the combustion air such that a homogeneous flame envelope, without a fuel rich centre or pockets, is established, with spacing between particles such that combustion energies will be shared throughout the combustion process.

iii) Maximum use should be made of radiant heat feedback, both during ignition and burning of the particles by such means as a properly shaped hot refractory in the system.

iv) Given a choice between the two, air is preferred to steam as the 'atomising' medium because it directly enters the combustion process, whereas steam will aggravate the already high water content.

v) Depending on availability and economics, gas fuel may be considered for the 'atomising' medium.

vi) Auxiliary fuel (gas or oil) should be incorporated in the burner for cold system startup, for supplemental heat input as may be necessary, and for takeover firing in the event of interruption of the coal-water slurry supply.

vii) Preheated air is desirable to offset the heat losses inherent in the fuel; we have found that in the absence of preheated bulk combustion air preheating the 'atomising' air improves combustion performance.

viii) All of the above should be orientated for a specific coal-water slurry formulation.

HYPOTHETICAL BURNER CONCEPTS FOR COAL-WATER SLURRY

In conclusion, rather than a written summary of the above, a pictorial display encompassing some of the salient features is presented. Again, specific design should be orientated for specific fuel formulations and, in all probability, a development programme will be necessary to achieve the optimum.

Figure 11 shows a Dual-Phase Atomiser, internally modified for coal-water slurry, and externally modified to accommodate ignition delay. Used in conjunction with a refractory lined combustion chamber, we feel that this would be the most likely approach for maximum coal-water combustion efficiency. By imparting swirl to the burning mass, the coal particles will follow a helical trajectory so that they can experience a longer burning time in a given longitudinal distance.

Figure 12 shows a Hyperbolic Atomiser used in conjunction with a refractory-lined air diffuser plate, mounted in an air-staged combustion chamber. Basically, the chamber is segregated into the ignition zone and the combustion zone.

Figure 13 shows a shaped combustion chamber designed to concentrate radiant heat at a focal point within the combustion process.

Figure 14 shows a Dual-Phase Atomiser installed in a "combustion can", in which the combustion air is sequentially metered into the combustion process.

REFERENCES

(1) Dixit G.S. "Combustion of Lignite Oil Mixtures using a Sonic Burner System" Master's Thesis, Dept.Of Mech.Eng. , Louisiana University, Aug. 1983.

(2) Yao S.C. Liu L., "Behaviour of suspended Coal-Water Slurry Droplets in a Combustion Environment", Combustion and Flame 51:335 1983.

(3) Cundy V.A., Maples D. and Buch T., "The Combustion of a Fuel Mixture Composed of Coal and Light Fuel Oil" paper presented at American Flame Research Committee, "International Symposium on combustion diagnostics from Fuel Bunker to Stack" Akron, Ohio, October 1983.

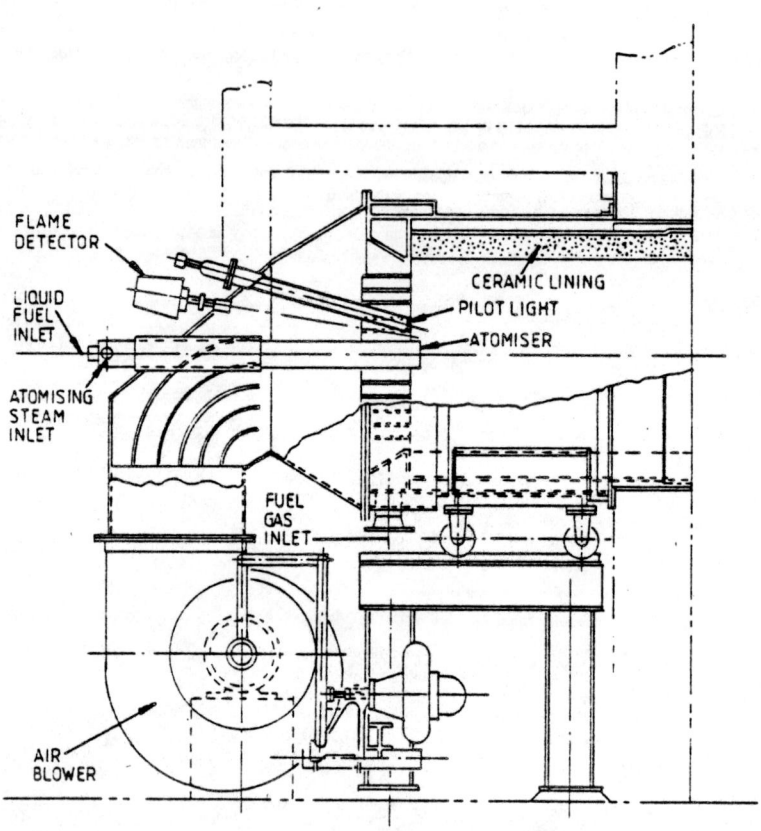

Figure.1. Configuration of a Natural gas/Fuel oil/Coal slurry
Test Burner

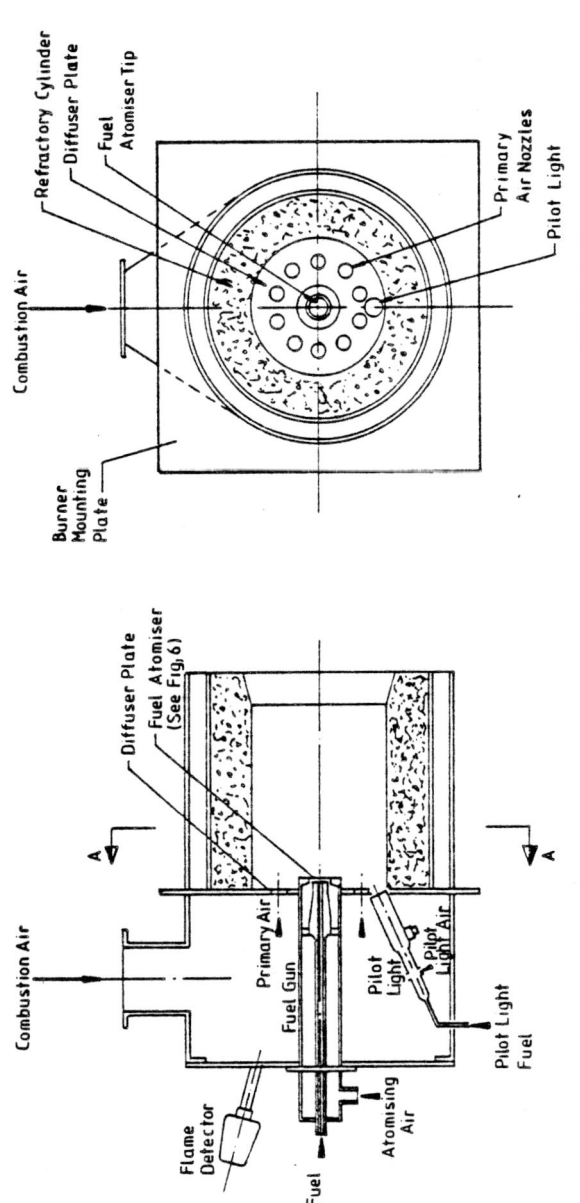

Figure. 2. Configuration of the Sonic Burner

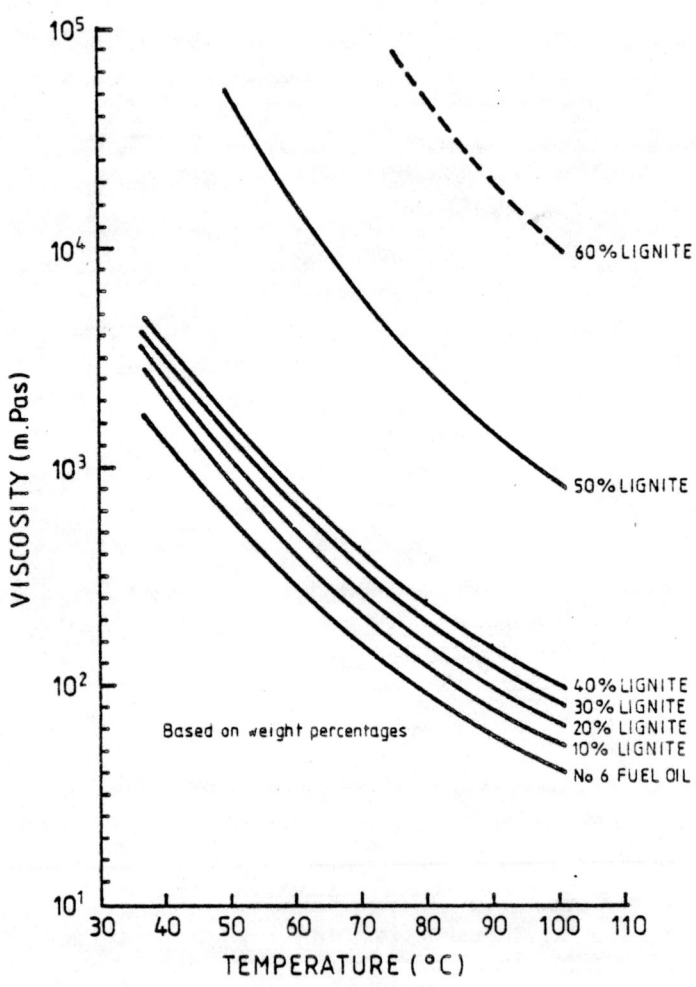

Figure. 3. L.O.M. Vicosity Data

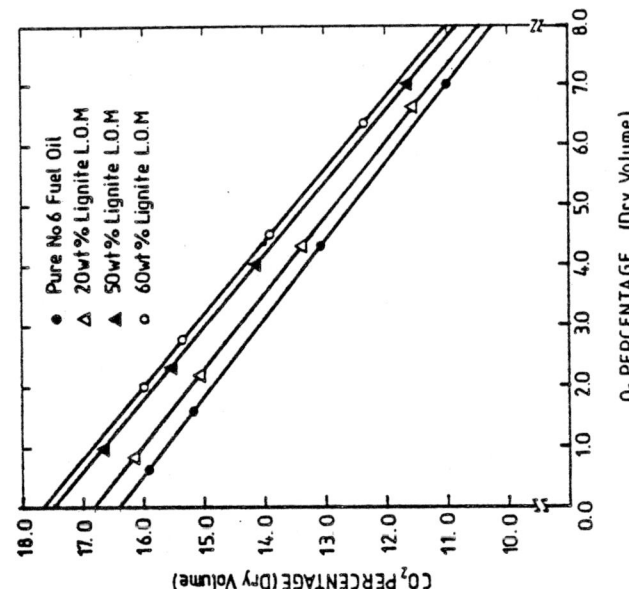

Figure 5. CO_2 versus O_2 with Sonic Burner

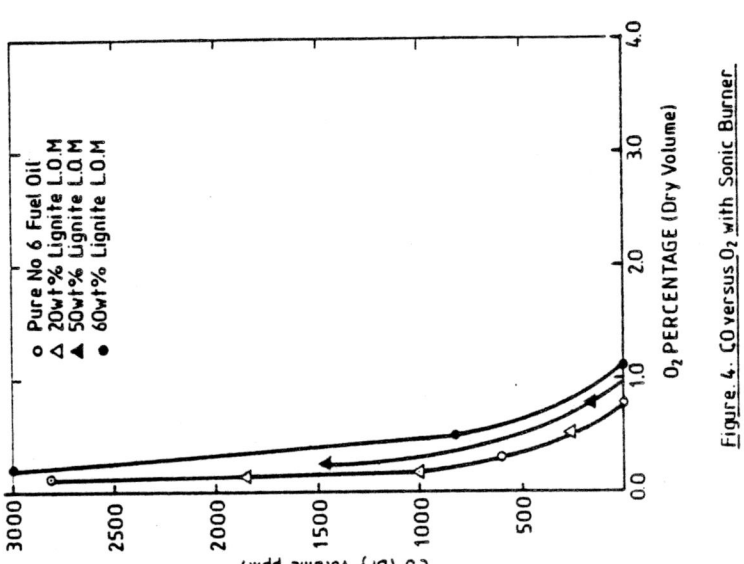

Figure 4. CO versus O_2 with Sonic Burner

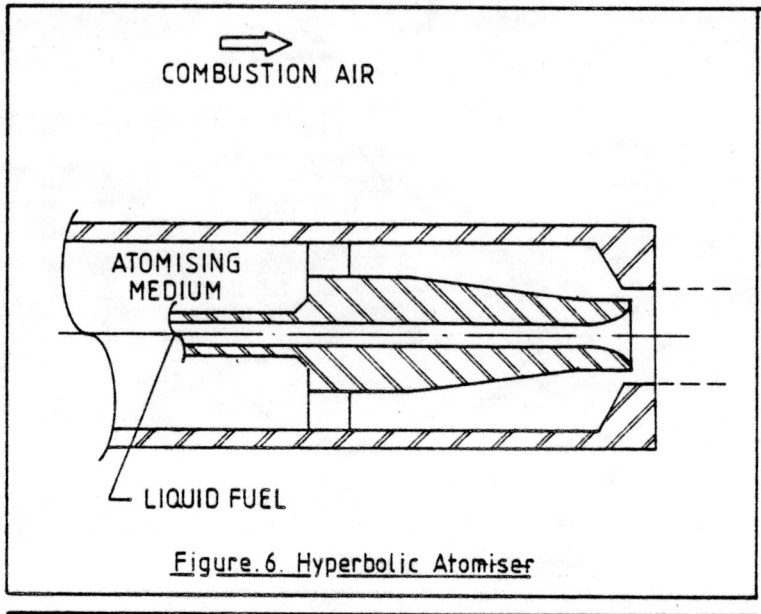

COMBUSTION AIR

ATOMISING MEDIUM

LIQUID FUEL

Figure. 6. Hyperbolic Atomiser

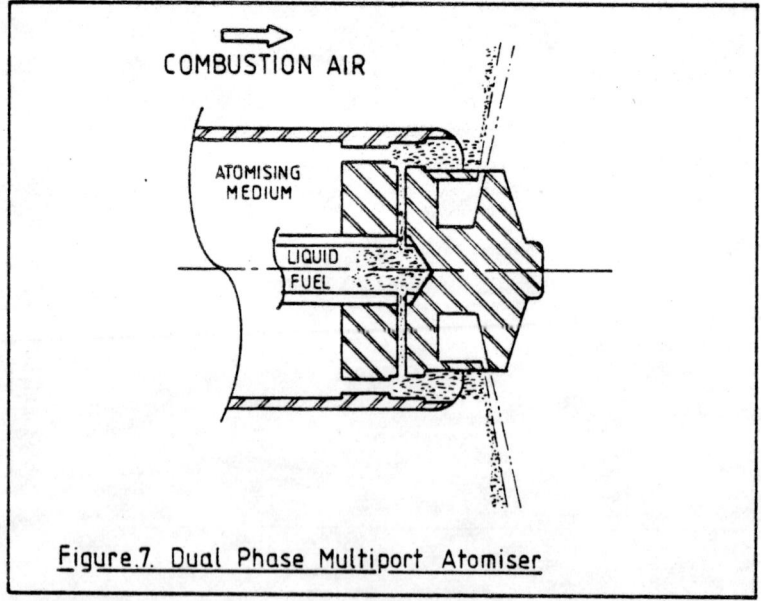

COMBUSTION AIR

ATOMISING MEDIUM

LIQUID FUEL

Figure.7. Dual Phase Multiport Atomiser

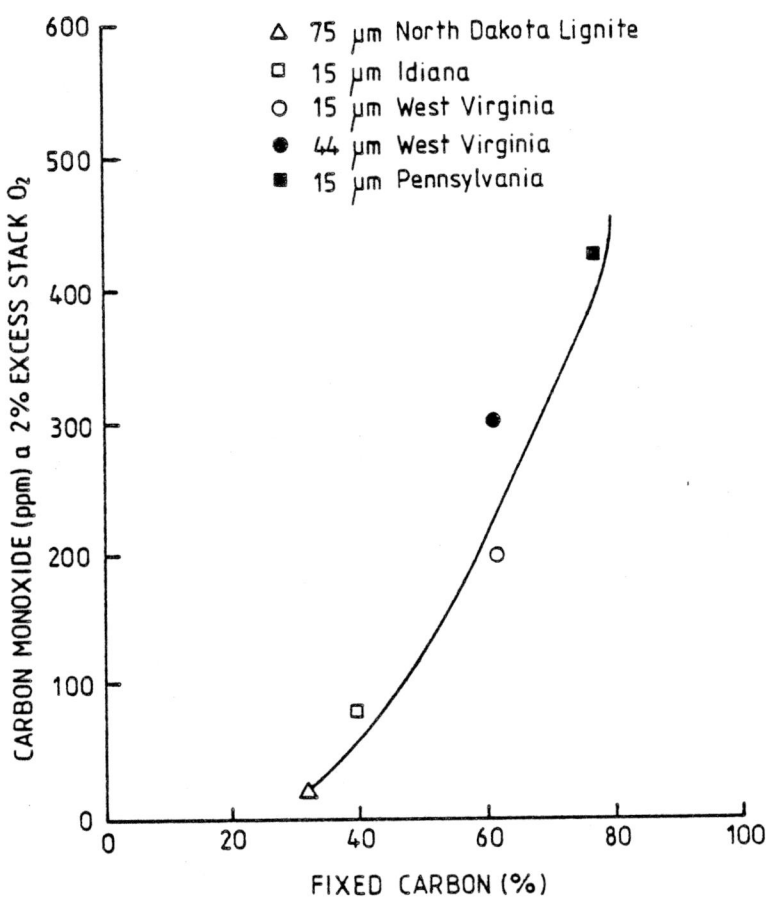

Figure. 8. Carbon monoxide as a function of fixed carbon
(in the parent coal) at 2% excess stack oxygen

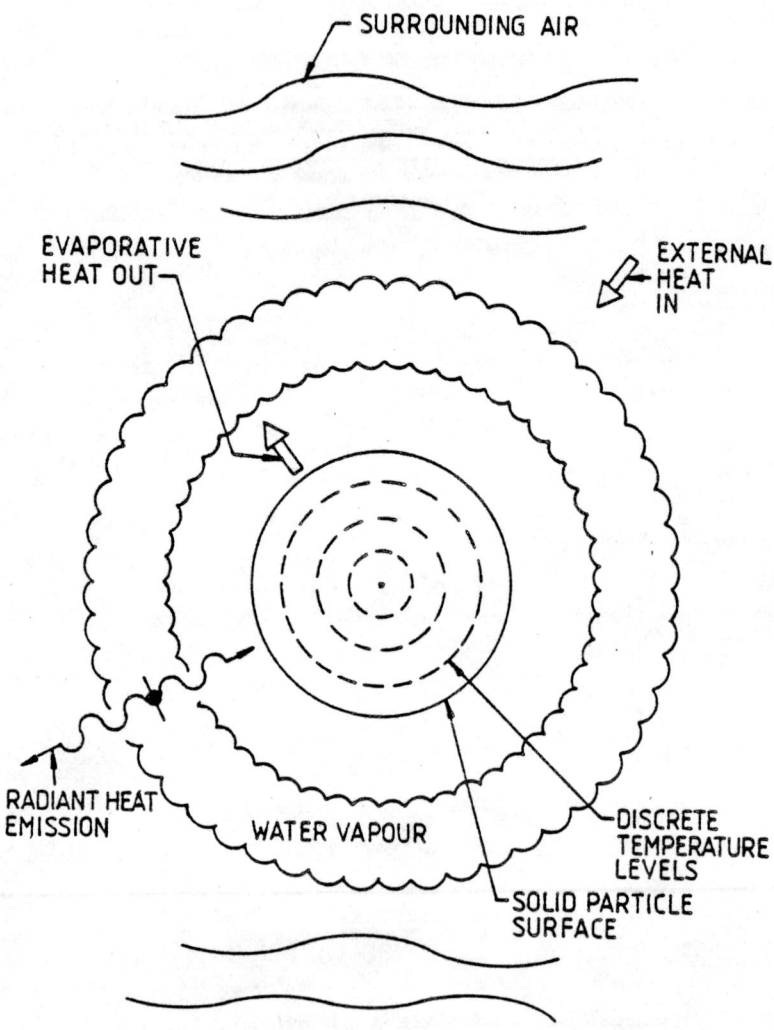

SURROUNDING AIR

EVAPORATIVE
HEAT OUT

EXTERNAL
HEAT
IN

RADIANT HEAT
EMISSION

WATER VAPOUR

DISCRETE
TEMPERATURE
LEVELS

SOLID PARTICLE
SURFACE

Figure.9 Coal/Water Droplet Model

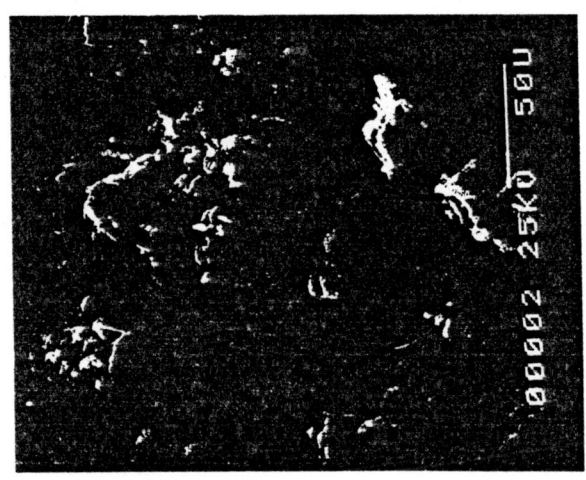

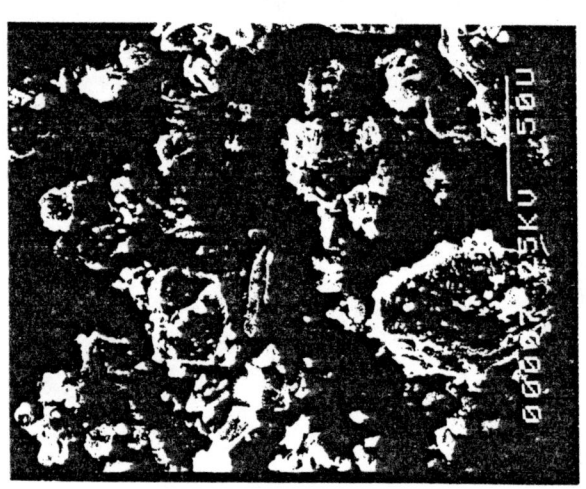

Figure 10. Coal Particle Comparisons

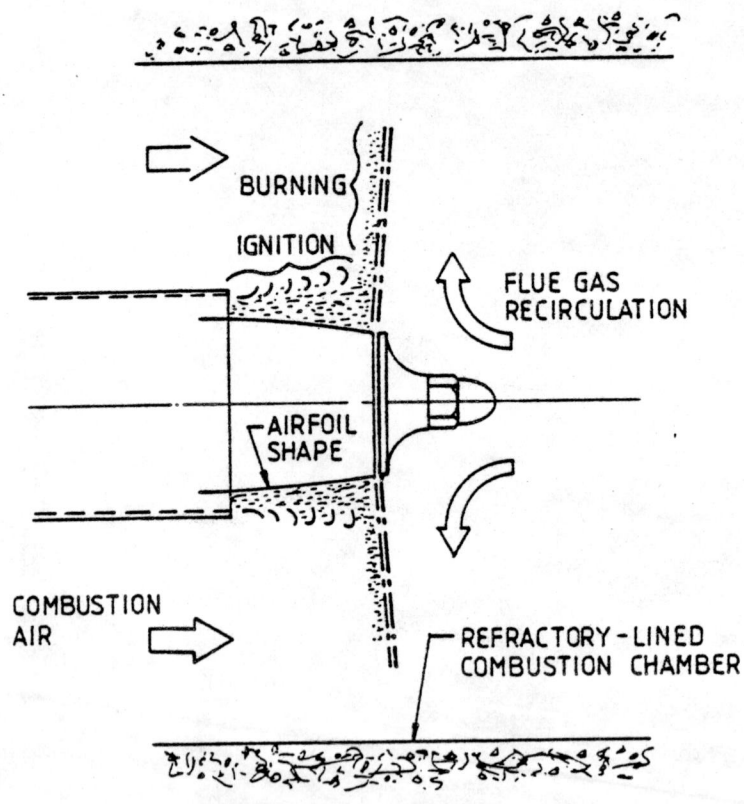

Figure.11. Dual Phase Atomiser for Coal-water slurry

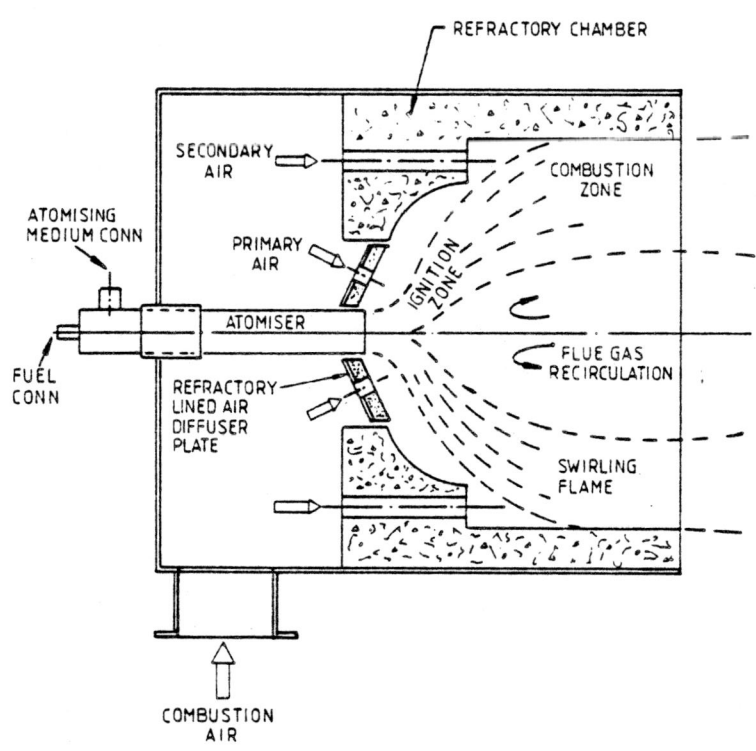

Figure.12. Hyperbolic Atomiser in an Air Staged Combustion Chamber

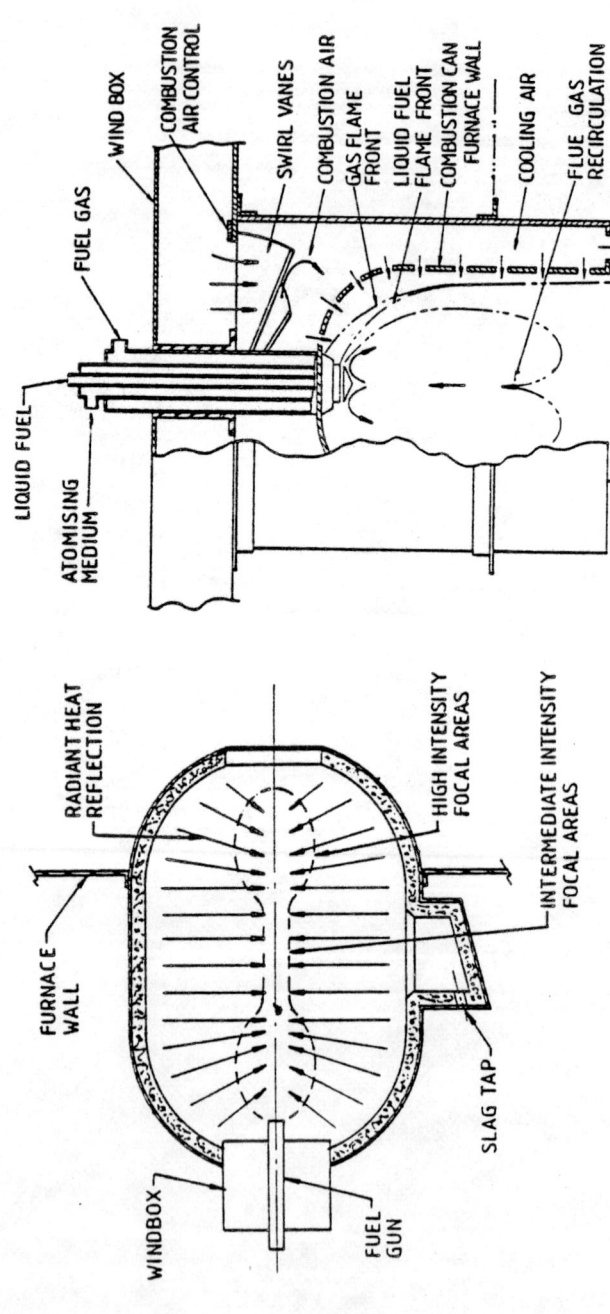

WIND BOX

COMBUSTION AIR CONTROL

SWIRL VANES

COMBUSTION AIR

GAS FLAME FRONT

LIQUID FUEL FLAME FRONT

COMBUSTION CAN

FURNACE WALL

COOLING AIR

FLUE GAS RECIRCULATION

FLAME SHAPE

FUEL GAS

LIQUID FUEL

ATOMISING MEDIUM

Figure 14. Duel Fuel Burner with Combustion Can

RADIANT HEAT REFLECTION

HIGH INTENSITY FOCAL AREAS

INTERMEDIATE INTENSITY FOCAL AREAS

FURNACE WALL

SLAG TAP

WINDBOX

FUEL GUN

Figure 13 Burner with Ellipsoidal Combustion Chamber

STATUS REPORT ON Co-AL FUEL

E.G.Atkins*

The recent renewed interest in coal water slurries has
been brought about by the need to have an alternative
fuel to oil. Oil is expensive and in the long term its
reliability of supply is questionable. From an
examination of the economics of slurry preparation
incorporating beneficiation, boiler conversion and fuel
price differential it has been concluded that payback
could be acceptable for the conversion of oil fired
plant to coal water firing.

Introduction

Coal water slurries have been produced as a waste product from coal
washeries for many years and were usually deposited in slurry ponds and
on spoil heaps. Babcock designed and supplied fuel handling and boiler
plants in the 1950's to utilise some of this type of fuel.

Slurries have been specially produced for transporting coal by
pipeline, with dewatering systems at the users plant to enable the fuel
to be fired in boilers. Political and economical problems have
prevented this application from being widely developed.

The substantial increase in the price of fuel oils in the 1970's gave
an incentive to convert oil fired plant to coal firing by several
means. Coal oil mixtures were developed, several production plants
built and a number of power station and industrial boilers were
converted. However, the production and conversion costs were too high
and the differential between oil and coal price too low for this
technology to be an economic success.

Many efforts were made to produce a coal water mixture which is stable
and can be transported, stored, pumped and fired in equipment similar
to that required for heavy fuel oil.

A small number of coal water mixtures (CWM) have been developed to the
stage where they appear to be capable of being produced at an
attractive price, with present day differentials between the price of
oil and coal. They have been demonstrated in combustion rigs and
experience is now being obtained in a series of boiler trials.

*Babcock Product Engineering Ltd., 11 The Boulevard, Crawley, Sussex RH10 1UX

There is considerable confidence among the producers of these CWM's that reliable and economic coal water alternatives to heavy fuel oil will soon be available for many types of conversion from oil to coal.

The advantages of low ash low sulphur coals for conversion and for new plants has accelerated developments in coal cleaning technology which can be incorporated into the design of a CWM production plant. The availability of low ash low sulphur CWM will substantially reduce conversion and operating costs and its price level should make it an economic choice for new installation requiring low sulphur emission levels.

Early Experiences With Coal Water Slurries

Coal mine slurry ponds, some over 100 years old, and consisting of wet coal fines from the mine washeries and filtration units have long been a feature of UK mining areas. These slurries have no chemical additives and random particle size distribution.

Probably one of the first occasions when coal water slurries were used as a boiler fuel was in the early 1930s at the Mine Orange Nassau, Heerlen plant, in the South of Holland.

In 1950 a survey indicated there was sufficient slurry in ponds in Ayrshire, Scotland to design and build a 60 MW Power Plant to utilise these slurries as fuels. Accordingly in 1951, Babcock was awarded the contract to design and supply the fuel handling and steam generation units for a 60 MW(E) power station at Barony, Ayrshire.

The fuel had a GCV of not less than 14.09 KJ/Kg, a moisture content up to 28%, ash content up to 22%, volatile content not less than 32% and at least 50% of the fuel larger than 3mm.

This slurry does not flow and had to be grabbed from wagons to specially designed rotating hoppers which discharged into a beater mill supplied with hot gas from the furnace for drying the fuel. The dried fuel and vapour was transported in an air stream through exhauster fans to pulverised coal burners. (Fig.1)

The four 150,000 lb/hr boilers at Barony operated for 27 years. Two 275,000 lb/hr boilers with similar fuel handling plant were supplied by Babcock in 1965 for Methil Power Station, Fife, and are still operating with coal water slurry.

Pipeline Slurries

Several long distance pipelines have been constructed in USA for transporting coal. A coal water mixture with 50-60% coal is prepared at the coal mine. Chemical additives are not normally used but particle size distribution is important. If the particle size is too large sedimentation can occur in the pipeline yet if the particle size is too fine dewatering prior to combustion can be a problem.

The Black Mesa pipeline, 275 miles long, supplies 2 x 755 MW boilers at Mohave Generating Station, Arizona. The slurry was a 50/50 coal water mixture with coal sized 0-2% over 1200 micron, 0-15% 1200-1600 micron,

70-84% 600-50 micron and 16-20% below 50 micron. The coal water mixture is centrifuged at the plant to give a coal cake containing 25% moisture. This cake is fed to hot air swept pulverisers and a conventional PF burner system. Special measures are required for thickening and treating the effluent from the centrifuges.

Babcock and Wilcox USA successfully operated a 475,000 lb/hr cyclone fired boiler plant at South Amboy, New Jersey (Fig 2) with a pipeline coal water mix supplied directly to the boiler without intermediate dewatering. This was possible because of the very high air temperature available and the high temperatures within the cyclone combustion chambers.

Stable Coal Water Mixtures

In the late 1970's after many efforts had been made to economically produce satisfactory coal oil mixtures the attention of several research and development programmes was concentrated on producing a stable coal water mixture which could be transported, stored, pumped and fired in equipment similar to that used for heavy fuel oil.

Stable low density homogeneous suspensions were produced, the dispersion being achieved by chemical stabilisers and wetting agents. Particle size distribution was found to be very important in the preparation of mixtures with a high coal content.

Professor James Funk at Alfred University, New York has developed one of the most successful coal water mixtures, now known as Co-Al, and available through Slurrytech Inc and its associates.

Co-Al is a stable homogeneous combination of coal particles, carrier water and additives wherein the particle sizes are controlled in accordance with a distribution formula for providing a coal compact with minimum porosity and maximum solids containing sufficient colloidal (less than 3 micron) size particles. These features, combined with the dispensing effect of the chemicals, provide a maximum electro-kinetic potential to the particles and a low viscosity to the mixture. The result is a pipeline pumpable stable coal slurry.

The quality of coal and ash, ash content and the nature of the water do not pose any constraint. Some 30 different coals from USA, Australia and UK, plus petroleum cokes and solvent refined coal have been successfully converted into coal water slurries.

(a) Particle Size Distribution

Alfred University have produced a guidance formula for optimum size distribution as follows:-

$$CPFT = \left[\frac{D^n - D_s{}^n}{D_1{}^n - D_s{}^n} \right] \times 100$$

CPFT	-	cumulative weight % of particles finer than a selected particle size in microns.
D_1	-	largest particle size, microns.
D_s	-	smallest particle size, microns.
n	-	0.33 to 0.50 dependent on D_1 and D_s.

When this formula is followed, optimum particle packing with minimum void space is obtained and enables the preparation of a coal water mixture with minimum water content and maximum solids loading.

Co-Al production is aimed at achieving:-

(i) Minimum number of particles near the largest size to minimise collision energy required to shear the system.

(ii) Maximum packing efficiency in order to minimise the inter-particle void volume which is first filled with water before additional water separates the particles and allows shear to occur easily.

(iii) Sufficient colloidal content (3-5% less than 3μm) and total surface area to react to additives to effectively lubricate the larger particles, fill the larger interstital voids, maintain a constant inter-particle distance between all particles and provide the substrate for a gel structure with controlled strength.

(b) Solids Loading

Solids loading is achieved by the control of particle size distribution. This is in accordance with the equation above, with the largest size of D_1 up to 300μm, the smallest $D_s = 0.1$μm and n = 0.37.

A value for n of 0.37 enables the particles to be most densely packed. In the first step, a distribution as close to this and as high a solids loading as possible is prepared without the addition of chemicals. The size distribution is obtained by grinding two or three batches of coal to varying degrees of fineness in water alone, as well as a few batches which are produced dry to different D_1 then calculating and mixing an optimum blend from these. The blending is carried out based on set correlations developed for calculating the minimum porosity and maximum solids loading to which a distribution can theoretically be packed.

(c) Stability

Stability is achieved by controlling the chemistry of the product. The desired chemistry is one in which the slurry is completely deflocculated and when the zeta potential on the particle surfaces is at a maximum. These parameters are achieved by the selection of the chemical of optimum concentration most suitable for the above blend which produces a slurry of minimum viscosity and maximum zeta potential.

(d) Handlability

The properties desired for pumpability and atomisation characteristics of the slurry are achieved by developing a suitable yield stress for the product promoted by the zeta potential so that the viscosity decreases with increase in shear rate.

(e) Final Preparation

The final step is to prepare a batch in which the test coal, water
and the chosen chemicals are mixed to produce an approximate solids
loading of 75 to 78%. Fine tuning of the chemistry to optimise
slurry properties is carried out by preparing additional batches with
slight variation of the chemicals.

(f) Co-Al Application Testing

A full programme of storage, handling, pumping, transportation and
freezing/thawing trials has been satisfactorily completed.

In 1980 Babcock & Wilcox successfully fired Co-Al self-sustained at
their Alliance Research Centre.

Since that time refinements have been made to the burner/atomiser
system to obtain higher carbon utilisation, increase stability,
increase turndown, increase wear life on the atomiser, simplify the
burner design and reduce pressure drops, air preheat and atomisation
medium required.

Trials have been successfully completed on a 5 MM B.Th.U/h furnace,
on a 22 MM B.Th.U/hr furnace and a 35 MM B.Th.U/hr multiple burner
furnace.

Babcock-Hitachi in Japan have received substantial quantities of
Co-Al which was transported from Pennsylvania to Kure, Japan, and
which was used without recirculation. Stable combustion without
auxiliary fuel consumption was achieved, with unburnt losses lower
than obtained firing the parent coal alone.

Babcock Power have conducted small scale trials on Co-Al at Renfrew
Research Centre and found that stability, pumpability and combustion
were satisfactory. Further trials will begin shortly with coal water
mixtures made from UK coals and boiler tests will be made on a 60,000
lb/hr water tube boiler at Renfrew next year.

Babcock & Wilcox USA are currently making a series of trials with
Co-Al on a specially converted 60,000 lb/hr water tube package boiler
originally designed for oil firing. The T-jet burners each 35-50 MM
B.Th.U/hr rating fire either oil or CWM (see fig 4).

The coal water mixture is fed through the centre tube and exits at
right angles into the atomising medium stream. Mixing takes place in
the swirl chamber before entering the furnace. The original T-jets
were manufactured in mild steel to determine the wear characteristics
of mild steel exposed to coal water mixtures. Specially developed
ceramic wear resistant materials are now incorporated at those points
where wear rates were shown to be high in the preliminary tests.

Another series of boiler tests sponsored by EPRI are being currently
carried out on Co-Al and ARC fuels on a 70,000 lb/hr Babcock & Wilcox
F type boiler designed for oil/gas firing.

Commercial Application

By Spring 1985 a semi-commercial demonstration plant producing 100,000 tons/year Co-Al will be in operation in USA and a series of large scale demonstration trials on industrial and utility boilers will be carried out.

A plant of similar capacity is planned to be built in the UK by Babcock Power and several partners and should be in operation in 1985.

The basic process of 100,000 ton per year plant is as follows. (Fig.3)

Coal is received from a stockpile by mechanical shovel delivering to a reclaim hopper. By means of a vibrating feeder and bucket elevator the coal is then deposited in a coal feed storage bin. A magnetic separator removes any tramp iron.

The coal is then fed via weigh feeder to a transfer chute where water and chemicals are added in controlled proportions. A fine product crusher (to $-\frac{1}{4}"$) may be employed after the weight feeder to reduce the crushing duty in the wet ball mill, which is fed by a screw feeder from the transfer chute.

Final grinding and slurry preparation is carried out in a conventional Kennedy Van Saun horizontal rotating wet ball mill. The CWM produced inside the mill overflows via a weir discharge to an integral sump.

CWM from the sump is pumped to a wet high frequency vibrating screen which separates any large particles from the CWM. Such oversize particles are recycled back to the wet mill for further processing, undersize material from the vibrating screen passes to a CWM holding tank. From the holding tank the CWM is pumped to the plant storage facilities from whence it is pumped via a loading arm to drums or road rail or barge transport facilities.

The proportions of coal and water in the CWM are generally 70% - 30% respectively.

Two chemicals are added to the water feed at the entry to the wet mill: caustic to control the pH of the CWM and a surfactant to control the electrochemistry of the CWM. A softener or wetting agent may be required to facilitate dissolving the surfactant.

A plant operation monitoring system will be installed with on-line data acquisition equipment for continuous monitoring and control of the plant. parameters. These parameters (process variables) are particle size distribution, solids content, viscosity, density and pH. The system will record and alarm excessive deviations of all monitored signals and permanent logs will be printed out at prescribed intervals.

It is anticipated that the 100,000 t/pa demonstration plant would be manually controlled to minimise capital cost and allow for process development, but commercial plants would be fully automated.

If a change in CWM characteristics is indicated, the plant operator will be able to take either or both of the following corrective actions:-

- increase/decrease mill retention time by varying coal/water/chemicals feed rate;
- increase/decrease mill retention by adjustment of recirculation rate.

To minimise the number of variables to be controlled on a manual basis, and hence operating/capital cost, local laboratory facilities will be utilised to monitor:-

(a) coal size/type;
(b) surfactant required for specific coals;
(c) pH analysis;
(d) particle size;
(e) fuel analysis of slurry.

A design of a commercial pit head production plant of 500,000 tons/year CWM capacity has been studied using data available from the operation of the 50 ton/day Co-Al plant at Danville, Pennsylvania.

The coal feed was assumed to be froth flotated fines which is available in adequate quantities from the Yorkshire, Derbyshire and Nottinghamshire coal fields. The coal was assumed to have 6.5% ash, 28% moisture GCV 9,660 B.Th.U./lb (22.83 GJ/ton) and a pithead price of £37.6/ton. The CWM produced was assumed to have the same GCV and 28% water content.

The estimated cost of the production plant site, services and initial coal stock was £20 million.

Annual running costs would be:-

	£K
Power	1,155
Labour	440
Chemicals, services etc	925
	2,520

Capital charges based on 10 year life	2,720
Coal cost	18,800
Profit, overheads, contingency	2,460
	23,980
Running costs	2,520
	26,500

Selling price of CWM at pithead	£53/ton
Delivered price to customer	£58/ton
	(£2.54/GJ)

Present day delivered prices of a heavy fuel oil having a GCV of 43.72 GJ/ton is £130/ton. The equivalent quantity of oil to 500,000 tons/year CWM, assuming CWM operates at 5% lower efficiency is:-

$$500,000 \times 0.95 \times \frac{22.83}{43.72} = 248,038 \text{ tons/year}$$

Assuming that the boiler plant is capable of burning either 500,000 tons/year CWM or 248,038 tons/year oil:-

Cost of oil	£32,245,000/year
Cost of CWM	£29,000,000/year
Annual fuel cost saving	£3,245,000

A plant using 500,000 tons/year CWS would supply 4 x 200,000 lb/hr boilers. Conversion costs would be £8-10m, depending on the individual plant and site. The customer may be able to claim a grant from the DoT&I of 25% of the conversion cost.

In addition to the cost of conversion the customer will evaluate the time required for conversion, the increased labour, maintenance and operating costs with CWM and any reduction in availability and maximum output. Dual fired burners would normally be added in the conversion to ensure that the customer could obtain full rating on oil alone.

The economics of the conversion are very largely affected by the relative prices of coal used to produce the CWM and the price of oil.

The approximate calculations above are based on current commercial prices and the differential between oil and coal prices is expected to increase.

Future Development

Coals commercially available at a price level from which CWM can be profitably produced at a price which will make conversion from oil to CWM economically attractive are the froth flotated fines having ash contents of 6-10%.

Boiler conversions using CWM produced from these coals must be designed to include furnace ash removal and ash and dust collecting and disposal systems sized for 10% ash in the fuel.

Developments in the cleaning of fine coals will be incorporated into the design of new CWM production plants so that preparation of a 1% ash CWM can be economically made from the froth flotated fines feedstock.

The proposed flow diagram of the CWM production plant incorporating beneficiation is shown in Fig.5. In this process the coal is pulverised in a hot air swept dry mill. The fines fraction is collected in the cyclone separators and fed to the wet mix beneficiation plant where the necessary chemicals are added. The beneficiated slurry is pumped to the primary tank and the normal additives are incorporated before final conditioning.

Using as a basis the study of the 500,000 ton/year CWM plant it should be possible to produce a 1% ash CWM at about £2.7/GJ. The boiler conversion and operating costs will be reduced and the maximum load obtainable increased.

Coal cleaning to 1% ash will usually reduce sulphur content and in cases where limits are imposed on sulphur emissions would reduce the costs of clean up equipment.

Conclusion

The gradual development from the firing of coal water mixtures won from slurry ponds, and the firing of pipeline slurries to the present day stable fuels such as Co-Al has produced an economic alternative to oil. Provided the differential between the price of oil and coal delivered to the customer does not decrease there is expected to be a growing demand for conversions of oil fired plant to coal water mixture firing. Further development of CWM production and firing equipment technology, coupled with economic deep cleaning processes will make this type of fuel attractive for a wide variety of applications, even extending to new plant specifically designed to fire low ash low sulphur CWM.

The author thanks the Directors of Babcock Power Ltd., for permission to publish this Paper.

The author also thanks the following for the assistance received in the preparation of this Paper:

Mr.D.Heyburn - Senior Vice President & Group Executive, Babcock & Wilcox, Ohio, U.S.A.

Mr.D.Dunlop - Chairman - Slurrytech Inc., Florida, U.S.A.

Mr.D.Z.Richards - Vice President, Sales & Contracts,
                  Kennedy Van Saun Corporation, Danville,
                  Pennsylvania, U.S.A.

BIBLIOGRAPHY

Barony Power Station   - Babcock & Wilcox Publication No.1691
Slurry for Steam Raising   - Babcock & Wilcox Publication No.1590/1

Mohave Generating Station   - Proceedings of the American   Power
Design Features   Conference 1969. Vol.31.

Coal Water Slurry at
Werner Station   - 'Electrical World' Nov.20 1961.

Slurry Burning Demonstration   - Babcock & Wilcox PublicationOctober 1961.

**Methil Power Station
Section through Boiler and P.F. System**

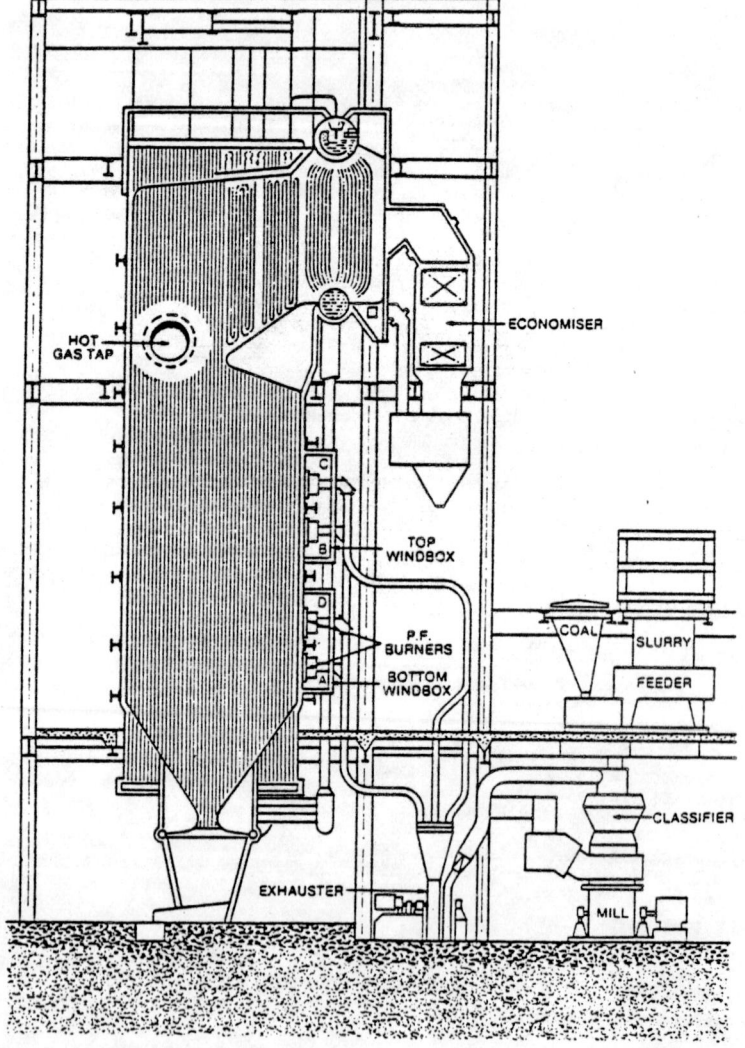

Fig 1

E. H. Werner Station
Section through RB Boiler with Cyclone Furnace

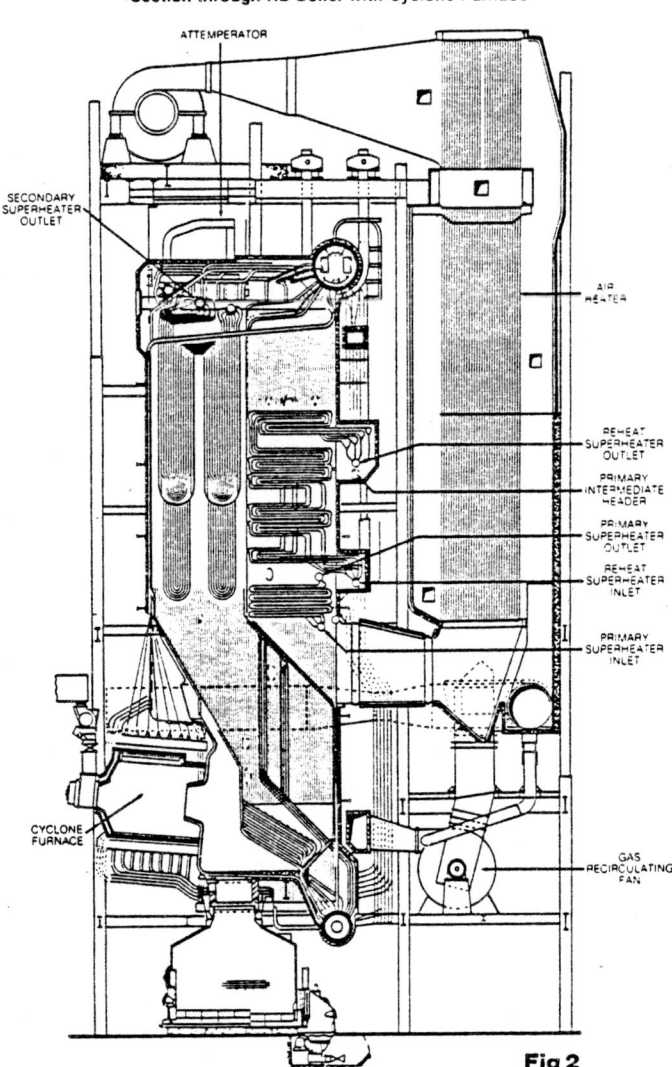

Fig 2

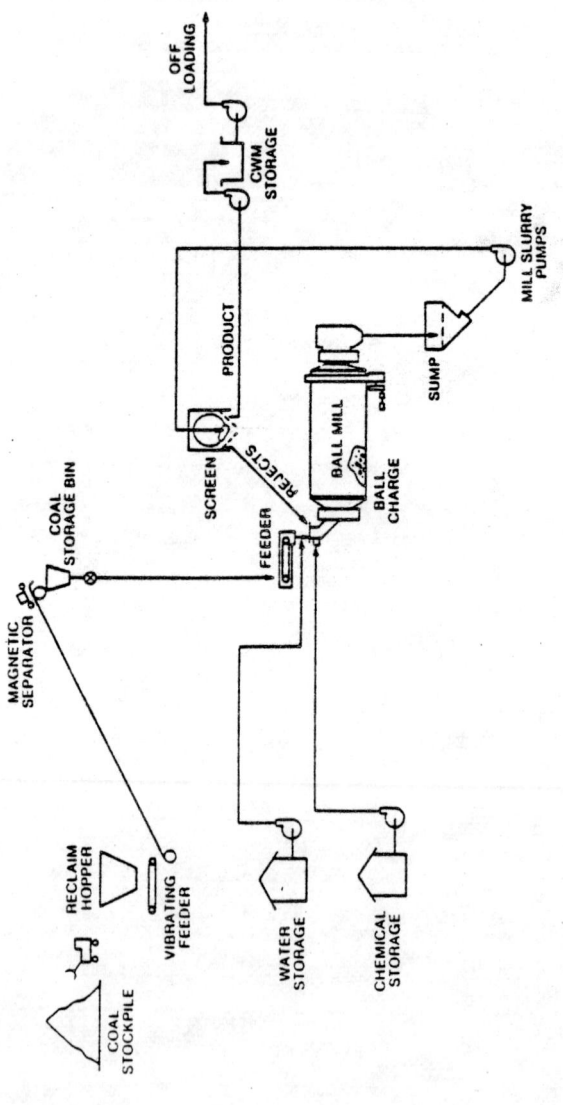

Fig 3

Basic Flow Diagram
100,000 Tpa CWM Preparation Demonstration Plant

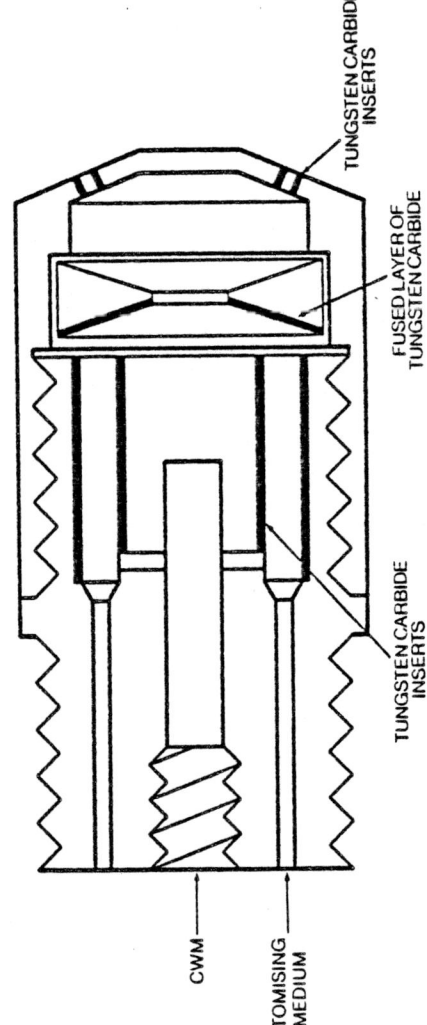

T–Jet Sprayer Plate
W/wear combating provisions

TUNGSTEN CARBIDE INSERTS

FUSED LAYER OF TUNGSTEN CARBIDE

TUNGSTEN CARBIDE INSERTS

CWM

ATOMISING MEDIUM

Fig 4

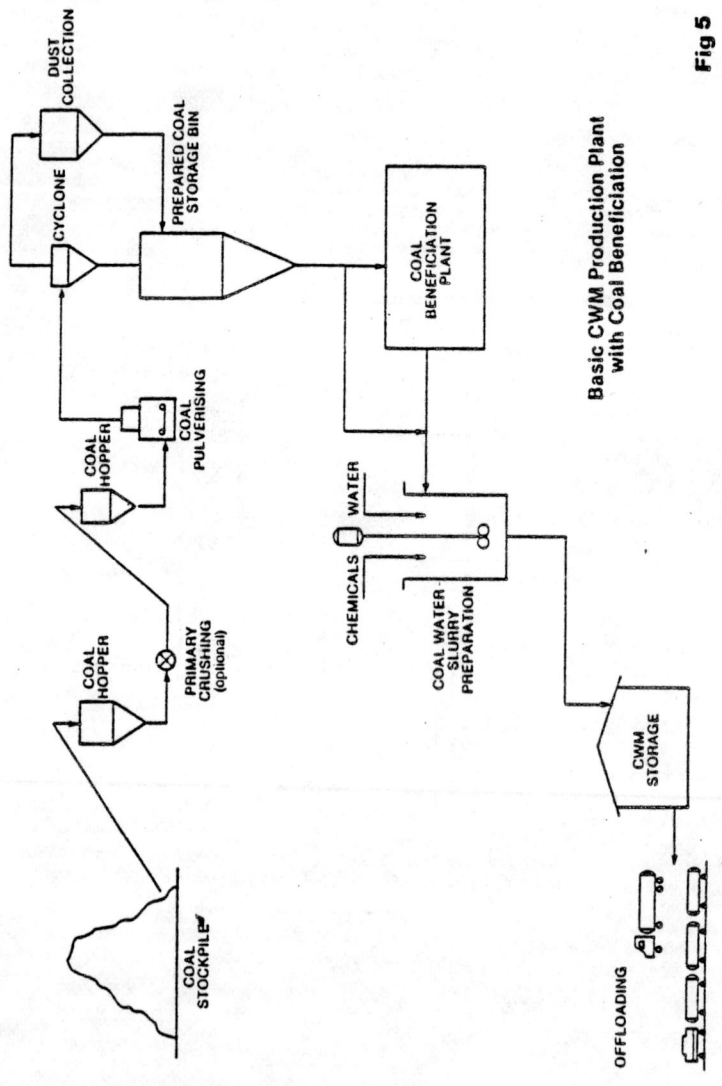

Fig 5

**Basic CWM Production Plant
with Coal Beneficiation**

PREPARATION AND ECONOMICS OF A BENEFICIATED COAL WATER FUEL BASED ON THE CARBOGEL PROCESS

Eric H Beckhusen*, John W Groel**, Michael J Shires***

SYNOPSIS

This paper describes some technical and economic aspects of commercial scale plants for the production of Coal Water Fuel (CWF) by the Carbogel process.

In addition to producing a coal-water slurry, the Carbogel process reduces both ash and sulphur by froth flotation, thus producing a high quality fuel. Typically, it is possible to recover between 93% and 98% of the feed coal in the CWF product, while achieving around 50% ash removal.

The Carbogel additives consist of dispersants and stabilisers which together with controlled size reduction produce a stable, mobile slurry at solids loadings as high as 75%. The product resembles No. 6 Fuel oil.

The paper gives the economics for two plant capacities, at two locations. The first location being a grassroots site and the second being at a mine mouth washery. Calculations show that the cost of producing Carbogel CWF (excluding the cost of coal and transportation) will be in the range $12-$26 per short ton.

1.0 INTRODUCTION

On a thermal basis, coal is cheaper than oil or gas.

One of the new techniques for taking advantage of this favourable cost difference is through the use of coal water slurries. By properly combining a finely ground coal with water and suitable additives it is possible to produce a stable, oil-like, coal water fuel (CWF). Further, once the coal has been fine ground to produce the slurry it can be beneficiated at a reasonable additional cost to produce a high quality fuel.

Often there is an initial concern that the presence of 25 to 30% water in the slurry will reduce the heating value due to vaporisation of the water. In fact in combustion only about 4% of the available heat is lost to vaporisation.

*   Foster Wheeler Synfuels Corporation, USA
** Foster Wheeler Energy Corporation, USA
*** Foster Wheeler Synfuels Limited, UK

Some of the outstanding advantages of CWF are:

- It is a low priced fuel
- It is handled as a liquid similar to fuel oil
- It is pumpable, storable, and atomisable
- It can be substantially cleaner (lower ash and sulphur) than run of mine coal.

The particular type of CWF considered here, Carbogel, has a unique feature of being a low sulphur and low ash fuel because it has been intensely beneficiated, principally by froth flotation. This process has shown its ability to reduce ash to well below 4% and sulphur to under 1% on many coals while maintaining thermal recoveries in the 93 to 99% range. These good results are achieved by very fine grinding, which releases many discrete mineral particles which can then be removed in the flotation step.

This high quality low cost fuel should be particularly attractive to those owners of industrial and utility boilers considering conversion from one of the higher cost fuels because it avoids the high capital cost of conventional coal-feed boilers and associated storage, handling and pulverisation equipment and possibly in some cases of flue gas desulphurisation plants.

It should be emphasised that each conversion of a boiler installation is site specific and must be individually studied to determine the overall economics of any change.

2.0 PLANT DESCRIPTION

Figure 1 is a block flow diagram showing the principal elements in the Carbogel CWF process.

The basic process blocks are:

- Coal Receiving, Storage and Crushing
- Wet Grinding to proper size distribution
- Froth Flotation Benefication for ash and sulphur removal
- Filtration and Dewatering
- Mixing
- Storage and Loading

These elements are further explained in the following section.

There are two scenarios for CWF production plants, these are:

a) A "grassroots" plant, to which run-of-mine coal is delivered by rail.

b) A plant associated with a "mine-mouth washery", and supplied with washed coal.

These facilities are illustrated on tentative plot plans, Figures 2 and 3.

Coal Receiving and Crushing

It has been assumed that coal for the "grassroots" plant would arrive by rail in a unit train from 70 to 100 cars, each containing 100 tons of run of mine coal, 2" x 0". These cars would have manually released doors for unloading through a grizzly to an undertrack hopper.

Coal from the hoppers is taken by feeders and conveyor belts to either a live storage pile or directly to the crusher tower. Once through the crusher where it is reduced to 3/4" x 0", it is put in silo storage awaiting subsequent transfer to the grinding mills.

For a "mine mouth washery" location, it has been assumed that the coal would be received directly via a conveyor belt and taken directly to the crusher tower.

Wet Grinding

Wet grinding is accomplished in a closed circuit configuration using primary and secondary ball mills.

The slurry leaving the secondary ball mills is picked up by a slurry pump and circulated through a separator which separates any coarse particles. These coarse particles are then recycled for further grinding. The fine particles pass on to the froth flotation for washing and cleaning.

Froth Flotation

In the production of any coal water fuel it is necessary to grind to controlled fine particle sizes to produce a fuel having desirable stability, handling and flow characteristics. Having expended the energy to produce the fine particles the Carbogel process then applies froth flotation as an effective means of removing liberated ash and sulphur constituents, thereby producing a high quality low cost coal based fuel.

The system for cleaning used in this process is multiple stage flotation wherein the ground coal is processed first through a bank of rougher cells and then finally through a bank of cleaning cells.

The froth flotation process relies on differences in the surface chemistry between the particles of clean coal and its impurities in order to make a separation. During the process, chemical agents are added, some of which act as frothers to prolong the life of the air bubbles and others which are collectors to increase the hydrophobicity of the coal and improve its movement into the froth.

Filtering and Dewatering

Dewatering is necessary to achieve the high concentration of solids required in the final fuel. This is accomplished by a combination of thickeners and vacuum filters producing a relatively dry clean filter cake.

Both the tailings stream and the coal concentrate stream are dewatered and the clear water recovered is recycled to the front end of the process.

Mixing

The filter cake of clean coal that is discharged from the filters is sent to a continuous intensive mixer where the stablising chemicals are added to produce the desired fuel characteristics.

Upon leaving the mixers the product fuel is pumped to storage for final loading and shipping or immediate use.

Storage and Loading

On site storage in conventional vertical steel tanks is provided for 15 to 30 days of plant output. This large storage capacity assures continuous supply of fuel to the user.

At the grassroots location, facilities are included for loading of railroad tank cars and river barges. The mine mouth location does not include barge facilities.

Support Facilities

Support facilities include buildings and utilities.

The following utilities and facilities are required for the production plant; they are shown on the plot plans.

Steam — 15,000 lb/hr steam generator delivering 90 psig steam for space heating, tracing and heating coils is included (Carbogel must be kept above 0°C). Current projections foresee using the tailings from the froth flotation as fuel for a fluid bed boiler as an alternative steam supplier.

Water — City water is available.

Air — A compressor in the utility building is included to supply air at 90 psig for plant air and a small dry stream for instrument air.

Power — Power is available at 13,200 volts. A substation on the plant site is included. An emergency generator is not considered necessary and is not included.

Firewater — A fire loop and monitors around the coal storage area are included. No additional facilities such as a foam system, fire trucks or emergency fire pump were included.

Sewers — Sanitary sewage is to be handled in a septic tank system.

Uncontaminated surface drainage will go to a nearby waterway. Contaminated drainage and process effluent will go to a waste water effluent pond for treatment. Outfall from this treatment will be recycled to the plant or conducted to a nearby waterway.

Environmental   -   Fugitive dust from coal handling is controlled by sprays at the coal unloading area and around the live coal storage and dust collectors (baghouses) on conveyor systems. Once the coal is mixed with water there is no serious problem with dusting or fugitive losses.

3.0 PLANT PERFORMANCE

The mass balance for a plant producing 2,750,000 short tons/year of Carbogel CWF is given in Figure 4. All flowrates are in short tons per hour, (short ton = 2000 lbs). It can be seen that 1 ton of Carbogel slurry containing 25% water requires 0.91 tons of coal feed. The quantity of Carbogel additive required is 0.5% of the product.

Feed Coal

This study is based on reference coal of the following analysis:

Proximate Analysis

Moisture	-	6.0%
Ash	-	10.3%
VM	-	29.9%
FC	-	53.8%
		100.0%

Ultimate Analysis

	Dry	As Received
C	74.5	70.03
H	4.7	4.42
N	1.4	1.32
S (pyrite)	0.7	0.66
S (organic)	0.8	0.75
O	6.9	6.49
Ash	11.0	10.33
Moisture	0.0	6.00
	100.0	100.00

Yields and Ranges

The effectiveness of mineral matter removal and thermal recovery that is achievable with the Carbogel process has been studied in laboratory tests and plant scale production. Typical results have been as follows:

Ash Removal 50 to 75%
Pyritic Sulphur Removal 40 to 90%
Thermal Recovery 93 to 98%

The ranges are broad since the process is sensitive to differences between various coals. Each coal is unique and must be individually tested on the small scale to gain some indication of its behaviour in a full size plant.

How the various constituents (i.e. combustibles, ash, sulphur, etc.) divide as they proceed through the processing steps is influenced by specific coal properties.

Preliminary washability tests are required to gain some insight into each coals behaviour.

The specific figures used in this study are as follows:

Ash Removal	60%
Pyrite Removal	80%
Thermal Recovery	93%

Additives

The preparation of Carbogel CWF is based on selected surface active agents and dispersants blended in a proprietary formulation to achieve desired fluid properties.

Agents used in the froth flotation step include coal flotation promotors, frothers and depressants for mineral constituents.

Additives for final slurry formulation are chosen from surface active water soluble dispersants, stabilisers, and possibly bacteriostats for long term storage (more than 6 months).

Normally the dispersants are added in amounts to a final concentration of 0.3 to 0.5% and other chemicals are used in concentrations below 0.1%.

The addition of these additives can affect the fuel behaviour and they must be carefully formulated and blended with ground coal of proper size distribution to achieve the desired final slurry properties. It is critical therefore for the user to carefully define and specify the product desired and to assure himself that the supplier has sufficient quality control awareness and capability to continuously manufacture an acceptable product.

Product Properties

Coal water fuels are pourable liquid mixtures of coal resembling No. 6 fuel oil in appearance. Viscosities and rheologies vary with coal loadings, particulate sizes and additive systems chosen. Normal utility boiler fuels display working viscosities in the 1000-1500 cp range at 50 to 100 S^{-1} shear rate.

The rheologies of these liquids are non-Newtonian, increasingly so with increased coal loading. At very low to moderate shear, pseudoplasticity dominates (up to 400-500 S^{-1}). At very high shear, slight tendencies towards dilatant behaviour are apparent, particularly with highly loaded slurries (above 75-80% solids). It is desirable for slurries to exhibit a slight initial shear stress before initiating flow (similar to a Bingham plastic). This initial resistance to deformation indicates a stable fuel formulation.

These slurries have been extensively tested to determine physical properties (most specifically their rheology and viscosity) and have been subject to successful test burns that evaluate atomisation and combustion characteristics.

Settling can be controlled by means of proper selection of slurry additives and particle size distribution. However, provision should be made to agitate tanks that are used for long term storage. Agitation is particularly important where slurries are to be continually replenished to assure an even fuel quality. This agitation can be provided by side entering mixers.

The calorific value of the CWF product is lower per lb than that of the parent coal. However the drop in calorific value due to the additional water, is partly offset by the reduction in ash levels. For the reference coal the calorific values lay in the following ranges:

HHV, BTU/lb

	HHV, BTU/lb
Parent Coal	12,000 - 12,800
CWF	9,800 - 10,800

For this study an HHV of 10,000 BTU/lb has been assumed for the CWF product.

4.0 ECONOMICS

Coal Water Fuels (CWF) offer the attraction of getting the low price of coal while simultaneously getting the ease of handling of oil.

Elements of the cost of producing CWF are the following:

- Capital Costs (Debt Service and Depreciation)
- Raw Materials (Coal and Chemicals)
- Utilities (Power, Fuel Oil, Ash Disposal)
- Processing Cost (Operating Labour, Maintenance, Supervision, Overhead, etc.)

Assumptions and Basis

Costs were estimated for two plant sizes having the following capacities:

2,750,000 ST/yr CWF
770,000 ST/yr CWF

Two locations were chosen. One was a "grassroots stand alone" facility at a location approximately 500 miles from a mine and 100 miles from the market area. It would have an extensive rail siding to provide holding space for a unit train and also provision for loading of rail cars and river barges with CWF.

The other plant was located at a "mine mouth washery" receiving coal directly to the crusher and sharing some existing support facilities. Rail shipment to CWF was included but no provision was provided for river barge shipping.

Capital Cost Estimate

The details of the capital cost estimates for the different plants are shown in Table 1. They can be summarised as follows:

Slurry Production short tons/year	770,000	2,750,000
Grassroots Plant	$58,000,000	$107,000,000
Mine Mouth Plant	$44,000,000	$81,000,000

Raw Materials

Raw material requirements which are determined from the plant material balance consist of coal, additives and water. It is to be noted that although the final slurry may contain 75% coal, it requires 0.91 lbs of feed coal to make 1 lb of slurry. This occurs because this process rejects a considerable quantity of the incoming material as ash and sulphur and they do not appear in the final product.

Raw materials costs per lb of slurry are insensitive to plant size and will be constant regardless of plant size.

Utility Cost

Power Cost has been estimated at $0.05/KWH. City water is available at $0.25 per 1000 US gals.

Utility costs have been determined from plant scale operations and confirmed with equipment manufacturers information. The general relation is as follows:

$$\text{Utilities (\$/ST)} = \left(\frac{259.13}{(ST/Yr)^{0.08}} \right) \times .05 \text{ \$/KWH} + 0.32$$

This includes power for both the Slurry Production and Support Facilities.

The figure of $0.32/ST accounts for plant fuel oil and miscellaneous utilities.

To provide for Ash Disposal, it is proposed that ash and tailings from a "grassroots" plant would go to secure ponds which are built over membrane liners with drainage systems. Periodically, these would be dredged and the solid sent to a landfill. A cost of $15/ST of ash (dry basis) has been assigned to this in the production cost.

For a plant at the mine mouth only $5/ST (ash) has been included for the ash disposal.

Processing Cost

Processing costs are those charges for operating labour, supervision, maintenance, supplies, overhead and insurance. They are based on specified percentages of the plant payroll and the Total Plant Cost.

Production Costs

These are shown in detail in Table 2 for the different plant configurations. The results can be summarised as follows:

	Slurry costs, $/ston	
Slurry Production short tons/year	770,000	2,750,000
Grassroots Plant	79.71	69.98
Mine Mouth Plant	72.11	64.71

These results are based on a mine mouth coal priced at $30/short tons, and include transportation costs of CWF to the user.

Sensitivity Studies

The variation in the cost of slurry caused by changes in plant cost and plant size was studied for the mine mouth location.

In the case of the plant size study the cost of slurry was evaluated for three different size plants. Results are summarised below:

Size Plant ST/Yr Slurry	Slurry Cost Change
770,000	Base
2,750,000	-11%
4,000,000	-13%

As can be seen from the above, the change from 770,000 ST/Yr to 2,750,000 ST/Yr is significant in that the slurry costs drop by 11%. However, as the plant size goes over 2,750,000 ST/Yr the price only drops by an additional 2%.

In the case of holding the plant size constant and varying the construction cost we find the following results for a 2,750,000 ST/Yr slurry plant at a mine mouth location.

Plant Cost	Slurry Cost Change
$67,500,000	Base
$81,000,000	+1.6%
$97,200,000	+3.6%

Hence it is seen from the above that a large increase in cost of plant (+44%) produced only a 3.6% increase in the slurry cost.

The effect of coal price on production cost of fuel is shown in Figure 5. This illustrates the strong impact that the cost of coal has on the final coal water fuel. The Figure also gives the cost of CWF in US dollars per million BTU. This is based on CWF with a higher heating value of 10,000 BTU/lb.

5.0 COMMERCIAL EXPERIENCE

Pilot Plant

Carbogel CWF has been prepared in test quantities in a pilot plant. These quantities have been used in combustion, rheological, and transportation stability tests.

Overall approximately 50 tons have been produced in this facility.

Production Plant

As a next step in scaling up, an ore plant was converted to CWF manufacture. The plant was started on Harbour Seam coal from Nova Scotia and it operated at a rate of 5 to 10 tons/hour. By the end of 1982, approximately 8000 tons of various coals had been processed through this plant.

The CWF produced was used in a variety of test burns, rheological, pumping and transportation tests. All of the tests indicated that the material could be handled and burned utilising fairly conventional equipment.

Commercial Demonstration

The first complete commercial demonstration programme is currently underway in Nova Scotia Canada. The programme starts with the production of coal water fuel using technology developed by A.B. Carbogel and continues through testing in commercial boilers. This fuel production plant, which will be producing in mid-1983, is part of the largest and most extensive testing to date using CWF's.

The project is sponsored by the Canadian government (Energy Mines and Resources Canada) and involves the Cape Breton Development Corporation, the New Brunswick Electric Power Commission and the Nova Scotia Power Corporation.

Two boilers, one a front wall fired 11 MW unit and the other a corner fired 22 MW unit, have been converted for CWF firing. Test firing is currently underway at these plants which are located in Chatham, New Brunswick.

REFERENCES

1. Coal-Water Slurry Systems for Oil-Design Power Plants, Combustion Processes Inc. EPRI FP-1164, Sept 1979

2. Clean Air and Coal Water Slurry, Leon Green, Mech. Eng., November 1982, p. 38.

3. Some Rheological Data and Atomisation Behaviour of CWM's Containing 68 to 83% Coal, AB Carbogel, L. Gillberg et. al.

4. Proceedings of the Coal Water Fuel Technology Workshop, March 19 and 20, 1981 at Pittsburgh, Pa. Sponsored by Pittsburgh Energy Technology Center COE Contract No. DE AC0276CH00016.

5. Development of Coal Water Slurries as Boiler Fuel, Atlantic Research Corp., Henderson, Scheffee & McHale, Coal Technology 81, Vol. 1, Pg. 251.

6. Direct Firing of Coal Water Slurries - State of the Art, Brookhaven National Lab, Dr. P. Marnell, Coal Technology 80, Vol. V, Pg. 485.

7. Changing Times and Units to Fire Coal, Foster Wheeler Corp., Raskin & Freidrick, Coal Technology 80, Vol. IV, Pg. 501.

8. Feasibility of Power Plant Conversions to Slurries, Richard C. Furman, Consulting Engineer, Coal Technology 81, Vol. V, Pg. 199.

9. Combustion Tests of Coal-Water Slurry, The Babcock & Wilcox Co. EPRI CS-2286 March 1982.

10. Coal-Water Slurry as Utility Boiler Fuel, Atlantic Research Co., EPRI CS-2287 March 1982.

11. Seminar Proceedings - The Use of Coal in Oil-Design Utility Boilers, EPRI, Palo Alto, California, EPRI WS-80-141 July 1981.

12. Coal Desulphurisation During the Combustion of Coal/Oil/Water Emulsions Dr. J.P. Dooher, Adelphi University, Garden City, N.Y. DOE/PC/10328-TI, November 15, 1979.

13. A Cost/Benefit Analysis for Evaluating Coal Cleaning, Envirotech Coal Services, Richard Terry, Coal Technology 81, Vol. II, Pg. 259.

14. Beneficiating Coal for Utilisation, Kaiser Engineers Inc., M.C. Albrecht, Coal Technology 81, Vol. II, Pg. 295.

15. Cost Effects of Coal Cleaning on Power Generation, Bechtel and EPRI, Bunder and Clifford, Coal Technology 79, Vol. I, Pg. 265.

16. Evaluation of Chemical Coal Cleaning, Bechtel and DOE, Hullenham and Taylor, Coal Technology 80, Vol. II, Pg. 207.

17. Coal and Coal Mixture Fuel Conversion, N. Raskin and J. DeAnna, Foster Wheeler Corporation, 1982 International Steam Generator Conference.

Table 1
Capital Cost Estimates
(1982 East Coast, USA Costs)

	Grass Roots Plant		Mine Mouth Washery	
Plant Capacities short tons/year	2,750,000	770,000	2,750,000	770,000
Coal Handling				
Receiving & Storage	7,000,000	6,000,000	-	-
Crushing & Storage	7,800,000	6,000,000	7,800,000	6,000,000
	14,800,000	12,000,000	7,800,000	6,000,000
Slurry Preparation				
Wet Grinding	16,000,000	7,500,000	16,000,000	7,500,000
Froth Flotation	13,400,000	6,250,000	13,400,000	6,250,000
Filtering & Dewatering	12,500,000	5,750,000	12,500,000	5,750,000
Mixing	7,500,000	3,500,000	7,500,000	3,500,000
	49,400,000	23,000,000	49,400,000	23,000,000
Storage & Load Out				
Tank Car Loading	450,000	300,000	450,000	300,000
Barge Facilities	2,900,000	2,900,000	-	-
Day Tanks	1,300,000	600,000	1,300,000	600,000
Storage Tanks	13,750,000	6,500,000	6,500,000	6,500,000
	18,400,000	10,300,000	8,250,000	7,400,000
(1) Buildings	14,000,000	6,500,000	11,550,000	5,100,000
(2) Utilities	10,400,000	6,200,000	4,000,000	2,500,000
Total	107,000,000	58,000,000	81,000,000	44,000,000

(1) Buildings include Administration, Maintenance, Gate/Change House, Slurry Preparation, Laboratory, Boiler House and Substation.

(2) Utilities include Boiler, Utilities, Ash Handling Ponds, Waste Water Treatment, Sewer and Sanitary, Power Distribution, Roads and Parking, Piping and Firewater, Fencing, Railroad Siding and Site Preparation.

Table 2
Production Costs

	Grass Roots Plant		Mine Mouth Washery	
Plant Capacity short tons/year	2,750,000	770,000	2,750,000	770,000
Total Plant Cost	$107,000,000	$58,000,000	$81,000,000	$44,000,000
	Production Cost, $/ST Slurry			
(1) Raw Materials	31.85	31.85	31.85	31.85
(2) Coal Transportation	18.20	18.20	-	-
Utilities	4.14	4.55	4.14	4.55
Ash Disposal	2.34	2.34	0.78	0.78
Processing	4.73	9.62	3.85	7.99
(3) Capital Charges	4.72	9.15	3.58	6.94
	65.98	75.71	44.20	52.11
(4) Slurry Transport	4.00	4.00	20.00	20.00
Total	69.98	79.71	64.71	72.11

(1) $30.00/ST for coal at the mine

(2) $20.00/ST for coal transportation to Carbogel plant

(3)   20 years @ 10.5%

(4)   Transportation of Slurry from Carbogel plant to user.

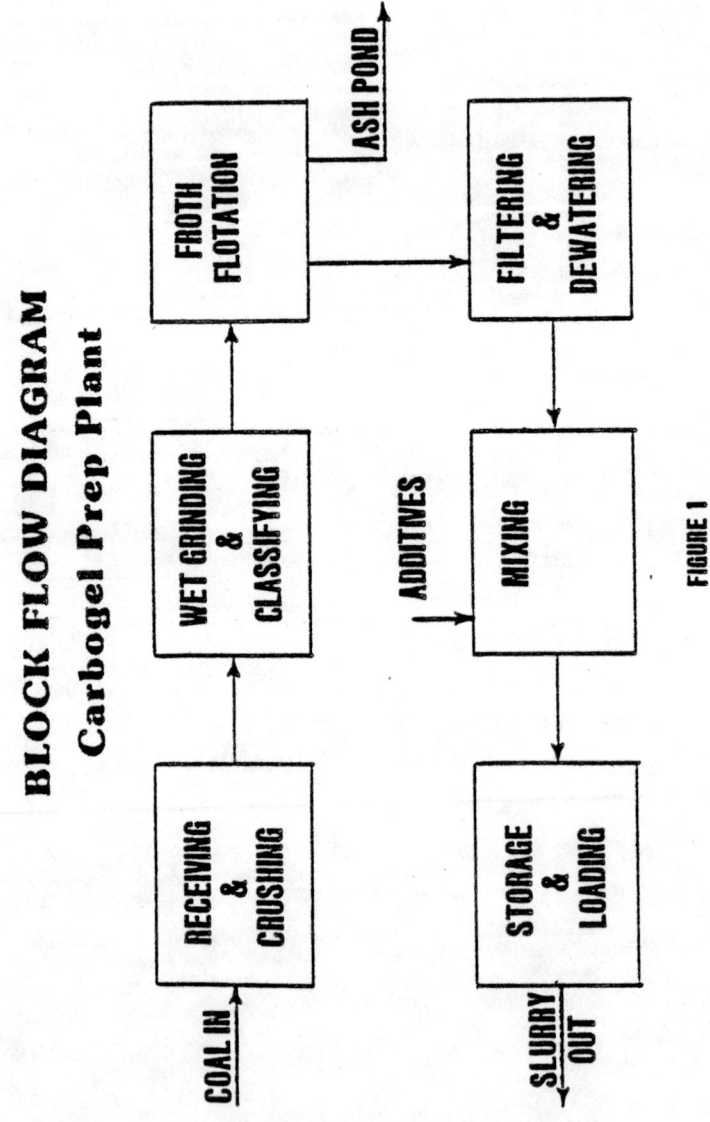

BLOCK FLOW DIAGRAM
Carbogel Prep Plant

FIGURE 1

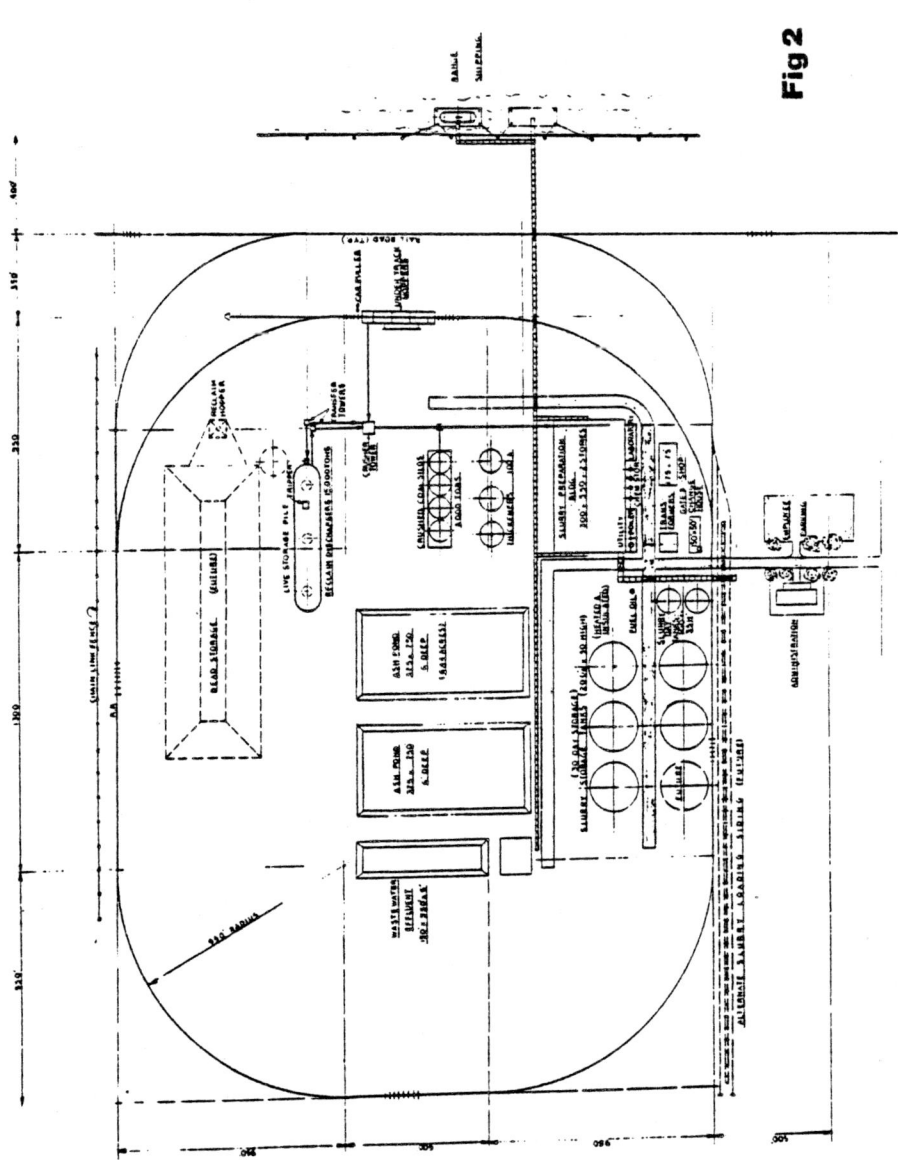

Fig 2

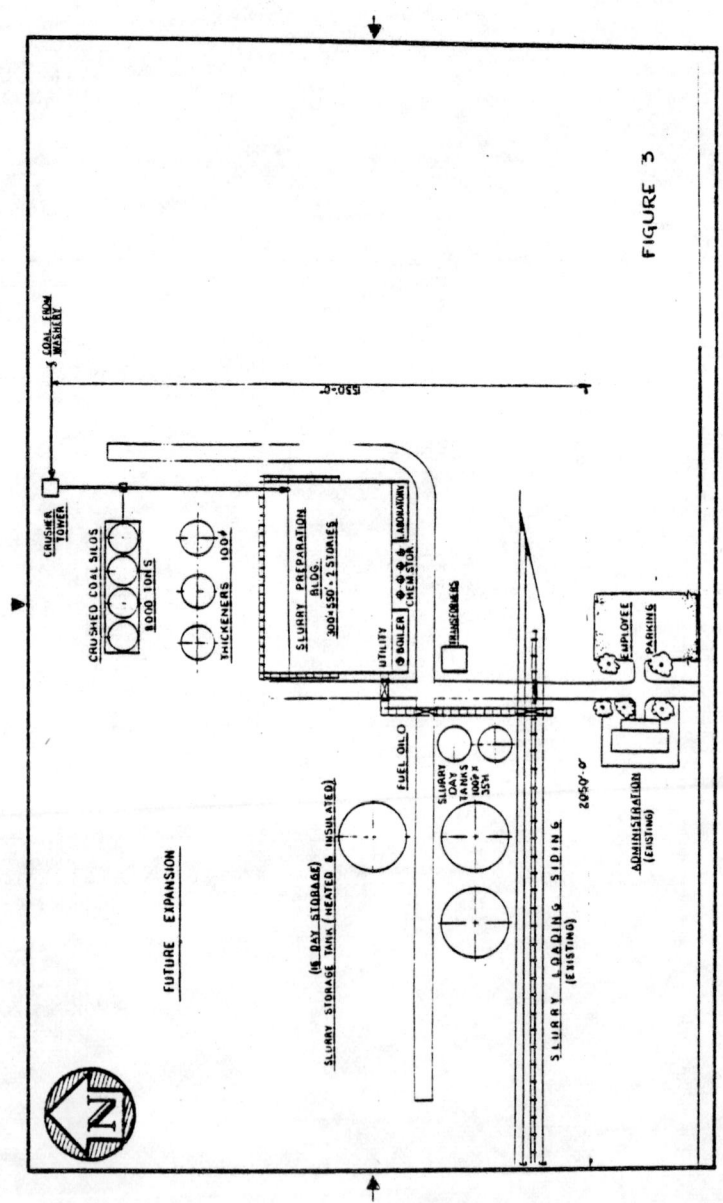

FIGURE 3

Figure 5 EFFECT OF COAL PRICE
ON
COST OF COAL WATER FUEL

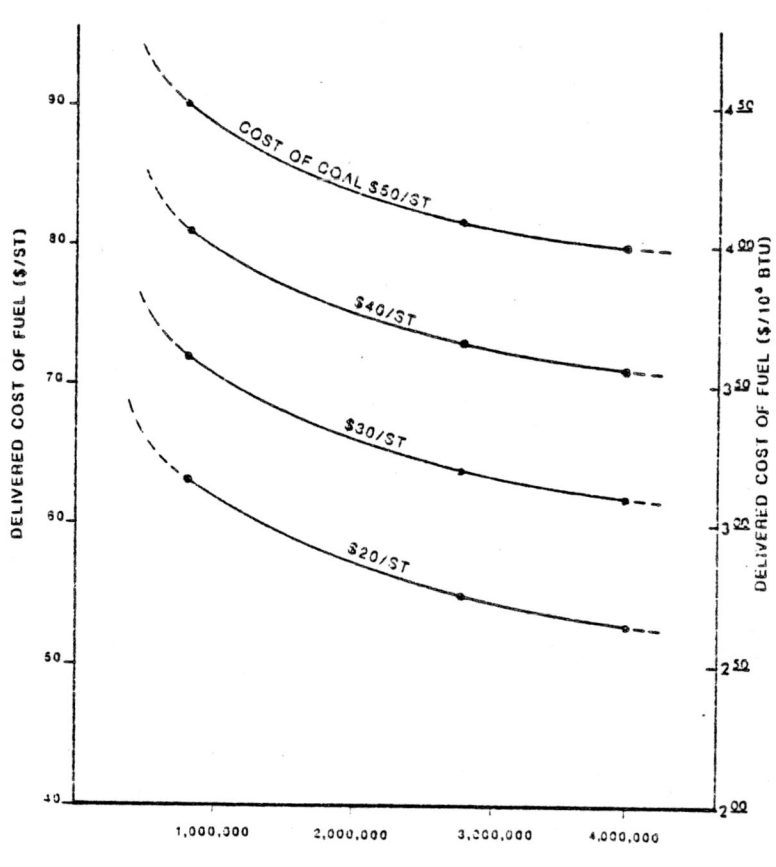

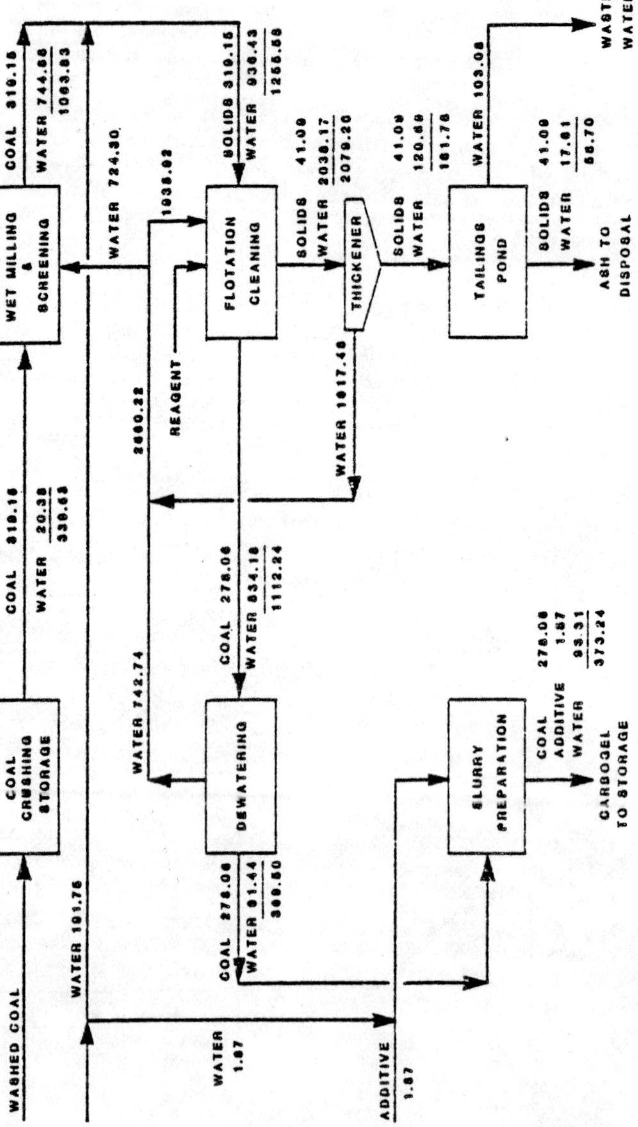

Figure 4. CARBOGEL PREPARATION PLANT – OVERALL MATERIAL BALANCE
BASIS: 2,500,000 TONNES/YR – 2,750,000 ST/YR. OPERATING FACTOR – 307 DAYS/YR.
ALL FLOWS IN SHORT TONS (2,000 LBS) PER HOUR

CORRIGENDUM

PAPER 4.3

Page 234.   Line 3. Correct 4 Mt p.a. to 3 Mt p.a.
Page 248.   Table 4. Boiler efficiencies.   Revise as below:
Page 251.   Revise Figure 1 (c) as below:

TRIAL	PADIHAM 1		PADIHAM 2		PADIHAM 3		
Fuel	COD	RFO	COD	RFO	COD	COD	COD
Load MW$_e$	94	86	117	120	94	120	120
Excess Oxygen %	0.9	0.5	1.0	0.6	1.2	1.7	3.0
LOSSES*(%):-							
Dry Flue Gas	5.24	4.30	5.88	4.47	5.65	5.56	6.73
Flue Gas Moisture	5.70	6.37	5.51	6.21	5.95	5.92	5.96
Unburnt Carbon	0.67	0.07	1.45	0.05	0.60	0.76	0.11
Unburnt CO	0.06	0.11	0.03	0.06	0.01	0.01	0.01
Air Moisture	0.03	0.07	0.09	0.09	0.09	0.09	0.11
Radiation etc. Assumed.	0.64	0.70	0.5	0.5	0.64	0.5	0.50
Boiler Efficiency %	87.6	88.4	86.5	88.6	87.1	87.2	86.6

*   To facilitate comparisons, data for all the COD trials has been
     normalised to the degree of boiler fouling typical of the
     Padiham 3 series.

TABLE 4:   BOILER EFFICIENCIES.

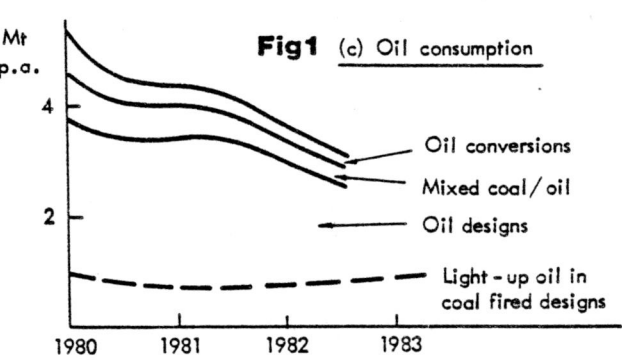

Mt p.a.

Fig1 (c) Oil consumption

Oil conversions
Mixed coal/oil
Oil designs
Light-up oil in coal fired designs

1980   1981   1982   1983